CRC Handbook
of
Electrical Resistivities
of
Binary Metallic Alloys

Editor

Klaus Schröder, Dr. rer. nat.
Professor of Materials Science
Department of Chemical Engineering
and Materials Science
Syracuse University
Syracuse, New York

CRC Press, Inc.
Boca Raton, Florida

Library of Congress Cataloging in Publication Data

Main entry under title:

Handbook of electrical resistivities of binary metallic alloys.

   Bibliography: p.
   Includes index.
   1. Alloys -- Electric properties -- Handbooks, manuals, etc. 2. Binary systems (Metallurgy) -- Handbooks, manuals, etc. I. Schröder, Klaus, 1928-
TN690.4.H27   669'.94   82-1199
ISBN 0-8493-3520-5   AACR2

Direct all inquiries to CRC Press, Inc., 2000 Corporate Blvd., N.W., Boca Raton, Florida, 33431.

© 1983 by CRC Press, Inc.

International Standard Book Number 0-8493-3520-5

Library of Congress Card Number 82-1199
Printed in the United States

THE EDITOR

Klaus Schröder is Professor of Materials Science at Syracuse University. He received his Dr. rer. nat. from the George August University at Gottingen in 1954. He worked for 3 years as a Research Officer at Tribophysics, C.S.I.R.O. University of Melbourne, Australia. From 1958 to 1960 he was a Research Associate and from 1960 to 1961 he held the position of Research Assistant Professor at the University of Illinois. From 1961 to 1962 he worked as Wissenschaftlicher Mitarbeiter at the University of Gottingen. In 1962 he became an Associate Professor at Syracuse University and in 1968 he became a Professor.

Professor Schröder studied deformation and fracture characteristics of single crystals and polycrystalline materials, and electronic properties of transition element alloys. He is presently concerned with magnetic memories and uses Barkhausen effect type measurements for nondestructive testing.

Professor Schröder has published more than 60 research papers. He holds several patents and has written a book, entitled *Electronic, Magnetic, and Thermal Properties of Solid Materials* (Marcel Dekker).

PREFACE

During the last decades the low temperature resistivity of metallic alloys was studied in detail and deviations from simple rules were reinvestigated. The effect of different types of second components, and changes in ordering in binary alloys on the scattering phenomena of conduction electrons are generally understood but it is not possible to calculate the resistivity of specific alloys from first principles. Therefore, a review of experimental data should be of general interest.

ACKNOWLEDGMENTS

The editor gratefully acknowledges the assistance of the many people who helped in the preparation of this manuscript. First, to Professor Zuckermann, Department of Physics, McGill University, Montreal, Canada, who read critically the contribution of Dr. Gratz. The staff of the Physics and Engineering Library at Syracuse University helped in the collection of references. Dr. Miller, head of the Engineering and Life Science Library, provided a computer search for abstracts. Mr. Henry Taubenfeld checked the science abstracts.

Special thanks are due to Mr. Rolf Ziemer, who was responsible for the drawings, and to Mrs. Helen Turner, who typed the various drafts of the manuscript; without her help this manuscript would not have reached completion. I would also like to thank Dr. R. Evan-Iwanowski and Mr. D. Suconick for translations.

ADVISORY BOARD

CONTRIBUTORS

John Orehotsky, Ph.D.
Professor of Engineering
Department of Engineering
Wilkes College
Wilkes-Barre, Pennsylvania

Klaus Schroder, Dr. rer. nat.
Professor of Materials Science
Department of Chemical Engineering
 and Materials Science
Syracuse University
Syracuse, New York

With a section on "Rare Earth Metal Alloys" by:

Ernst Gratz, Dr., Dipl. Ing.
University Assistant
Institut für Experimentalphysik
Technische Universität
Vienna, Austria

TABLE OF CONTENTS

Introduction

INTRODUCTION

J. Orehotsky and K. Schröder

As technology advances, a wider variety of demands are frequently imposed upon our device materials. Material design engineers often require a tabulation of physical properties in order to select the appropriate material for use, while developmental engineers need to be aware of basic principles and systematic trends that can be utilized as a guide for developing the next generation of materials and alloys for future use. The electrical resistivity of metallic alloys is a particularly relevant property since many applications require conductive materials that need to match given specifications in other physical properties such as strength or environmental stability. The trade-off in conductivity that results when metals are replaced by alloys for strength and stability purposes becomes an important consideration and a tabulation of electrical resistivities is needed to provide some insight into the magnitude of the trade-off. While a number of physical properties are readily available in graphical or tabular form, an extensive updated comprehensive compilation of the electrical resistivity of metallic alloys does not exist, and the object of this monograph is to provide both the necessary background and the desired tabulation of the electrical resistivity of metallic alloys.

CONDUCTIVITY OF PURE METALS

An adequate presentation of the factors that influence the conductivity of alloys requires a background knowledge of the conductivity of pure metals. Historically, the modern treatment of the electrical conductivity of metals originated with the Drude theory.[1] The conduction electrons in metals were assumed to behave like a gas of weakly interacting particles. Under the influence of an applied electric field, ξ, the electrons with charge, q, experience a force

$$F = q \xi \qquad (1)$$

that results in an acceleration

$$a = q \xi / m \qquad (2)$$

in the direction of the field. The accelerated electron is subjected to collisions with the fixed ions of the crystal lattice where τ is taken to be the average time between collisions. The resulting average drift velocity of the elctrons, v_d, in the direction of the field would be

$$v_d = a\tau = q\tau / m \qquad (3)$$

assuming that the motion of the electron is completely terminated at each collision event. The mobility of the electron is defined by

$$\mu \equiv \frac{v_d}{\xi} \qquad (4)$$

and becomes

$$\mu = q\tau / m \qquad (5)$$

where m is the mass of the electron. Since current density J is expressed as

$$J = nqv_d \tag{6}$$

where n is the number of electron carriers per unit volume, the conductivity becomes

$$\sigma \equiv J/\xi = nq^2\tau/m \tag{7}$$

The result of the Drude model contains two attractive features. First, it gave a reasonable order of magnitude value for the conductivity of metals at room temperature if n is taken to be equal to the number of atoms per unit volume, assuming each atom contributes one conduction electron, and if the relaxation time, τ, is estimated from the formula

$$\tau = \ell/v_t \tag{8}$$

where ℓ, the mean free path of the electrons, is taken to be equal to the separation between atoms in the crystal lattice and v_t is the thermal velocity of the electron as given by the equipartition law

$$(1/2)\, mv_t^2 = (3/2)\, k_B T \tag{9}$$

The drift velocity resulting from the electric field is superimposed on the thermal velocity. The second attractive feature of the Drude approach is that it was used to determine a relationship for the thermal conductivity, K. The ratio of the thermal to electrical conductivity yields for the Lorenz number

$$K/\sigma = 3(k_B/q)^3 \tag{10}$$

which is in modest agreement with experimental results. There are inherent difficulties with the Drude model. This free electron approach does not provide a physically realistic mechanism for conduction in metals since the influence of the periodic lattice atoms was ignored and since the electrons in the metal were assumed to act independently of each other with no mutual interactions. Lorentz[2] found that the Drude theory, when refined by the application of Boltzmann statistics, does not yield a Lorenz number that is reasonably consistent with experimental values. Another failure of the simple Drude model is that it does not account for the observed conductivity differences among the metallic elements. The model also cannot easily provide for the enormous increase in the mean free path needed to explain the rapid increase in conductivity that results with decreasing the temperature. Another gross deficiency of the model is that the mean free path and therefore the conductivity of a metal is expected to decrease with increasing pressure while experimentally the conductivity normally increases with pressure. The final drastic failure is that the model predicts an electronic specific heat, C_e, per mole of free electrons as:

$$C_e = (3/2)R \tag{11}$$

which is considerably larger than the experimentally determined values. In spite of its obvious limitations and failures, the strength of the Drude theory lies in the fact that both its unrealistic assumption of essentially noninteracting electrons and the derived expression for the conductivity (see Equation 7) is contained in the subsequent modern theory which emerged from quantum considerations and which formulates both a physically attractive conductive mechanism in solids and a relationship for the temperature dependence of the resistivity that is reasonably consistent with experimental results.

The quantum approach started with Sommerfeld,[3] who treated the electrons as a quantum gas. He employed the Pauli exclusion principle and Fermi-Dirac statistics to show that only a fractional number of the conduction electrons are affected by thermal agitation. This fractional number is $k_B T/E_F$, where E_F is the Fermi energy, and this result provides for the experimentally observed magnitude of the electronic specific heat showing that the electrons must be treated as a quantum gas rather than be represented as a Drude electron gas. The Sommerfeld theory also yields a Lorenz number that is in remarkably good agreement with experiment which again emphasizes the inadequacy of the Drude approach and suggests that the theory of conduction will be contained in quantum mechanical considerations.

This agreement between theory and experiment for the Lorenz number and for the electronic specific heat is impressive, but a simple application of the theory does not yield an experimentally verifiable result for the conductivity. The thermal velocity of the electron in the Drude model is replaced in the quantum gas model by the Fermi velocity upon which the drift velocity is superimposed. The Fermi velocity, v_F, is given by

$$v_F = (2E_F/m)^{1/2} \tag{12}$$

When the Fermi velocity is cast into the formalism of the Drude equations, the conductivity becomes relatively temperature insensitive:

$$\sigma = nq^2 \tau/m = nq^2 \ell/mv_F$$

$$= nq^2 \ell/(2mE_F)^{1/2} \tag{13}$$

which is not experimentally acceptable if the mean free path is restricted to be the separation between atoms. Also, the descriptive Sommerfeld quantum gas does not specify a mechanistic model for the electrical conduction process in a crystalline lattice.

Bloch[4] provided a framework to resolve these basic difficulties. In an elegant treatment, he conceived the basic concepts for conduction in the periodic atomic structure of a crystalline solid. The conduction electrons in metals were considered as waves rather than particles. As waves, the electrons are now considered in accordinace with the newly emerged quantum theory and their mean free path is no longer restricted to atomic dimensions. The wave function description of the electron, $\psi_k(r)$, in the crystal lattice was suggested to be a plane wave, $\exp(i\vec{k}\cdot\vec{r})$, modulated by a periodic function, $U_k(r)$,

$$\psi_k(r) = \exp(i\vec{k}\cdot\vec{r})\, U_k(r) \tag{14}$$

where $U(r)$ contains the periodicity (R) of the lattice

$$U(r) = U(r + R) \tag{15}$$

In Cartesian coordinates,

$$\psi_k(x,y,z) = \exp[i(k_x x + k_y y + k_i z)]\, U(x,y,z) \tag{16}$$

The wave vector, k, designates the direction of propagation, the energies, and the wavelength, λ, of the electron

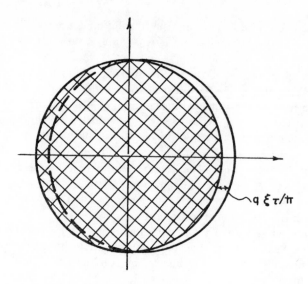

FIGURE 1. Equilibrium and steady state, distribution functions electron distribution.

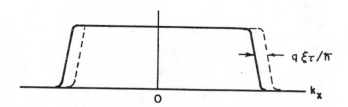

FIGURE 2. The electron distribution function for equilibrium (solid line) and steady state current (dashed line).

$$k = 2\pi/\lambda \tag{17}$$

Within the structure of this quantum mechanical wave vector representation of the electron, Bloch proposed a mechanism for the behavior of metallic electrons when subjected to an applied electric field in the presence of scattering centers due to thermally induced atomic vibrations in a crystal lattice. The analysis is presented in detail in many fine articles,[5-8] but briefly the analysis first involved a consideration of the drift movement in k-space of a wave packet composed of superimposed Bloch electrons centered at k when an electric field is applied. The drift rate or acceleration of the wave packet dk/dt is given by

$$\frac{dk}{dt} = -\frac{q\xi}{\hbar} \tag{18}$$

which is the quantum analogue of the Drude acceleration given by Equation 2. Bloch associated an electron occupational probability function, f_k, with the wave packet drift, and in thermal equilibrium, this function was subsequently shown to be equal to the statistical Fermi-Dirac distribution function, f_k^o,

$$f_k^O = 1/\left[1 + \exp[(E_k - E_F)/k_B T]\right] \tag{19}$$

The probability function, f_k, is drift displaced (see Figures 1 and 2) when an electric field is present and changes in time according to

$$\frac{df_k}{dt}\Bigg|_{dr} = \frac{df_k}{dk}\frac{q\xi}{\hbar} \tag{20}$$

so that in a time interval, δt, the distribution shifts in k-space by an amount

$$\delta k = (q\xi/\hbar)\delta t \tag{21}$$

As a first approximation, f_k can be replaced by f_k^o and the drift rate term becomes

$$(df_k/dt)_{dr} = (df_k^o/dt)_{dr} = (q\xi/\hbar)(df_k^o/dk) \tag{22}$$

$$= (q\xi/\hbar)(df_k^o/dE)(dE/dk) \tag{23}$$

In thermal equilibrium, the change in the distribution due to the electric field is balanced by the change resulting from a scattering of the electrons by thermally induced lattice vibrations giving the Bolzmann equation:

$$(df_k/dt)_{dr} + (df_k^o/dt)_{sc} = 0 \tag{24}$$

This phonon scattering term can be evaluated from the probability $\phi(k,k')$ that an electron scatters from state k to k' with the Pauli exclusion restriction that k' must be unoccupied. The result is

$$(df_k/dt)_{sc} = (1/2\pi^3)\int \left\{ \Phi(k',k)f_{k'}\,[1-f_k] - \Phi(k,k')f_k[1-f_{k'}] \right\} dk \tag{25}$$

For elastic, isotropic scattering and a spherical Fermi surface where the scattering probability is dependent only on the angle θ between the k and k' wave vectors, the scattering term becomes

$$(df_k/dt)_{sc} = (f_k - f_k^o)/\tau \tag{26}$$

Equating the drift and scattering terms,

$$(q\xi/\hbar)(dE/dk)(df_k^o/dE) - (f_k - f_k^o)/\tau = 0 \tag{27}$$

or

$$f_k = f_k^o + (\tau/\hbar)(dE/dk)(df_k^o/dE)q\xi \tag{28}$$

Along with the density of states, this result can be used to calculate the electron concentration in k-space that contributes to the conductivity. The density of electron states per unit volume, D(k), with wave vectors in the interval δk, is given by the free electron model to be:

$$D(k) = \delta k/8\pi^3 \tag{29}$$

The number, dn, of electrons per unit volume with wave vectors in the δk interval is then the product of the density of states multiplied by the probability function, f_k, that a state in this interval is occupied,

$$dn = 2f_k(\delta k/8\pi^3) = f_k \delta k/4\pi^3 \qquad (30)$$

where the factor of two is included to account for the fact that two electrons of opposite spin can occupy each state. The electrical current density is determined from a form of the familiar formula (see Equation 6) or

$$J = \int q v_k \, dn \qquad (31)$$

where the velocity, v_k, would be given by the group velocity of the wave packet which, from physical optics, is expressed as

$$v_k = d\omega/dk \qquad (32)$$

where

$$E = \hbar\omega \qquad (33)$$

giving

$$v_k = (1/\hbar)(dE/dk) \qquad (34)$$

In contrast to the Drude velocity, v_k does not depend on scattering from fixed ions, but rather is governed by the form of the energy relationship to the wave vector, k. Unless the slope of the functional relationship between E and k is zero, the conductivity of a perfectly periodic lattice will be infinite. Combining Equations 28, 30, 31, and 34, the current density becomes

$$J = (1/4\pi^3)\int (q/\hbar)(dE/dk)\left[(\tau/\hbar)(dE/dk)(df^0/dE)q\xi\right] dk \qquad (35)$$

where the leading team f_k^0 in the distribution function (see Equation 28) makes no contribution to the current and is not included in the expression for the current density. Since

$$J = \sigma\xi = \xi/\rho \qquad (36)$$

where σ is the conductivity and ϱ is the resistivity, then

$$\sigma = 1/\rho = (q^2/4\pi^3\hbar^2)\int_E (dE/dk)\left[\tau(dE/dk)(df^0/dE)\right] dk \qquad (37)$$

$$\sigma = (q^2/4\pi^3\hbar^2)\int_E (dE/dk)^2 (\tau dS/grad_k E) \qquad (38)$$

Employing the result of free electron theory,

$$E = \hbar^2 k^2 / 2m^* \tag{39}$$

where m^* is the effective mass, the conductivity becomes

$$\sigma = (q^2 \tau_{E_F} / 4\pi^3 m^* \int k dS \tag{40}$$

if τ_{E_F} is the relaxation time evaluated at the Fermi energy and does not depend on the electron distribution. Since

$$\int k dS = 4\pi^3 n \tag{41}$$

the conductivity becomes:

$$\sigma = nq^2 \tau_{E_F} / m^* \tag{42}$$

showing that the Drude result is recoverable from the Boltzmann-Bloch approach to the conductivity in the relaxation time approximation.

This simple Drude expression for the conductivity is not particularly useful for metals and alloys, but the equation realizes a considerable amount of strength in the fact that it adequately describes the temperature dependence of the conductivity of intrinsic and extrinsic semiconductors and insulators up to elevated temperatures, since the value of n at any temperature is obtained by a relatively straightforward calculation involving a simple consideration of the density of states and the Fermi-Dirac statistics applied to a band gap material. The asset of the Boltzmann-Bloch approach for obtaining the conductivity is that it provides a realistic model from which the more complicated problem of the conductivity of metals and alloys can be treated.

A treatment of the temperature dependence of the conductivity of both metals and alloys requires a detailed consideration of the scattering mechanisms involved within the framework of the Bloch model. Scattering results when a deviation from the perfect periodicity of the pure atomic arrangement is introduced into the crystal structure. These deviations result from displacement or replacement of atoms from fixed periodic lattice positions by such effects as:

1. Atomic vibrations due to thermal agitation
2. Defects in the crystal, such as vacancies, dislocations, grain boundaries, Frenkel and Schottky defects
3. A random substitution of impurity atoms for pure metal atoms on the pure metal lattice sites.

To obtain the conductivity of metals and alloys, these deviations must be incorporated into the Boltzmann-Bloch analysis utilizing Equations 24 and 25, but the integral-differential nature of these equations presents considerable difficulty in finding a solution, and it is easier to employ the relation time approximation characterized by Equation 26 where the relaxation time is given by

$$1/\tau = \int \left\{ 1 - \cos\theta \right\} \Phi(\phi) d\Omega = \int [1 - \cos\theta] \Phi(\phi) \sin\theta \, d\theta \tag{43}$$

where $\Phi(\theta) d\Omega$ is the probability per unit time that an electron will be scattered through

θ into the solid angle, $d\Omega$. To use Equation 43, $\Phi(\theta)$. must be determined from the probability per unit time, $\Phi(k,k')$, that an electron will be scattered to its final state, k', from its initial k-state:

$$\phi(\theta) = 1/8\pi^3 \int \Phi(k,k') \, k^2 \, (dk/dE)dE = (k^2/4\pi^2\hbar) \, (dk\langle k'|U|k \rangle^2 /dE)$$

(44)

where θ is the angle between k and k' and $\langle k'|U|k\rangle$ is the matrix element involving both the wave functions in the initial, $\psi_k(r)$, and final, $\psi_{k'}(r)$, states and the interaction potential, $U(r)$, responsible for the scattering mechanism. The theory of conductivity is then dependent on selecting the appropriate interaction potential for representing the scattering process in this relaxation time approximation approach.

The electron-phonon interaction potential for determining the temperature dependence of the conductivity in pure metals is frequently obtained by considering the potential, V_r, at a distance, r, from a simple ion in the latter:

$$V_r = (Zq^2/r) \exp(-r/r_c)$$

(45)

The shielding of the ion of charge, Z_q, by the surrounding conduction electrons is contained in the potential equation by the exponential term containing the effective screening radius, r_c. The electron-phonon interaction potential in the crystal lattice at any temperature can be obtained from the difference between the potential, V_r, at a point in the lattice due to the undisplaced n^{th} ion which is a distance, r, away when there is no thermal excitation and the potential $V_{r+\delta r}$, at this point due to the nth ion which has suffered a δr vibrational displacement due to thermal effects. The electron-phonon interaction potential is equal to the difference between these two potentials summed over all n ions in the lattice

$$U = \sum_n \left[V_{r_n + \delta r_n} - V_{r_n} \right]$$

(46)

Employing this relationship, the relaxation time is found to be inversely related to the temperature divided by the Debye temperature (Θ_D) squared

$$1/\tau \sim T/\Theta_D^2$$

(47)

The Debye temperature appears in the final expression because it characterizes the vibrational response of the lattice and, as a result, the electron-phonon scattering probability at a given temperature. The resistivity from the Drude equation in this relaxation time approximation would then be given by a linear temperature dependence:

$$\rho \sim T/\Theta_D^2$$

(48)

This equation is applicable for the conductivity at high temperatures. At low temperatures $(T \ll \Theta_D)$, the analysis for the conductivity becomes more complex. The $\hbar\omega$ quantization of the lattice waves relative to the thermal energy, $k_B T$, of the lattice places restrictions on the phonon energies that can be absorbed or emitted by electrons. The Boltzmann equation cannot be solved by the relaxation time approximation. In a complicated analysis, several years after he derived the linear temperature dependence for the resistivity at high temperatures, Bloch[9] showed that the resistivity at low temperatures was proportional to the fifth power of T:

$$\rho \sim T^5 \qquad\qquad (49)$$

Bloch's solutions were only applicable at the high- and low-temperature extremes, and he sugested that the solution at intermediate temperatures would prove to be a more difficult problem. Gruneisen[10] suggested that Bloch's low-temperature expression could be used for all temperatures, and he formulated the temperature dependence of the resistivity as:

$$\rho(T) = 4[T/\Theta_R]^5 \rho(\Theta_R) \int_0^{\Theta_R/T} (x^5/[e^x - 1][1 - e^{-x}])dx \qquad (50)$$

which yields a T^5 response at low temperature and a linear temperature dependence at high temperatures. Θ_R is a characteristic temperature that is chosen to fit the experimental data to this Bloch-Grüneisen function. In general, the temperature dependence of the resistivity of many simple metals were found to be consistent with the universal Bloch-Grüneisen equation over a fairly wide temperature range. The characteristic temperature, Θ_R, selected to provide the best fit of experiment to the Bloch-Grüneisen function usually differs slightly from the Debye temperature (see Table 1). Table 2 gives the Debye temperatures of the characteristic temperature according to Gerritsen.[11]

The agreement between the Bloch-Grüneisen equation and experimental results for these metals is remarkably good considering the assumptions involved in the Bloch analysis, but detailed investigations for many of these same metals show that the Bloch-Grüneisen relationship is not rigorously obeyed by experimental results, particularly at the temperature extremes.[12,13] This is shown for high temperature in Figure 3.

The proposed T^5 law at low temperatures is often violated for many reasons. Umklapp scattering[14] may be sizeable, particularly for metals whose Fermi surface is in close proximity to the Brillouin zone. Electron-electron scattering,[15] particularly in transition metals, spin density fluctuation scattering[16] in nearly magnetic materials, and magnon scattering[17] in magnetic materials are all manifest as a T^2 contribution to the resistivity, and this fact has been used to account for the experimental inability to verify the expected Block T^5 behavior at low temperatures.

Systematic departures from the predicted linear ϱ-T behavior at elevated temperatures is evident in many metals. The departures in the noble metal are suggested to result from thermal expansion effects, while the more sizeable departures in some transition metals are attributed to s-d electron-electron scattering. The hybridized s-d orbitals associated with the transition metals presents several unique considerations in the analysis of the resistivity response at high temperature. Since the s electrons are very mobile while the d electrons are not, electron-electron scattering transistions occur from the s to the d band. The transitional probability is related[18] to the density of states in the d-band, and depending upon the density of states, these transitions can be the predominate contributing mechanism to the resistivity even at room temperature. The large density of d-band states at the Fermi level and the extremely rapid change in the density of states with increasing energy has been employed to account for the nonlinear resistivity response at elevated temperatures in Pd and Pt.[18]

Another exhibition of anomalous behavior of the resistivity occurs around the critical temperature of the magnetic materials, as shown in Figure 4 where the normalized resistivities of the sister elements Ni (ferromagnetic below 350°C and paramagnetic above) and Pd (paramagnetic at all temperatures) are compared. The obvious influence that magnetic ordering exerts on the resistivity is even more convincingly demonstrated

Table 1

CHARACTERISTIC TEMPERATURE (Θ_R) AND DEBYE TEMPERATURE (Θ_D) OF SELECTED METALS[6]

	Metal							
	Li	Na	Cu	Ag	Au	Pb	Al	W
Θ_R	363	202	333	203	175	86	395	333
Θ_D	340—430	159	310—330	212	168—186	82—88	385	305—357

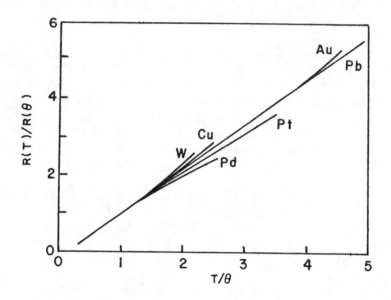

FIGURE 3.   High-temperature resistivity of elements.

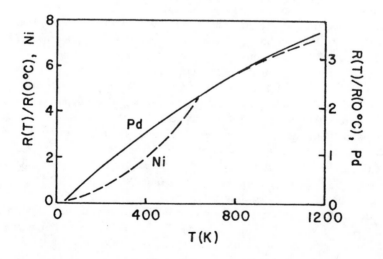

FIGURE 4.   Resistivity of Ni and Pd. Curves are matched near Curie temperature.

Table 2

VALUES FOR Θ_D OR Θ_R, CONDUCTIVITY σ AT 273°K, RESISTIVITY ϱ AT 273°K, R_T/T_{273} AT TEMPERATURE T, AND THE TEMPERATURE COEFFICIENT OF THE RESISTIVITY NEAR ROOM TEMPERATURE[11]

		Θ (°K)	σ_{273} 10^6 $(\Omega m)^{-1}$	ϱ ($\mu\Omega cm$)	T (°K)	RT/R_{273}	α $10^{-3}(°C)^{-1}$
Aluminum	Al	419	40	2.50	77.7	0.1008	4.67
Antimony	Sb	201	3.12	32.1	88.4	0.2441	5.1
Arsenic	As	100	4.5	26	85	0.195	—
Barium	Ba	133	2.8	36	78	0.215	6.1
Berylium	Be	1160	36	2.78	86.1	0.2419	—
		652	—	—	—	—	—
Bismuth	Bi	120	0.93	107	81	0.3561	—
Cadmium	Cd	300	14.8	6.73	81	0.2522	4.26
Cesium	Cs	219	5.12	19	82	0.265	5
		150	—	—	—	—	—
Calcium	Ca	—	28	3.6	83.6	0.4582	~4
Chromium	Cr	403	6.6	15.0	80	0.134	—
Cobalt	Co	445	19	5.2	86.9	0.1929	6.58
Copper	Cu	335	64.5	1.55	81	0.152	4.33
Gallium	Ga	125	7.30	13.7	84.3	0 2204	4.1
Gold	Au	165	49	2.04	84.9	0.2480	3.98
Indium	In	109	12.2	8.2	88.9	0.2567	5.1
Iridium	Ir	316	21.1	4.74	77.7	0.1905	4.33
Iron	α-Fe	462	11.5	8.71	78.2	0.0741	6.57
Lead	Pb	90	5.17	19.3	81	0.2634	4.22
Lithium	Li	363	11.8	8.5	86.3	0.1514	4.37
Magnesium	Mg	330	25.4	3.94	82.2	0.2006	4.2
Manganese	α-Mn	—	0.14	710	—	—	0.17
	β-Mn	410	1.1	91	—	—	1.4
	γ-Mn	—	4.3	23	—	—	6.4
Mercury	Hg	69	4.71	21.2	90.1	0.258	
Molybdenum	Mo	425	19.9	5.03	86.9	0.1701	4.7
Nickel	Ni	413	15.2	6.58	87	0.1179	6.75
Niobium	Nb	250	4.28	23.3	90	0.403	2.28
Osmium	Os	—	1.05	95	—	—	4.2
Palladium	Pd	275	10.2	9.77	81	0.1963	3.8
Platinum	Pt	233	10.2	9.81	81	0.2060	3.92
Potassium	K	100	—	—	—	—	4.57
		163	16	6.3	87.8	0.1635	5.4
Rhenium	Re	310	5.30	18.9	—	—	3.1
Rhodium	Rh	370	23	4.33	87.4	0.1466	—
Rhuthenium	Ru	426	15.0	6.67	81.7	0.1734	4.5
Rubidium	Rb	—	8.62	11.6	87.8	0.3043	5.3
Silver	Ag	—	66	1.5	82	0.2109	4.1
Sodium	Na	160	—	—	—	—	—
		202	23.4	4.27	81.8	0.2304	5.5
		148	—	—	—	—	—
Strontium	Sr	100	5.0	20	86.3	0.3313	~5
Thallium	Tl		6.7	15	86.1	0.2785	5.2
Tantalum	Ta	247	8.06	12.4	88.3	0.2511	3.6
Thallium	Tl	89	—	—	88.6	0.2721	—
Tin	Sn	160	9.9	10.1	88	0.2457	4.63
Titanium	Ti	278	2.38	42	78.5	0.2150	5.5
Tungsten	W	380	20.4	4.89	87.4	0.1565	4.83
Vanadium	V	300	0.54	18.2	90	0.457	—
Zinc	Zn	190	18.3	5.45	82.5	0.2141	4.20
Zirconium	Zr	270	2.47	40.5	88.2	0.2380	4.0

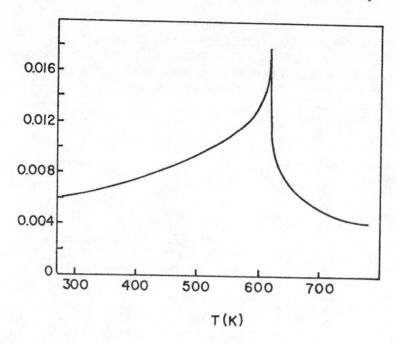

FIGURE 5. $(d\varrho/dT)/\varrho$ of Ni. $T_c = 617°K$.

in the temperature coefficient of resistivity which displays a λ-peak discontinuity at the Curie temperature (Figure 5) of nickel. The temperature dependence of $d\varrho/dT$ around the Curie temperature, T_c, for the nickel group ferromagnetic transition metals is expected[20-23] to behave like the specific heat anomaly near T_c and obey the relationship

$$(d\rho/dT) = A\left\{[(T - Tc)/Tc]^{-\lambda} - 1\right\} \tag{51}$$

where A is a constant and λ is the characteristic critical exponent. Experimental results for nickel are in reasonable accord[20-23] with this predicted behavior.

The anomalous behavior of the resistivity around the critical temperature is not just confined to the nickel group ferromagnetic transition metals. Most metals and alloys experiencing the ferromagnetic-paramagnetic, the antiferromagnetic-paramagnetic, or the atomic order-disorder phase transitions will usually exhibit anomalies in the resistivity and in its temperature derivative around the critical point. Depending on the nature of the material being investigated, the behavior of the anomaly can be significantly different from that observed in nickel.

These general considerations suggest that the high-temperature resistivity of the monovalent metals where electron-electron and Umklapp scattering effects are minimized should be in reasonable agreement with the Bloch-Grüneisen formula, while the high-temperature resistivity of the divalent, trivalent, and transition metals are not expected to be in agreement. This is amply demonstrated in Table 3 where the measured resistivity at room temperature is compared with theory for several metals.

While the Boltzmann equation may be an adequate but not the most rigorous approach for deriving the electrical conductivity and while the experimental results for the resistivity of the monovalent metals are not in detailed agreement with the predicted Bloch T^5 and T^1 dependencies at low and high temperatures, refinement of the Bloch treatment still remains at the heart of most modern theories of conductivity. After 50 years, this fact alone is a tribute to the originality, power, and insight of his pioneering work.[24]

Table 3
RESISTIVITY AT 273°K
OF SELECTED
ELEMENTS AND
COMPARISON WITH
THEORY[6]

Element	Resistivity at 273°K ($\mu\Omega$cm)	$\varrho_{exp}/\varrho_{theor}$
Li	8.5	1.40
Na	4.27	0.77
K	6.3	0.80
Rb	11.6	1.34
Cs	19.0	1.50
Cu	1.55	1.87
Ag	1.50	1.16
Au	2.01	3.26
Be	2.78	7.33
Cd	6.73	8.05
Mg	3.94	3.23
Zn	5.45	5.75
Al	2.50	3.62
In	8.2	8.2
Co	5.2	14.7
Cr	15.0	34.2
Fe	8.71	26.0
Mo	5.03	25.2
Ni	6.58	9.9
Pd	9.77	13.0
Pt	9.81	17.2
Ta	12.4	36.9
Ti	42	81.0
V	18.2	48.6
W	4.89	35.2

CONDUCTIVITY OF DILUTE ALLOYS

The Bloch model suggests that the electrical resistivity is a consequence of disturbances in the atomic periodicity in a crystal structure, and the Bloch-Grüneisen relationship characterizes the temperature dependence of the resistivity in pure metals when the disturbance is a result of atomic displacements due to vibrations associated with thermal excitation. As was suggested previously, disturbances in the periodicity of atoms on a pure metal lattice can result from numerous other causes and perhaps the most prominent cause is the random substitutional replacement of the pure metal atoms on their lattice sites with impurity atoms to form an alloy. The resistivities of an alloy would then be expected to be larger than that of its pure metal solvent at the same temperature.

With some approximations, the resistivity of dilute random solid solution alloys can be easily determined. If the alloy is dilute enough so a conduction electron interacts with only one impurity at any instant and if, for simplicity, the solute impurity atoms dissolve randomly and substitutionally in the solvent lattice, the alloy can be treated as containing two predominate and independent scattering mechanisms: impurity scattering and phonon scattering. The electron-electron scattering contribution to the resistivity is neglected in this analysis. Since the impurity and phonon scattering mechanisms are assumed to act independently, the resistivity, $\varrho_a(c,T)$, of an alloy as a function of concentration and temperature can be expressed as the sum of two terms, $\varrho_p(T)$ and $\varrho_i(C)$:

$$\rho_a(c,T) = \rho_p(T) + \rho_i(c) \tag{52}$$

where $\varrho_i(C)$ is the temperature-insensitive, concentration-dependent residual resistivity due to the impurity scatterers, while $\varrho_p(T)$ is the temperature-dependent contribution to the alloy resistivity due to phonon scattering which is taken to be equal to the resistivity behavior as a function of temperature for the pure metal solvent. This equation is known as Matthiessen's rule after Matthiessen and Vogt,[25] who observed experimentally that the temperature coefficient of the resistivity of dilute alloys and the pure solvent metal at comparable temperatures are approximately equal. The resistivity as a function of temperature of a dilute alloy obeying Matthiessen's rule is then a shifted replica of the resistivity-temperature response of the pure metal solvent where the amount of the shift would be equal to the residual resistivity. Alloy systems rarely obey Matthiessen's rule and considerable interest has been directed toward the deviations from Matthiessen's rule (DMR) for various alloy systems. The deviations are frequently expressed as $\Delta(C,T)$ which is given by:

$$\Delta(c,T) = \rho_a(c,T) - \rho_p(T) - \rho_i(c) \tag{53}$$

where $\varrho_a(c,T)$, $\varrho_p(T)$, and $\varrho_i(c)$ are the experimentally measured resistivities of the alloy at any temperature, of the pure solvent metal at the same temperature, and of the alloy at 0°K, respectively. Deviations from Matthiessen's rule for various alloy systems have been exhaustively studied and are presented in several comprehensive review articles.[26,27] In general, the deviations are usually small relative to the magnitude of $\varrho_i(c)$ or $\varrho_p(T)$, and as a fair approximation for application purposes, it is often reasonable to treat the resistivity of a dilute alloy as if it did obey Matthiessen's rule and is the simple sum of the resistivity of the pure host metal at the specified temperature plus the residual resistivity of the alloy. In order to formulate a numerical equation for the resistivity of a dilute alloy that is receptive to rapid and easy use, it is necessary to consider Matthiessen's rule in detail.

Although Matthiessen's rule was formulated from experimental data many years before the advent of modern theory, the rule as a first approximation can be obtained from the Boltzmann-Bloch approach. In the relaxation time approximation, the justification of Matthiessen's rule is particularly simple. In a material containing independently distinct phonon and impurity scattering mechanisms, the total collision rate, R_T, will be the sum of the individual collision rates due to phonons, R_P, and impurities, R_i,

$$R_T = R_i + R_P \tag{54}$$

Since the individual collision rates are expected to be inversely related to the separate relaxation times due to phonons, τ_p, and impurities, τ_i, the total relaxation time would be

$$1/\tau_T = 1/\tau_p + 1/\tau_i \tag{55}$$

The resistivity of an isotropic alloy material is a scalar given by

$$\rho_a = (m/n_a q^2)\left[1/\tau_{p_a}\left(E_{F_a}\right) + 1/\tau_i\left(E_{F_a}\right)\right] \tag{56}$$

where n_a is the number of conduction electrons per unit volume in the alloy and $\tau_{P_a}(E_{F_a})$ and $\tau_i(E_{F_a})$ are the relaxation times in the alloy evaluated at the Fermi energy level of the alloy for the phonon and impurity scattering mechanisms, respectively. The relaxation time due to phonons is dependent only on the temperature

$$\tau_{P_a} = F(T) \tag{57}$$

and the relaxation time due to impurities is a function only of the impurity concentration, c:

$$\tau_i = f(c) \tag{58}$$

The resistivity of the pure metal is

$$\rho_M(T) = (m/n_M q^2) / \tau_{P_M}\left(E_{F_M}\right) \tag{59}$$

where n_M is the number of conduction electrons per unit volume in the pure metal and $\tau_{P_M}(E_{F_M})$ is the relaxation time in the pure metal evaluated at the Fermi energy of the metal. If the solute and solvent have the same valence, then

$$n_a = n_M \tag{60}$$

and if

$$\tau_{P_M} = \tau_{P_a} \tag{61}$$

then a simple manipulation of Equations 56, 59, 60, and 61 will yield

$$\rho_a = \frac{m}{n_M q^2} \left[\frac{1}{\tau_{P_M}\left(E_{F_M}\right)} + \frac{1}{\tau_i\left(E_{F_a}\right)} \right] \tag{62}$$

which is Matthiessen's rule (see Equation 52), where

$$\rho_a = (m/n_M q^2) / \tau_{P_M}\left(E_{F_M}\right) \tag{63}$$

and

$$\rho_i(c) = m/n_M q^2 \tau_i\left(E_{F_a}\right) \tag{64}$$

A simple examination of these equations shows that deviations from Matthiessen's rule results when alloying changes: the number of electrons per unit volume (i.e., $n_a \neq n_M$), the Fermi energy, and τ_p which is related to changes in the phonon vibrational spectra of the host lattice.

The impurity contribution, $\varrho_i(c)$, to the alloy resistivity, $\varrho_a(c,T)$, can be estimated from a relaxation time approximation. The relaxation time due to impurities is given by[28]

$$1/\tau_i = 2\pi n_i v_F \int_0^\pi I(\theta)[1 - \cos\theta]\ \sin\theta\, d\theta \tag{65}$$

where n_i is the number of impurity scattering centers per unit volume, v_F is the electron velocity at the Fermi level (see Equation 12), and $I(\theta)$ represents the scattering difference between impurity and host atoms. The impurity contribution to the resistivity of the alloy would be

$$\rho_i(c) = (mv_F n_i/n_a q^2)2\pi \int_0^\pi I(\theta)[1 - \cos\theta] \sin\theta \, d\theta \qquad (66)$$

where the bracketed term is the total scattering cross section of the impurity. This equation has been used frequently to evaluate the residual resistivity of dilute alloys. When $I(\theta)$ is calculated in the Born approximation, the scattering cross section of an ion is found[28] to be proportional to the square of the ionic charge $(Zq)^2$. This result can be applied to dilute impurity atoms in a solvent metal lattice by employing the simple expedient that scattering cross section of the impurity can be represented by the square of the difference between the ionic charges of the solvent and solute atoms $(\Delta Zq)^2$. The residual resistivity is then proportional to

$$\rho_i \sim n_i[\Delta Zq]^2 \qquad (67)$$

This predicted behavior has been employed with some success in analyzing experimental data of dilute alloys.

The resistivity of dilute alloys can also be derived from the Boltzmann equation without resorting to the relaxation time approximation, and this approach has been used to calculate the DMR. The essential details of the procedure are presented adequately elsewhere[26] and will only be summarized here. The expression for the Boltzmann scattering term is proposed by Ziman[5] to be

$$(df/dt)_{sc} = (1/k_B T) \left[\int [\Omega_k - \Omega_{k'}] \Phi_i(k,k')dk + \right.$$
$$\left. \int [\Omega_k + \Omega_{k'}] \Phi_p(k,k')dk' + \int [\Omega_k - \Omega_{k'}] \Phi_p(k,k')dk' \right] \qquad (68)$$

where Ω_k is a function that is defined by

$$f_k - f_k^0 = -\Omega_k \left[df_k^0/dE_k \right] \qquad (69)$$

Employing the variational principle,[29] the resistivity of a dilute alloy can be shown.[5,6,29] to be given by

$$\rho_a(c,T) = (1/k_B T) \iint \sqrt{\kappa} = \kappa \div^{-1^2} \kappa' \kappa \div \phi \kappa \phi \kappa \div$$

$$\rho_a(c,T) = (1/k_B T) \iint [\Omega_k - \Omega_{k'}]^2 \Phi(k,k')dkdk' / |qv_k\Omega_k(\partial f_k^0/\partial E_k)dk|^2 \qquad (70)$$

An exact expression for the resistivity is difficult to obtain from this equation, since the function Ω_k must be obtained from the Boltzmann equation once $\phi(k,k')$ is known. Instead, an approximate expression for the resistivity can be obtained by the variational principle where, for a given $\phi(k,k')$, the correct Ω_k will yield the smallest values for $\varrho_a(c,T)$. The resulting functional relationship for the resistivity is often separable into temperature- and concentration-dependent terms which then can be used to calculate the deviations from Matthiessen's rule. Attempts have been made to calculate the $\Delta(c,T)$ deviation for a number of different possible operative mechanisms, such as inelastic scattering, anisotropic scattering, changes in the phonon spectrum with alloying, electron-electron scattering, and phonon drag. One noteworthy result of the calculations for inelastic electron scattering by phonons is the prediction of a hump or maximum in the temperature dependence of $\Delta(c,T)$. A thorough treatment of the mechanisms responsible for deviations from Matthiessen's rule is reviewed in the literature.[18]

Several of the predicted features, both of the residual resistivity and of the calculated deviations $\Delta(c,T)$, are worthy of specific consideration since they have a measure of experimental support. The residual resistivity in the Mott treatment of dilute alloys is directly proportional to concentration of impurities n_i and to the square of the difference between the ionic charge of the solute and solvent (see Equation 67). The predicted linear relationship between ϱ_i and n_i has been verified for many dilute alloy systems,[30] and the linearity often extends to concentrations as large as 5%. From the linear region, the residual resistivity per atomic percent solute, K, for various solvent host metals has been determined and the results are presented in Table 3.

The Mott analysis for the $[\Delta Zq]^2$ dependence for ϱ_i is consistent with the experimental behavior observed by Linde[31] in Cu, Au, and Ag alloys containing dilute B group impurities of various valencies. The Linde data suggests that the residual resistivity of these alloys obeys the equation:

$$\rho_i = n_i[A + B(\Delta Z)^2] \tag{71}$$

where A and B are constants that depend on the solvent element and on the row in the periodic chart containing the B group impurity element. The observation that the residual resistivity is dependent on the row position in the periodic chart of the B group impurity is also evident when the impurity element is a transition metal in Cu or Au (Figures 6 and 7).[26] In general, the residual resistivity per atom is larger for noble metals containing 3d rather than 4d transition metal impurities when sister elements in the same column of the periodic chart are compared (see Figures 6 and 7). The resisitivity associated with transition metal impurities in noble metals does not systematically obey Equation 71. The behavior of the residual resistivity per iron group transition metal impurity atom in Zn and Al is shown in Figures 8 and 9. The general form of this behavior for the Fe group impurities in Al is consistent with the predictions of Friedel's[34] partial wave analysis for the scattering of conduction electrons of the host metal lattice which shows that the residual resistivity associated with transition metal impurities is

$$\rho_i = n_i 10[m/n\pi q^2 \hbar N(E)] \sin^2(\pi n_d/10) \tag{72}$$

where $N(E)$ is the density of electron states for the host metal and n_d is the number of unpaired d electrons in the transition metal impurity. Allowing for a partial population of s states, this equation predicts that ϱ_i will first increase as the transition metal impurity goes across the 3d row from Ti ($n_d = 2 - 3$) to Mn ($n_d = 5 - 6$) and then decrease as the impurity goes from Mn to Ni ($n_d = 8 - 9$), provided the impurity behaves nonmagnetically. This predicted single peak response whose maximum value occurs when Mn is the impurity is verified by the observed behavior (see Figures 8 and 9) in Zn- and Al-based alloys where the transition metals are nonmagnetic, but is not verified when Cu is the host metal. In a Cu host lattice, the transition metal impurities retain their magnetic character and the residual resistivity, measured at room temperature to avoid errors due to the Kondo effect, displays a double peak that can be crudely accounted for by an unequal population of spin subbands by the electrons.

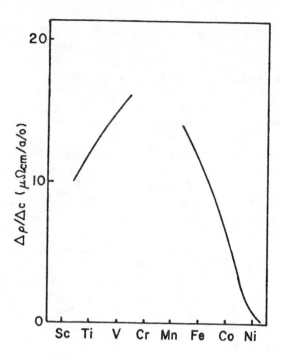

FIGURE 6.   Resistivity due to 1 a/o 3d transition element impurities in a Au matrix at 1.5 to 4.2°K.[32]

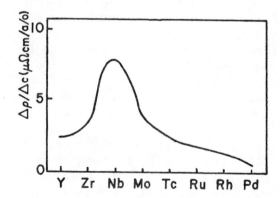

FIGURE 7.   Resistivity due to 1 a/o 4d-transition element impurities in a Au matrix at 1.5 to 4.2°K.[32]

While a comparison of theory and experiment was relatively fruitful for the residual resistivity, a similar comparison between theory and experiment for the deviations from Matthiessen's rule at low temperatures as a function of temperature is not as rewarding. Theory for nonmagnetic alloys suggests that $\Delta(c,T)$ as a function of temperature can exhibit a peak or hump at relatively low temperatures (i.e., $T \simeq \Theta_D/4$). The experimental data of several dilute alloy systems[26] do display this general form, but the magnitude of the experimental results is significantly larger than predicted.

Dilute alloys sometimes show a resistance minimum (see Figure 10) at low temperatures in these alloys. This effect has been attributed by Kondo[36] to an s-d exchange interaction involving the conduction electron and impurity spins that introduces a logarithmic term into the thermal dependence of the resistivity:

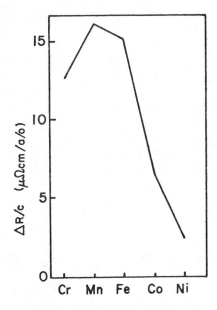

FIGURE 8. Resistivity increase due to 1 a/o 3d transition elements in a Zn matrix.[33]

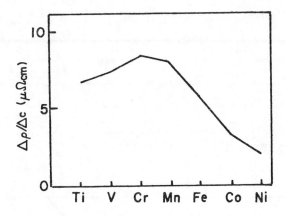

FIGURE 9. Resistivity increase due to 1 a/o 3d transition elements in Al-rich alloys.[33]

$$\rho(c,T) = \rho_i(c) - A \ln T + \rho_p(T) \tag{73}$$

A similar resistance minimum effect is also evident[37] in spin glass alloys like $\underline{Pt}Mn$ where the mechanism responsible for the minimum is associated with spin orientation quenching below the spin glass transition temperature. The resistivity of $\underline{Pt}Mn$ can be described with an equation of the form

$$\rho_{\underline{Pt}Mn} = \rho_{Pt} + c[A + B \ln(T^2 - \Theta^2)^{1/2}]$$

$$= \rho_{Pt} + \Delta\rho(T) \tag{74}$$

where c is the impurity concentration, A and B are constants, and ρ_{Pt} is the resistivity of pure Pt. Θ reflects the life span of spin fluctuation. Depending on the magnitude

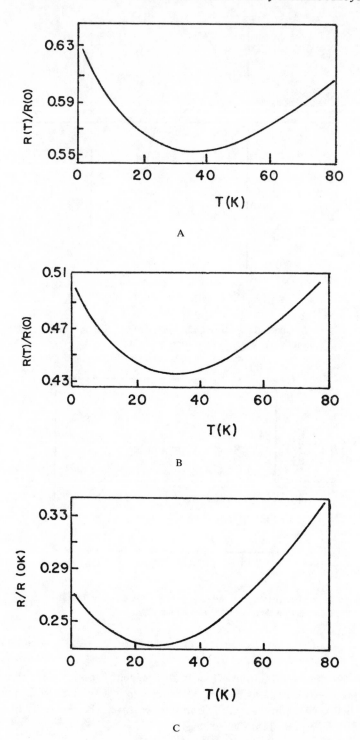

A

B

C

FIGURE 10.   (A) Resistivity of Cu + 0.2% Fe; (B) electrical resistivity of Cu + 0.1% Fe, (C) electrical resistivity of 0.05% Fe in Cu.[35]

Table 4A

VALUES OF K IN $\mu\Omega$CM/ (A/O)[6,31] FOR EVALUATING EQUATION 75

Host element / Impurity	Cu	Ag	Au	Al	Mg	Pb	Ti	Fe	Ni	Pd	Pt
Ag	0.2		0.36	1.1	0.75	0.11				1.5	
Al	1.25	1.95	1.87		2.0		12	5.8			
As	6.7	8.5	8.0								
Au	0.55	0.4						4.9		1.0	
Be	0.62										
Bi		7.3	6.5		8.8	0.96					
Ca	0.3			0.3							
Ce					9.1						
Cd	0.3	0.4	0.63	0.5	0.68	3.02					
Co	6.4		6.1		0.13			0.6			
Cr	4.0		4.3	8.5				5.0	6.4		
Cu		0.1	0.45	0.75			15	6.8	1.5		
Fe	9.3		7.9						6.8		
Ga	1.4	2.3	2.2	0.3							
Ge	3.8	5.5	5.2	0.8							
Hg	1.0	0.8	0.44			2.3					
In	1.1	1.8	1.4		2.0	1.13					
Ir	6.1										1.5
Li				0.94	0.75						
Mg	0.65	0.5	1.3	0.45		4.0					
Mn	2.9	1.6	2.4	6.5	3.8			5.9	2.5		
Mo								5.8			
Nb							1.8				
Ni	1.2	1.1	0.8	0.1				3.2			
P	7.0							6.0			
Pb	3.3	4.6	3.9	1.3	6.3						
Pd	0.9	0.45	0.4								0.6
Pt	2.0	1.5	1.0						0.7		
Rh	4.4		4.15								0.7
Sb	5.5	7.25	6.8			1.24					
Si	3.1			0.7				6.9			
Sn	3.1	4.3	3.36		4.8	0.29	17				
Te	8.0				3.2						
Ti	16		13	5.5							
Tl		2.2	1.9			0.635					
V				8.0				4.6			
W								4.8			
Zn	0.3	0.6	0.95	0.22							
Zr				4.5			1.8				

Note: See individual alloy systems for more details.

Table 4B

VALUE OF THE CONSTANT K′ IN
EQUATION 76 FOR EVALUATING
THE PURE METAL RESISTIVITY AT
TEMPERATURES IN RANGES
$2\,\Theta_R/3 < T < 250°C$

	Cu	Ay	Au	Al	W
K′ ($\mu\Omega$cm K)	530	226	208	1230	1780

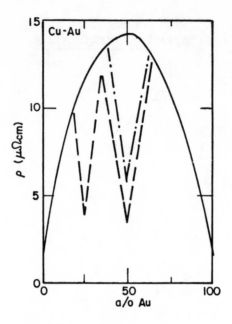

FIGURE 12.   Electrical resistivity of Au-Cu.[39]

interesting dilute alloy systems exhibit some solubility at room temperature. For concentrated alloys, complete substitutional solubility between the constituent atoms occurs relatively infrequently and most binary alloy systems are characterized by the presence of miscibility gaps at room temperature in their phase diagrams. Depending on the equilibrium phase diagram, the temperature, and the composition, any alloy will exist either as a single-phase solid solution or as a mixture of two or more phases, each with their own unique and separate resistivities. To specify the resistivity of an alloy at a given temperature and composition, it is necessary to know if the alloy exists as a single-phase or as a two-phase material at that temperature and composition.

Single-Phase Disordered Alloys

The resistivity of a single-phase concentrated alloy is most easily presented when the two atomic components form a complete series of solid solutions at a given temperature. The isothermal residual resistivity of the solid solution alloys as a function of the fractional atomic composition x_A and x_B for a series of binary A-B solid solution alloys was derived by Nordheim[38] to be

$$\rho_i = Cx_A x_B = Cx_A[1 - x_A] \tag{79}$$

where C is a constant. This proposed relationship is adequately verified by the electrical resistivity of the disordered Cu-Au and Au-Ag binary alloy systems where the predicted symmetry maximum occurs at the equiatomic composition axis (see Figure 12). Allowing for the different resistivities of the pure metal components, the resistivities of the In-Pb, Pt-Pd, and K-Rb binary alloy systems are in general agreement with Nordheim's rule, even though the maximum in the isothermal resistivity does not appear to occur[39] at the equiatomic composition (see Figure 13). The resistivities of the Ag-Pd, Au-Pd, and Cu-Pd system[40,41] are not in reasonable agreement with Nordheim's rule (see Figure 14).

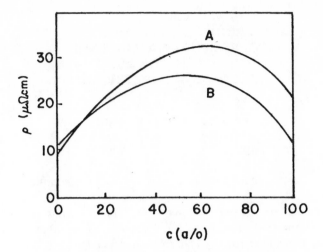

FIGURE 13. Resistivity of In-Pb (A) and Pt-Pd (B) at 75°C.[40]

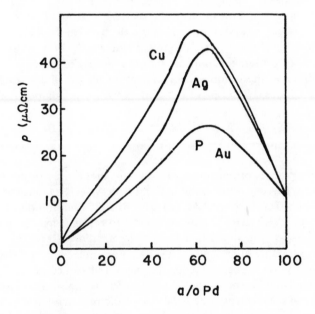

FIGURE 14. Resistivity of Cu-Pd, Ag-Pd, and Au-Pd alloys.[41,42]

These results show that Nordheim's rule cannot be used indiscriminately to determine the residual resistivity of concentrated single-phase alloys. Solid solution alloy systems containing a transition metal constituent generally do not obey Nordheim's rule, while solid solution alloys whose constituents are in the same column of the periodic chart are usually in reasonable agreement with the rule. For those alloy systems obeying the rule, the total resistivity at room temperature is approximately equal to the residual resistivity calculated from the rule, since the phonon contribution to the total resistivity is relatively small at room temperature

$$\rho_a \simeq \rho_i = Cx_A[1 - x_A] \tag{80}$$

Table 5

**APPROXIMATE VALUE OF THE
CONSTANT C IN NORDHEIM'S
RULE (EQUATION 80) FOR
SELECTED ALLOY SYSTEMS
AT 25°C**

	Ag-Au	Cu-Au	Pt-Pd	In-Pb
$C(\mu\Omega cm)$	44	56	104	128

The constant, C, for calculating the resistivity at room temperature of these alloys is given in Table 5.

Several of the solid solution alloy systems involving the transition metals Ni or Pd as one component and the noble metals Cu, Ag, or Au as the other component display a similar asymmetric shape for the composition dependence of the resistivity where the maximum resistivity occur at a transition metal rich composition. This characteristic response can be explained by a consideration of the behavior of the density of states in these alloys as a function of composition. The density of states at the Fermi level enters into the calculation because the scattering probability is proportional to the density of empty energy states into which the electron can be scattered. Nordheim's rule does not allow for changes in the density of states with composition. Low-temperature specific heat measurements indicate that the density of states is a complex function of composition in these noble metal-transition metal solid solution alloys. The density of d-band states, N(E), in Ag-Pd alloys varies with composition according to the relationship

$$N(E) \sim [x_{Pd} - 0.6]^2 \qquad 0.6 < x_{Pd} < 1 \tag{81}$$

where x_{Pd} is the fractional concentration of Pd atoms in the alloy. For Pd compositions less than 60 atomic percent, the d-band is filled which suggests that s-d electron transition does not occur for $x_{Pd} < 0.6$. Armed with this information on the density of d-band states, the total resistivity can be assumed[43] to result from the sum of individal contributions from s-s transitions and from s-d transitions (see Figure 15), and the sum of the two individual contributions yields an asymmetric resistivity-composition behavior that is in good agreement with the measured values of Pd-Ag alloys. The composition dependence of the isothermal resistivity in concentrated alloys involving just transition metal components, the behavior of the resistivity as a function of composition can also be approached by this density of states model provided the alloys are not magnetic. Unfortunately, this approach cannot be easily quantified into a mathematical relationship for calculating the resistivity of these alloys. More refined theories.[44,45] exist for addressing the resistivity of concentrated binary alloys, but they also have failed to give a detailed mathematical relationship upon which the conductivity of an alloy at any composition and temperature can be calculated with any degree of precision. The problem is even more complicated in concentrated ferromagnetic alloys since a two subband model[46] is often required in the analysis and since the resistivity frequently displays anomalies at the critical temperature which could introduce additional complexities into the composition dependence of the resistivity.

A consideration of the density of states was suggested previously to be responsible for the nonlinear resistance at high temperatures in Pd and Pt, and the same reasoning can also be used to explain the anomalous behavior observed in the resistivity as a function of temperature in several concentrated alloys containing transition metal components. The resistivity in these alloys will in some range of temperature decrease with

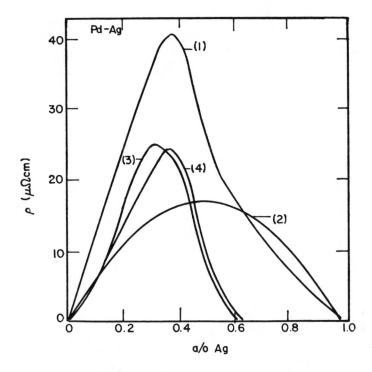

FIGURE 15.   Residual resistivities, ϱ_R, of Pd-Ag alloys. (1) Measured values; (2) K · c(1 − c) fitted to ϱ_R at c = 0.65; (3) calculated values of ϱ_R (s-d scattering); (4) ϱ_R − Kc(1 − c).

increasing temperature (see Figure 16). However, not all experiments show minima in these alloys. (See the discussion of individual alloys.) This negative temperature coefficient of resistance is reminiscent of the response in semiconductors and is not at all consistent with the Bloch predictions. The effect is observed in many concentrated alloys[48] and in some of these alloys may be attributed[44] to a rapid decrease in the density of d-band states near the Fermi level which results in a reduction of the average density of d-band states within $k_B T$ of the Fermi level as temperature increases.

In summary, the resistivity of most concentrated alloys for application purposes has to be obtained from experimental data and cannot, with some exceptions, be calculated from a standardized functional relationship. The exceptions are selected alloys usually involving elemental constituents from the same column in the periodic chart and not involving transition metal components. The resistivity of these alloys as a function of composition can be calculated from Nordheim's rule (see Equation 81), provided the constant C is known at the temperature of interest.

Order-Disorder Effects in Alloys

The previous discussion of concentrated binary alloys showed that the random mixture of two atomic species in a crystal lattice will result in a residual resistivity that displays a bell-shaped compositional dependence that is frequently centered near the equiatomic axis as suggested by Nordheim's rule. In many solid solution alloy systems, a random distribution of the component atoms on the crystal lattice is only realized at elevated temperatures. At lower temperatures in several alloys of specific stoichiometric compositions, the unlike atoms can preferentially populate distinct lattice sites creating an ordered arrangement of the atoms on the crystal lattice sites. The ordering processes occur below a critical temperature and result in a considerable reduction of

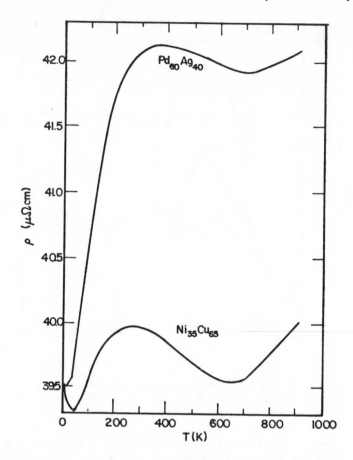

FIGURE 16.   Resistivity of $Pd_{60}Ag_{40}$ and $Ni_{35}Cu_{65}$.[47]

the residual resistivity of the alloy. These features are illustrated for the Au_1Cu_3 alloy system. Above the critical temperature of 380°C, Au_1Cu_3 is a disordered alloy with the Cu and Au atoms randomly arranged on a face centered cubic crystal lattice. The resistivity is large and reflects a large residual resistivity associated with the disordered lattice. Below 380°C, the Au and Cu atoms tend to preferentially arrange themselves on the lattice such that the Cu atoms occupy face-centered sites and the Au atoms reside on the corner sites of the fcc unit cell, creating a long-range order throughout the entirety of the crystal. The resistivity monitors the ordering phenomena by an abrupt drop in its continuity as a function of temperature at 380°C (see Figure 17). This observed behavior of the resistivity as a function of temperature in the order-disorder Au_1Cu_3 system is exactly similar to the response observed in (see Figure 4) pure Ni where the resistance decreases relatively rapidly just below the Curie temperature due to the asset of magnetic ordering.

Many binary disordered alloy systems contain specific compositions that exhibit the long-range order-disorder phenomena. Several examples are Fe_1Co_1, Fe_3Al_1, Mg_1Cd_1, Cr_1Pt_3, and Fe_1Ni_3.[46] Several alloy systems, Co-Pt, Cu-Pt, and Mn-Pt, are like the Au-Cu system which contains two compositions, Au_1Cu_3 and Au_1Cu_1, that order at low temperatures. Resistivity measurements in the equilibrium Au-Cu alloys would then obey Nordheim's rule only at temperatures exceeding the ordering temperatures for both alloys, but at lower temperatures, an appreciable departure from Nordheim's rule is apparent at the composition of the two ordered alloys (see Figure 12), provided the kinetics are sufficiently rapid to permit the atomic rearrangement necessary for

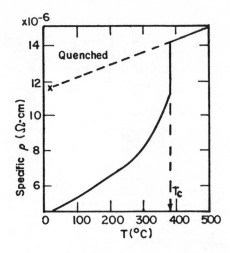

FIGURE 17.   Electrical resistivity of AuCu₃
as a function of temperature.

ordering. The change in resistivity during ordering is sufficiently large that resistivity measurements have been easily employed for monitoring the kinetics of both the long-range ordering process and the short-range ordering process in alloys not normally exhibiting long-range order.

The large resistivity change accompanying the ordering process shows that extreme caution must be exercised in specifying the resistivity of any alloy composition that is receptive to the ordering phenomena. Nordheim's rule cannot be used at all when the temperature is below the critical value unless the alloys have been rapidly quenched to a low temperature where the atoms are immobile. Even the use of established experimental data for the resistivity of an order-disorder alloy below the critical temperature is limited since the attainment of complete long-range order is a kinetic process that differs in each particular ordering alloy system. The resistivity will then be dependent on the state of order in the alloy which makes the specification of the resistivities of alloys exhibiting the ordering phenomena an extremely difficult task that cannot be given by a simple equation.

Lattice Defect Effects

The large increase in the resistivity when an impurity atom replaces a solvent atom in the crystal lattice can result from either the disturbance in the periodicity of the neighboring solvent atoms due to the strain field surrounding the impurity or from the change in potential at the lattice site that results by replacing the solvent atom with the impurity. If the strain field is instrumental in causing the large resistivity changes upon alloying, point and line defects with their accompanying strain fields would be suspected to contribute appreciably to the residual resistivity of an alloy which suggests that the concentration of the defects should be taken into account in any specification of the resistivity of alloys.

The effect of the vacancy concentration on the resistivity of a metal or alloy can be estimated using the results of several calculations for the resistivity of a vacancy[49-50] and a knowledge of the equilibrium vacancy concentration in any material. Depending upon the model and assumptions, the calculated value for the resistivity of a vacancy in a given metal can vary appreciably. Typical values for the vacancy resistivity for several selected metals from both theory and experiment are given in Table 6.

Table 6
RESIDUAL
RESISTIVITIES OF
VACANCIES ($\mu\Omega$cm/
(a/o)[49,50]

Metal	ϱ(Theory)	ϱ(Exp)
Li	1.84	—
Na	1.67	1.9—2.1
K	2.25	—
Rb	1.46	—
Cs	1.07	—
Cu	2.27	—
Ag	1.42	—
Au	1.94	—
Mg	0.87	—
Ca	0.84	—
Zn	0.59	—
Cd	1.06	—
Al	0.86	1.1—3.0
Pb	1.35	3.5

The equilibrium vacancy concentration, n_v, in a metal or alloy at any temperature depends exponentially on the activation energy for formation (Q_F) of a vacancy:

$$n_v/n_o \simeq \exp(Q_F/k_B T) \tag{82}$$

where n_o is the number of lattice sites so that the n_v/n_o ratio is the atomic percent vacancy concentration. Since the formation energy varies from metal to metal and often changes systematically with composition in alloys, it is necessary to know the particular values of this activation energy for calculating the vacancy concentration in a material. The Q_F for Cu is about 20,000 cal/mol[51] which is typical for many metals and alloys. Using this value, the vacancy concentration at room temperature in Cu is calculated from Equation 82 to be 4 (10^{-15}) a/o. Taking the vacancy resistivity in Cu to be 2.3 μ Ωcm/a/o, the vacancy contribution to the room temperature resistivity of Cu is 9.2 (10^{-15}) $\mu\Omega$cm which is insignificantly small compared to the 1.5 $\mu\Omega$cm experimental value for the room temperature resistivity of Cu. Since the vacancy concentration increases exponentially with temperature, vacancies are most likely to make a contribution to the resistivity only at elevated temperatures. The vacancy concentration in Cu near its melting temperature (1350°K) is approximately 10^{-3} a/o which results in a contribution of 2.3 (10^{-3}) $\mu\Omega$cm to the total resistivity 10.2 $\mu\Omega$cm of Cu near its melting temperature. Even near the melting temperature, vacancy effects can be neglected in the specification of the resistivity of a material.

A similar analysis can be employed to estimate the dislocation contribution to the residual resistivity. The dislocation is a line defect, and the nature of the line defect compared with the spherically symmetric nature of the vacancy point defect suggests that the resistance of a dislocation is maximized when the dislocation line axis is perpendicular to the direction of current flow and is minimized when the current flows parallel to the axis. Also, an edge dislocation is suspected to provide more resistance to current flow than a screw dislocation. With these features in mind, the resistivity per unit density of randomly oriented dislocations in Cu are estimated[52] to be

$$0.59 \, (10^{-20}) \, \Omega cm^3 \qquad \text{edge}$$

$$0.18 \, (10^{-20}) \, \Omega cm^3 \qquad \text{screw}$$

Table 7

DISLOCATION SPECIFIC RESISTIVITY (DSR) FOR SELECTED METALS[53]

Metal	K	Cu	Ag	Au	Be	Cd	Al	Zr	Ti	Pb	Bi	Mo	W	Pt	Fe	Ni	Rh
DSR($10^{-19}\Omega cm^3$)	4	1.3	1.9	2.6	34	24	1.5	100	100	1.1	10^5	5.8	7.5	9	10	10	36

Table 8

THE ROOM TEMPERATURE
RESISTIVITY OF ANNEALED,
ϱ_{an}, AND DEEP DRAWN, ϱ_{dd},
ALLOYS[30]

Cr composition (w/o)	ϱ_{an}	ϱ_{dd}
0	1.697	1.738
0.50	1.788	3.635
0.99	1.832	3.160
4.8	1.986	3.701
6.5	2.404	3.902
9.1	2.209	3.940
13	2.324	4.268
14.7	2.52	4.296

Experimentally, the dislocation specific resistivity (DSR) of metals[43] is usually 10^{-17} to 10^{-19} Ωcm^3 (see Table 7).

The dislocation contribution to the residual resistivity for Cu, as an example, can be calculated from this information and compared with the experimentally measured values. The dislocation density in annealed Cu is typically 10^6cm dislocation line per cubic centimeter. The resulting resistivity due to dislocations is 10^{-7} $\mu\Omega cm$ which is again insignificantly small compared to the measured resistivity of about 1.5 $\mu\Omega cm$ at room temperature. In highly cold worked Cu, the dislocation density will attain a maximum value of about 10^{+12} cm of line per cubic centimeter. The dislocation contribution to the resistivity of deformed Cu is then about 10^{-1} $\mu\Omega cm$ which will make a small contribution to the room temperature resistivity, but will contribute influentially to the resistivity of worked Cu at lower temperatures. These numbers suggest that the dislocation contribution to the resistivity of a pure material can be neglected for specification purposes except when the specification involves deformed pure materials employed at low temperatures. The resistivity of annealed and deformed Cu and Cu-Cr alloy at 25°C is presented in Table 8. The prediction that the resistivity will not change significantly at 25°C by the deformation process is confirmed by the experimental data for pure Cu, but is not true for the alloys where a significant increase in the resistivity results from the drawing operation. This data shows that the resistivity of alloys in particular can be influenced by its state of deformation.

Two-Phase Alloys

The resistivity behavior of two-phase alloys is a complex question that not only involves the resistivity and volume fraction of each phase, but also involves the distribution and shape of the phases in the alloy. If the alloy is directionally solidified such that the two phases, α and β, are greatly elongated and lie parallel to each other, the resistivity of the alloy might be expected to be that of two resistors in parallel:

$$1/\rho_a = V_\alpha/\rho_\alpha + V_\beta/\rho_\beta \tag{83}$$

if the current is passed down the elongated direction. The quantities ϱ_a, ϱ_β, V_a, and V_β are the resistivities and volume fractions of the α and β phases, respectively. If the phases lie in sandwiched layers perpendicular to the current flow, the resistivity of the alloy would be expected to be that of resistances in parallel

$$\rho_a = v_\alpha\rho_\alpha + v_\beta\rho_\beta \tag{84}$$

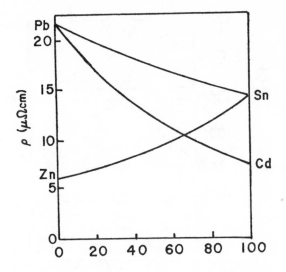

FIGURE 18. Resistivity of Pb-Sn, Zn-Sn, and Pb-Cd
as a function of volume percent.

In general, one phase is usually randomly dispersed as platelets or spherically shaped particles in the other phase, and the resistivity of the two-phase alloy is often a complex function of composition. A typical example of the resistance of a two-phase AB alloy system as a function composition is shown in Figure 18.

To predict the resistivity of a two-phase alloy at a given composition and temperature from any of the proposed equations (see Equations 83 and 84), it would be necessary to estimate the resistivity and volume fraction of each phase. The volume fraction is easily determined either metallographically or from the phase diagram if the density of each phase is known. The resistivity of each phase is more difficult to estimate. The difficulty is minimized if the evaluation is being done around 25°C, if the phase diagram shows a simple miscibility gap with very little mutual terminal solubility in the primary phases, and if the residual resistivity is linear with solute concentrations to the solubility limit. With these assumptions, each of the phases can then be considered as a dilute alloy, and in the vicinity of room temperature, the approximation of the preceding treatment of the dilute alloys (see Equation 77) can be used to determine the resistivity of each phase. For example, the resistivity of the primary phase in an $\underline{A}$B alloy system would be given by:

$$\rho_\alpha = K_B \, ni_B + \frac{K'_A T}{\Theta_{R_A}^2} \qquad (2/3)\Theta_R < T < 250°C \qquad (85)$$

where α is terminal solid solution in the A-rich end of the $\underline{A}$B phase diagram so A is the solvent and B is the solute, while K_B, K'_A, and Θ_{R_A} have their usual designations for pure metal A containing B impurities. If the phase shows even a modest maximum solubility of solute in the solvent lattice, this dilute alloy approximation is not acceptable, and the resistivity of the phase can possibly be estimated with extreme caution from the following treatment of concentrated single-phase alloys involving Nordheim's rule with its obvious limitations and deficiencies.

The effect of the multiphase structure can, however, be much more complicated than these simple equations indicate, as an investigation by Ho and Collings[54,55] showed. These authors studied Ti-Mo alloys with 5 to 20 a/o Mo. The phase diagram

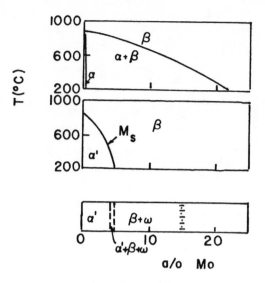

FIGURE 19.   Part of the equilibrium phase diagram for Ti-Mo.

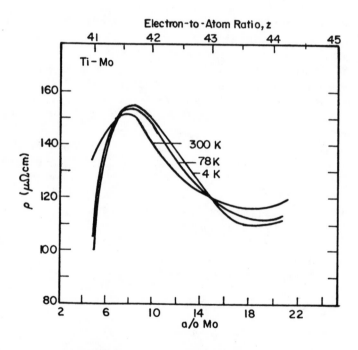

FIGURE 20.   Resistivity of Ti-Mo.[54,55]

of Ti-Mo is given in Figure 19. It shows a continuous series of solid solutions above 900°C from pure β-Ti to Mo. At low temperatures, a two-phase region ($\alpha + \beta$) exists which at equilibrium extends from ~1 a/o Mo to 28 a/o Mo at 200°C. Rapid cooling leads to a martensitic type transformation of β to α'. Additionally, a ω-phase is found in quenched and annealed samples. Figure 20 gives ϱ as a function of composition. The curve shows a maximum near 8 a/o Mo.

Ho and Collings[54,55] then tried to calculate the resistivity considering the resistivity of the individual bulk phases and that of static lattice defects. Figure 21 shows results

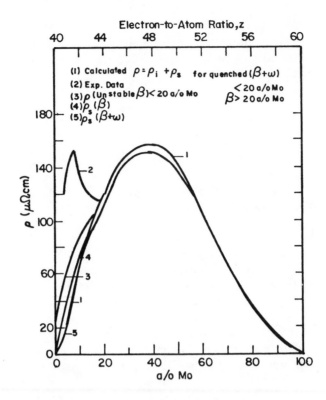

FIGURE 21.   Calculated resistivity of Ti-Mo.[54,55]

of the calculations and experimental data. The full line[1] represents the calculated values of $\varrho = \varrho_{lattice} + \varrho_{impurity}$ of the as quenched $\beta + \omega$-phase and of β-phase Ti-Mo. This curve clearly cannot explain the peak in ϱ_{exp} at 8 a/o Mo. Ho and Collings therefore assumed that an additional term due to matrix-precipitate interfacial scattering was required to explain the observations. This takes place at the surface of the ω-phase precipitates which form in a diffusionless process. The authors discussed these problems in-depth in a series of publications.

CONCLUSIONS

Pure Metals

The resistivity of the monovalent elements are in reasonable agreement with the Bloch-Grüneisen expression. Slight departures are evident at the high and low temperatures where the resistivity is predicted to obey a T^1 and a T^5 dependency. The multivalent elements and the transition metals in particular are often in substantial disagreement with the predictions of the Bloch-Grüneisen formula. The disagreement is most evident in the low-temperature region where the resistivity does not obey the T^5 behavior. The predicted linearity of ϱ vs. T at high temperatures is also violated particularly by the transition metals, yet in the vicinity of room temperature the resistivity of many metals obeys an approximately linear temperature dependence:

$$\rho_p = K'T/\Theta_R^2 \qquad (2/3)\Theta_R < T < 250°C \qquad (76)$$

which then can be employed to evaluate both the resistivity of pure metals and the pure metal contribution to the resistivity of dilute alloys in the vicinity of room temperature.

Dilute Alloys

The resistivity of dilute alloys near room temperature is reasonably represented by Matthiessen's rule at least for application specification purposes. The resistivity of many dilute alloys can be deduced from the proposed relationship shown in Equation 76:

$$\rho_a(c,T) = Kn_i + K'T/\Theta_R^2 \qquad (77)$$

provided $(2/3)\Theta_R < T < 250°C$ and provided the concentration does not in general exceed 1 to 5 a/o impurity, depending upon the alloy system. When the temperature and composition are outside these restrictions, the resistivity of a dilute alloy must be determined from existing experimental data.

Concentrated Alloys

The resistivity of random substitutional solid solution alloys as a function of composition cannot be easily represented by a simple equation, particularly if one or both of the components is a transition metal. The resistivity of these alloys must be determined experimentally or obtained from existing experimental information. For some solid solution alloys involving components that are from the same column in the periodic chart, the resistivity-composition relationship can be calculated from Nordheim's rule (see Equation 80), provided the constant, C, is known and provided the alloy composition is not receptive to the order-disorder phenomena. The resistivity of a random solid solution alloy will decrease appreciably if atomic ordering occurs, but will not be influenced to any great extent by the vacancy concentration. The influence of dislocation on the resistivity can be appreciable, particularly at low temperatures. The resistivities of two-phase alloys is a complex question of how to specify the resistance of each phase and then how to functionally combine the resistivities of both phases to give the resistivity of the composite alloy. Within the discussed limitations, the resistivity of a phase that represents only a very limited solubility of one component in the other can be taken from the dilute alloy approximation (see Equation 76), while the resistivity of a phase that exists with extensive terminal solubility can be, at best, taken from Nordheim's rule calculated at the solubility limit.

REFERENCES

1. Drude, P., *Ann. Phys. (Leipzig)*, 1 (4), 566, 1900.
2. Lorentz, H. A., *Proc. Acad. Sci. Amst.*, 7, 438, 1904.
3. Sommerfeld, A., *Z. Phys.*, 47, 1, 1928.
4. Bloch, F., *Z. Phys.*, 52, 555, 1928.
5. Ziman, J. H., *Electrons and Phonons*, Oxford, London, 1960.
6. Blatt, F. J., *Physics of Electronic Conduction in Solids*, McGraw-Hill, New York, 1968.
7. Ashcroft, N. W. and Mermin, N. D., *Solid State Physics*, Holt, Rinehart & Winston, New York, 1976.
8. Bardeen, J., *Handbook of Physics*, Uhler, E. U. and Odishaw, H., Eds., McGraw-Hill, New York, 1958, chap. 6.
9. Bloch, F., *Z. Phys.*, 59, 208, 1930.
10. Grüneisen, E., *Ann. Phys.*, 16, 530, 1933.
11. Gerritsen, A. N., *Encyclopedia of Physics XIX*, Flügge, S., Ed., Springer-Verlag, Berlin, 1956.
12. DeHaas, W. J. and van der Berg, G. J., *Commun. Phys. Lab. Univ. Leiden*, Suppl., 82a, 1936.
13. MacDonald, D. K. C. and Mendelssohn, K., *Proc. R. Soc. London Ser. A*, 202, 103, 1950.

14. Klemens, P. G. and Jackson, J. L., *Physica,* 30, 2031, 1964.
15. Barber, W. G., *Proc. R. Soc. London Ser. A,* 158, 383, 1937.
16. Schindler, A. I. and Rice, M. J,. *Phys. Rev.,* 164, 759, 1967.
17. deGennes, P. G. and Friedel, J. J., *Phys. Chem. Solids,* 4, 71, 1958.
18. Mott, N. F. and Jones, H., *Theory of Metals and Alloys,* Oxford Press, London, 1936.
19. Masing, G., *Lehrb. der allg. Metallkunde,* Springer-Verlag, Gottingen, 1950.
20. Kawatra, M. P. and Budnick, J. I., *Dynamical Aspects of Critical Phenomena,* Gordon and Breach, New York, 1972, 257.
21. Gerlach, W., *Phys. Z.,* 33, 953, 1932.
22. Craig, P. P., Goldburg, W. I., Kitchens, T. A., and Budnick, J. I., *Phys. Rev. Lett.,* 19, 1334, 1967.
23. Zumsteg, F. C. and Parks, R. D., *Phys. Rev. Lett.,* 24, 520, 1970.
24. Allen, P. B.and Butler, W. H., *Phys. Today,* 31, 44, 1978.
25. Matthiessen, A. and Vogt, C., *Ann. Phys. (Leipzig),* 122, 19, 1864.
26. Bass, J., *Adv. Phys.,* 21, 431, 1972.
27. Gimerle, M. R., Bodel, G., and Rizzuto, C., *Adv. Phys.,* 23, 639, 1974.
28. Mott, N. F., *Proc. Cambridge Philos. Soc.,* 32, 281, 1936.
29. Kohler, M., *Z. Phys.,* 126, 495, 1949.
30. *International Critical Tables,* Vol. 6, Washburn, E. W., Ed., McGraw-Hill, New York, 1929, 138.
31. Linde, J. O., *Ann. Phys.,* 15, 219, 1932.
32. Toyoda, T., *J. Phys. Soc. Jpn.,* 39, 76, 1975.
33. Boato, G., Buto, M., and Rizzuto, C., *Nuovo Cimento,* 45, 226, 1966.
34. Friedel, J., *Can. J. Phys.,* 34, 1190, 1956; *Nuovo Cimento Suppl.,* 7, 287, 1958.
35. Frank, J. P., Manchester, F. P., and Martin, D. L., *Proc. R. Soc. London Ser. A,* 263, 494, 1961.
36. Kondo, J., *Prog. Theor. Phys. (Kyoto),* 32, 37, 1962.
37. Van Vlack, C., *Elements of Materials Science,* 2nd ed. Addison-Wesley, Reading, Mass., 1964, 302.
38. Nordheim, L., *Ann. Phys.,* 9, 641, 1931.
39. Linde, J. O., Johannson, C. H., *Ann. Phys.,* 25, 1, 1936.
40. Kurnakow, N. S. and Zemczuzny, S. F., *Z. Aug. Chem.,* 64, 149, 1909.
41. Swenson, H., *Ann. Phys.,* 14, 699, 1932.
42. Geibel, W., *Z. Anorg. Chem.,* 70, 240, 1911.
43. Coles, B. R. and Taylor, J. C., *Proc. R. Soc. London Ser. A,* 267, 139, 1962.
44. Elk, K., Richter, J., and Christoph, V., *J. Phys. F,* 9, 307, 1979.
45. Levin, K., Velicky, B., and Ehrenreich, H., *Phys. Rev. B,* 2, 1771, 1970.
46. Brouers, F., Vedyayev, A. V., and Giovgino, M., *Phys. Rev. B,* 7, 380, 1973.
47. Ahmad, H. M. and Greig, D., *Phys. Rev. Lett.,* 32, 833, 1974.
48. Mooij, J. H., *Phys. Status Solidi A,* 17, 521, 1973.
49. Manninen, H. and Nieminen, R. M., *J. Phys. F,* B, 2243, 1978.
50. Mori, G. and Yoshioki, S., *Phys. Lett.,* 53A, 405, 1975.
51. Wright, P. and Evans, J. H., *Philos. Mag.,* 13, 521, 1966.
52. Hunter, S. C. and Nabarro, F. R. N., *Proc. R. Soc. London Ser. A,* 220, 542, 1953,
53. Brown, R. A., *J. Phys. F.,* 7, 1283, 1977.
54. Ho, J. C. and Collings, E. W., *Phys. Rev. B,* 6, 3732, 1972.
55. Collings, E. W., *Phys. Rev. B,* 9, 3989, 1974.

Binary Metallic Alloys

SILVER (Ag)

Ag-Al (Silver-Aluminum)

The phase diagram of Ag-Al shows that the Ag-rich α-phase can dissolve up to 20.34 a/o Al (6.0 w/o Al). At 200°C, up to 8.75 a/o Al may dissolve in Ag. The solubility of Ag in Al is small (0.8 a/o Ag at 300°C, 8.9 a/o Ag in Al at 500°C). The β-ε transformation of the Ag_3Al compound is studied with resistivity and dilatometric measurements between 20 and 700°C. The diffusionless β-ε transformation takes place at 610°C, the order-disorder transformation takes place at 300°C to 420°C, and the peritectic reaction takes place 450°C.

The resistivity of Ag-Al alloys has been investigated in References 1 to 27. References 1, 5, and 7 are concerned with the validity of Matthiessen's rule. Most of the publications deal with the precipitation process of metastable phases and the stable ξ-phase of Al-rich alloys.

Borelius and Larsson[26] measured the resistivity of one-phase and two-phase samples over the complete composition range. The alloys were prepared of 99.995% Al and the purest Ag from Kahlbaum. They were homogenized somewhat below the solidus line. The equilibrium, ϱ(T), values are shown in Figures 1, 2, and 3. The kink in the curves indicates the transition from two-phase conditions. Borelius and Larsson investigated in detail the time dependence of ϱ and the effect of quenching. Figure 3 shows that quenching of alloys with 18.0 a/o Ag leads to no transformation product. Only ϱ of the δ-phase is measured. This gives the top line in each of the figures. Annealing at lower temperatures leads to the formation of a second phase and a lowering of ϱ. The resistivity-composition diagram is shown in Figure 4. One should keep in mind that ϱ of a sample with two phases does not depend only on ϱ of each of the two phases, but also on the geometric shape of these phases. Therefore, only ϱ of the α-, γ,-, and β-phase are uniquely defined for the equilibrium state.

Studies on the effect of relatively small amounts of Ag in Al are usually concerned with alloys which are metastable at low temperatures. Sato et al.[28] prepared alloys from 99.999% pure Al and zone refined Al (RRR ≈ 15,000). Appropriate amounts of elements were melted in a graphite boat in a vacuum (≈ 2×10^{-6} Torr) for 12 hr and furnace cooled. After rolling to the final thickness, samples were annealed for 3 hr at 400°C. Table 1 gives the residual resistivities of these samples. $\Delta\varrho/\Delta c = 1.2~\mu\Omega cm/(a/o)Ag$. This is in good agreement with results by Fickett[29] who also obtained $\Delta\varrho/c = 1.2~\mu\Omega cm/(a/o)Ag$.

Turnbull et al.[21] studied the effect of clustering in AlAg alloys prepared from 99.996% Al and 99.994% Ag. Samples were vacuum cast, swaged, drawn to wires, and rolled to ribbons. The samples were quenched to −50°C from the homogenizing temperature and then placed into liquid N_2. The measurements were made while the sample heated slowly (∼20°C/hr).

Kubo et al.[30] measured the resistivity of Ag-Al alloys to determine, analogue to Ag-38 a/o Zn, the martensitic transformation. Powell and Evans[31] used 99.99% pure Al and pure Ag for an investigation of the complete Ag-Al system at room temperature. Alloys in the homogeneous AgAl phase with 0, 1.5%, 3%, 3.89, 4.92, 5.71 w/o Al had the following resistivity values: 1.51, 5.51, 15.40, 18.75, 22.78, 25.4 μΩcm at 0°C, respectively. Panova et al.[18] measured the resistivity of AlAg alloys with up to 0.58 a/o Ag from 1.2 to 300°K. Elements had a purity of 99.999%. After smelting, samples were extruded to wires and annealed at about 500°C. $[\Delta\varrho~(T) - \Delta\varrho(o)]/C \cdot \varrho_{Al}~(273°K)$ shows a maximum near 50°K. It is proposed that the sharp increase in the impurity resistance near 50°K is due to electron impurity atom scattering.

Dukin and Aleksandrov[32] gave a critical account of pseudopotential calculations and the resistivity increase due to 1% impurity. They gave $[\Delta\varrho/\Delta c]_{exp} = 1.1~\mu\Omega cm/(a/o)Ag$

in $\underline{Al}Ag$. Linde[33] gives for $\Delta\varrho/\Delta c$ for dilute alloys a value of 1.95 $\mu\Omega cm/(a/o)Al$ for $\underline{Ag}Al$.

The *International Critical Tables*[6] give the following room temperature resistivity data.

Al (w/o)	ϱ(Annealed) (μΩcm)	ϱ(Tempered) (μΩcm)
0	1.47	1.52
1.3	11.05	10.87
2.7	16.9	16.5
6.1	31.1	31.75
7.7	36.4	34.7
8.6	46.7	33.0
10.0	35.3	32.8
11.1	33.9	32.4
12.9	31.4	30.9
14.1	26.0	27.6
15.7	26.3	29.3
16.7	25.7	29.6
20.0	18.5	27.25
28.1	13.9	20.8
37.5	9.71	17.55
50.6	6.85	13.5
69.6	5.46	8.0
83.1	4.53	5.13
100.0	2.49	2.60

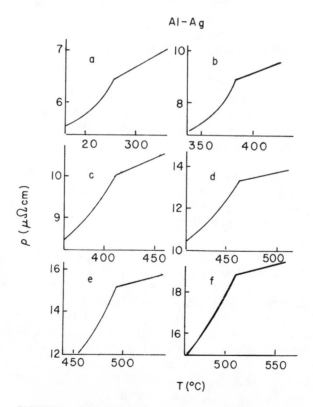

FIGURE 1.   Resistivity of Al-Ag alloys. Changes in slopes indicates phase transformation (a/o Ag). a = 0.64; b = 2.0; c = 2.7; d = 5.0; e = 7.5; f = 10.

45

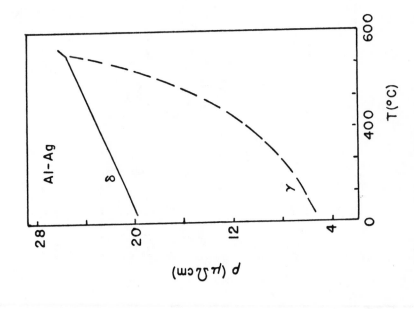

FIGURE 3. Resistivity-temperature diagram of Al + 18 a/o Ag. The solid line gives, at lower temperatures, the resistivity of the quenched (inelastable) δ-phase. The dashed curve gives the resistivity of the γ-phase + δ-phase.

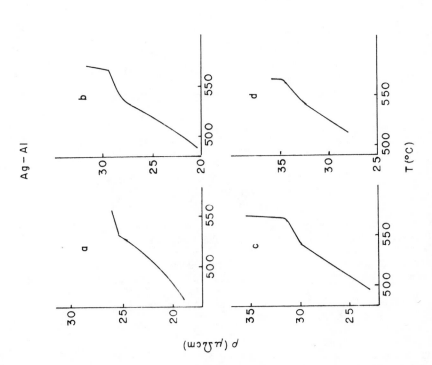

FIGURE 2. Resistivity-temperature diagram of Al-Ag alloys. Changes in slopes indicate phase transition (a/o Ag). a = 18; b = 22; c = 25; d = 30.

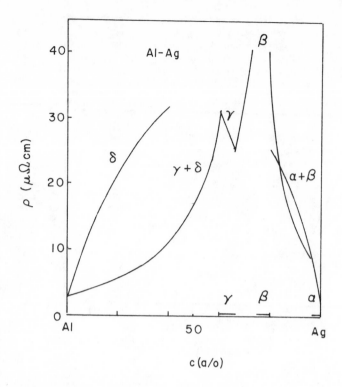

FIGURE 4.   Resistivity isotherms at 20°C for Al-Ag alloys.

Table 1

Al + Ag impurities	Residual resistivities
0 (a/o Ag)	$(1 \times 9 \times 10^{-4} \, \mu\Omega\text{cm})$
0.0096	0.01
0.032	0.036
0.073	0.096
0.12	0.16
0.17	0.23
0.30	0.32
0.42	0.41
0.61	0.48

REFERENCES

1. Hansen, M. and Anderko, K, *Constitution of Binary Alloys,* McGraw-Hill, New York, 1958; Mondolfo, L. F., *Aluminum Alloys, Structure and Properties,* Butterworths, London, 1976.
2. Fujita, Y. and Fukai, Y., *J. Phys. F,* 8(6), 1209, 1978.
3. Merlin, J., Merle, P., Fouquet, F., and Pelletier, J. M., *Scripta Metall.,* 12(3), 227, 1978.
4. Osono, H., Endo, T., Kawata, S., and Kino, T., *Trans. Jpn. Inst. Metals,* 19(2), 69, 1978.
5. Osono, H. and Endo, T., *Trans. Jpn. Inst. Metals,* 19(2), 69, 1978.
6. *International Critical Tables,* Vol. 6, Washburn, E. W., Ed., McGraw-Hill, 1929, 156.
7. Edwards, J. T. and Hillel, A. J., *Philos. Mag.,* 35(5), 1221, 1977.
8. Fujita, Y., Fukai, Y., and Watanabe, K., *J. Phys. F,* 7, L175, 1977.
9. Pawlowski, A., *Arch. Hutn.,* 22(4), 621, 1977.
10. Hiroaka, Y.,., Osamura, K., and Murakami, Y., *Jpn. Inst. Met.,* 40(12), 1223, 1976.
11. Dworschak, F., Monsan, T. H., and Wollenberger, J. *Phys. F,* 6(12), 2207, 1976.
12. Morinaga, M., Nasu, S., Murakami, M., Murakami, Y., Shingu, P. H., and Taoka, T., *J. Mater. Sci.,* 9(8), 1385, 1974.

13. Yoshida, H. and Kodaka, H., *Annu. Rep. Res. Inst. Kyoto Univ.,* 15, 721, 1926.
14. Yonemitus, K., *Phys. Status Solidi,* 13(1), 325, 1972.
15. Evsyukov, V. A., Postniko, V. S., and Sharshakov, I. M., *Fiz. Met. Metalloved.,* 32(2), 431, 1971.
16. Riviere, J. P. and Grilhe, J., *Phys. Status Solidi A,* 5(3), 701, 1971.
17. Murty, K. N. and Vasu, K. I., *Mater. Sci. Eng.,* 7(4), 208, 1971.
18. Panova, G. Kh., Zhernov, A. P., and Kutaitsev, V. I., *Zh. Eksp. Teor. Fiz.,* 56(1), 104, 1969; *JETP,* 29(1), 59, 1969.
19. Borchers, H. and Thym, G., *Z. Metallkd.,* 60(4), 303, 1969.
20. Raman, K. S., Dwarakadasa, E. S., and Vasu, K. I., *Curr. Sci.(India),* 38(6), 130, 1969.
21. Turnbull, D., Rosenbaum, H. S., and Treaftis, H. N., *Acta Metall.,* 8, 277, 1968.
22. Lasek, J., *Phys. Status Solidi B,* 3, K394, 1963.
23. Labusch, R., *Phys. Status Solidi B,* 3, 1661, 1963.
24. Labusch, R., *J. Phys. Radium,* 23, 823, 1962.
25. Federigi, T. and Passari, L., *Acta Metall.,* 7, 422, 1954.
26. Borelius, G. and Larsson, L. E., *Ark. Fys.,* 11(6), 137, 1956.
27. Powell, H. and Evans, E. J., *Philos. Mag.,* 34, 145, 1943.
28. Sato, H., Babauchi, T., and Yonemitsu, K., *Phys. Status Solidi B,* 89, 571, 1978.
29. Fickett, F. R., *Cryogenics,* 11, 349, 1971.
30. Kubo, H., Hamaba, A., and Shimizu, K., *Scr. Metall.,* 9, 1083, 1975.
31. Powell, H. and Evans, E. J., *Philos. Mag. Ser. 7,* 34, 145, 1943.
32. Dukin, V. V. and Aleksandrov, B. N., *Pseudopotential Calculations of the Residual Resistivities of Dilute Solid Alloys Based on Normal Metals, Physics of Low Temperatures,* Akademia Nauk Uk SSR Academy of Science, USSR, Kharkov, 1978, 1.
33. Linde, J. O., *Helv. Phys. Acta,* 41, 1013, 1968.

Ag-As (Silver-Arsenic)

Hansen and Anderko[1] report that up to 8.8 a/o As will dissolve in Ag at 595°C and up to 4.3 a/o As will dissolve in Ag at 300°C. The intermetallic phase, ε (HCP of the Mg (A3) type), probably has the composition $Ag_{89.5}As_{10.5}$. Its stability range is very narrow. Ag is practically insoluble in As.

Linde[2] determined $\Delta\varrho/c$ for AgAs with up to 2.05 a/o As. He used Kahlbaum Ag. A spectroscopic analysis revealed only Ca as impurity. As was also of Kahlbaum purity. He found that $\Delta\varrho/c = 8.5$ $\mu\Omega$cm/(a/o)As in measurements at +18, −100, and −180°C.

REFERENCES

1. Hansen, M. and Anderko, K., *Constitution of Binary Alloys,* McGraw-Hill, New York, 1958.
2. Linde, J. O., *Ann. Phys. Folge,* 5, 14, 353, 1932.

Ag-Au (Silver-Gold)

The phase diagram given by Mondolfo[1] and Hansen and Anderko[2] shows that Ag and Au form a complete series of solid solutions. They suggest that atomic order may be found near the equiatomic composition.

The resistivity of Ag-Au alloys has been studied extensively.[3-27] The effect of short-range order was especially investigated in References 7, 8, 10, 12, 13, 17 18, 20, and 26; the effect of pressure was investigated in References 13 and 18. Most of the studies deal either with only one or two alloys or with dilute alloy systems.

Davis and Rayne[19] measured the resistivity of Au-Ag alloys at 4.2, 77, and 295°K. Their data are shown in Table 1. They found that their resistivity values differed by up to 10% from values obtained by Beckman[32] at 20°C. Nordheim extrapolated from the data by Beckman the residual resistivity, $\Delta\varrho_R$, assuming $\Delta\varrho_R = \varrho$ (alloy, 20°C) − ϱ (pure element, 20°C) and concluded that $\Delta\varrho_R \approx c(1-c)$. Linde's data[31] in 1933 show that the addition of 1 a/o Au increases the resistivity of Ag by the same amount as 1 a/o Ag in Au, namely by 0.38 $\mu\Omega$cm/(a/o). An earlier paper by Linde[28] gives $\Delta\varrho/c$ for AgAu alloys with up to 3.02 a/o Au between −180 and +18°C. He found $\Delta\varrho/c$ = 0.36, 0.36, and 0.35 $\mu\Omega$cm/(a/o)Au for the temperatures +18, −100, and −180°C, respectively. More recent data obtained at liquid He temperatures[34,35] agree with those of Davis and Rayne; they are 0.298 $\mu\Omega$cm/(a/o) for Ag in Au and 0.375 ($\mu\Omega$cm/(a/o) for 1 a/o Au in Ag. Boes et al.[36] measured the resistivity of Ag-Au alloys and of Ag-Au-Yb with less than 0.15 a/o Yb to study the Kondo effect in these alloys. The result of their measurements on Ag-Au alloys at 4 and 77°K is given in Table 2.

Table 1[19]
ALLOY (a/o)

Au	$\varrho(\mu\Omega$cm)		
	4.2°K	77°K	295 ± 3°K
98.0	0.584	1.067	2.886
95.0	1.401	1.890	3.672
90.0	2.801	3.268	5.120
90.0	2.788	3.275	5.113
82.5	4.627	5.111	6.911
75.0	6.239	6.745	8.488
50.0	9.020	9.407	11.147
30.0	7.774	8.127	9.747
20.0	5.835	6.199	7.723
10.0	3.360	3.668	5.135
2.0	0.736	1.047	2.375
0	0	0.270	1.618

Table 2

Au (a/o)	ϱ ($\mu\Omega$cm)	
	(4°K)	(77°K)
0	0	0.28
10	3.32	3.66
20	5.82	6.22
30	7.65	8.05
40	8.82	9.17
50	8.98	9.41
60	8.47	8.95
70	7.70	7.72
80	5.29	5.81
90	2.83	3.36
100	0	0.45

Note: Köster and Störing[24] obtained at room temperature for a $Ag_{50}Au_{50}$ alloy: ϱ = 10.66 $\mu\Omega$cm. Older measurements give similar results,[37-39] as shown by Borelius.[35]

REFERENCES

1. Mondolfo, L. F., *Aluminum Alloys, Structure and Properties,* Butterworths, London, 1976.
2. Hansen, M. and Anderko, K., *Constitution of Binary Alloys,* McGraw-Hill, New York, 1958.
3. Barnard, B. and Caplin, A. D., *Commun. Phys.,* 2, 223, 1977.
4. Nakamura, F, Ogasawara, K., and Takamura, J., *J. Phys. F,* 6, L11, 1976.
5. Schule, W. and Crestoni, G., *Radiat. Eff.,* 29, 17, 1976.
6. Kus, F. W., *J. Phys. F,* 5, 1512, 1975.
7. Schule, W. and Crestoni, G., *Z. Metallkd.,* 66, 728, 1975.
8. Lucke, K. and Haas, H., *Scr. Metall.,* 7, 781, 1973.
9. Knobel, D. W., *Physica,* 69, 87, 1973.
10. Lucke, K., Haas, H., and Schulze, H. A., *J. Phys. Chem. Solids,* 37, 979.
11. Knoble, D. W., Josephson, W. D., and Roberts, L. D., *Physica,* 69, 87, 1973.
12. Lang, E., *Z. Metallkd.,* 64, 56, 1973.
13. Edwards, L. R., *Phys. Status Solidi,* 51, 537, 1972.
14. Schulze, A. and Lucke, K., *Acta Metall.,* 20, 529, 1972.
15. Giardina, M. D., Schule, W., Frank, W., and Seeger, A., *Radiat. Eff.,* 77, 20, 1971.
16. Davis, T. H. and Rayne, J. A., *Phys. Rev. B,* 6, 2931, 1972.
17. Giardina, M. D. and Schule, W., *1st Eur. Conf. on Condensed Matter,* European Phys. Soc., Geneva, 1971, 73.
18. Schule, W., Frank, W., and Seeger, A., *Radiat. Eff.,* 10, 123, 1978.
19. Davis, T. H. and Rayne, J. A., *Phys. Lett. A,* 36, 40, 1971.
20. Huray, P. G., Roberts, L. D., and Thomson, J. O., *Phys. Rv. B,* 4, 2147, 1971.
21. Makhyani, S. P. and Asimow, R. M., *Scr. Metall.,* 4, 63, 1970.
22. Crisp, R. S. and Rungis, J., *Philos. Mag.,* 22, 217, 1970.

23. Boes, J., van Dam, A. J., and Bigvot, J., *Phys. Lett. A*, 28, 101, 1968.
24. Köster, W. and Störing, R., *Z. Metallkd.*, 57, 34, 1966.
25. Asimov, R. M., *Philos. Mag.*, 9, 171, 1964.
26. van der Sijde, B., *Physica,* 29, 599, 1963.
27. Kloske, R. and Kahffman, J. W., *Phys. Rev.*, 126, 123, 1962.
28. Linde, J. O., *Ann. Phys.*, 5, 14, 353, 1932.
29. Auer, H., Riedl, E., and Seemann, H. J., *Physik,* 92, 291, 1939.
30. Broniewski, W. and Wesolowski, R., *C. R.,* 194, 2047, 1932.
31. Linde, J. O., *Ann. Phys.*, 10(1), 52, 1931.
32. Beckman, B., Thesis, Upsala, Sweden.
33. Linde, J. O., *Ann. Phys.*, 17, 52, 1933.
34. Roberts, L. D., Becker, R. L., and Obenshain, F. E., *Phys. Rev. A*, 137, 895, 1965.
35. Borelius, G., *Ann. Phys.*, 77, 109, 1975.
36. Boes, J., van Dam, A. J., and Bigvot, *Phys. Lett. A,* 28, 101, 1968.
37. Matthiessen, A., *Pogg. Ann.*, 110, 190, 1860.
38. Strouhal, V. and Barus, C., *Abh. K. Bohn Gesd. Wiss.*, 6, 12, 1883.
39. Clay, J., *Jahrb. D. Radioakt.*, 9, 383, 1911.
40. Schulze, A. and Lucke, K., *Acta Metall.*, 20, 529, 1972.

Ag-Bi (Silver-Bismuth)

Hansen and Anderko[1] report that 3 a/o Bi will dissolve in Ag. The limit of the solubility range is 900°C, 0.5_2 a/o Bi; 600°C, 2.6_7 a/o Bi; 200°C, 0.3_3 a/o Bi. The solubility of Ag in Bi was found to be negligibly small. The eutectic point lies at 94.92 a/o Bi at 262°C.

Linde[2] determined $\Delta\varrho/c$ for a AgBi alloy with 0.21 a/o Bi. Samples were prepared from spectroscopically pure Ag and "Kahlbaum" purity Bi. Samples were melted in an evacuated glass tube, drawn to wires, and homogenized at 500°C. $\Delta\varrho/c$ = 7.47, 7.55, and 7.61 $\mu\Omega$cm/(a/o)Bi at +18, −100, and −180°C, respectively. Table 1 gives resistivities near room temperature.[3]

Table 1

Bi (w/o)	ϱ ($\mu\sigma$cm)	T (°C)
0.0	1.54	0.0
2.2	3.22	22.9
16.0	19.05	20.3
49.0	33.0	22.4
65.8	46.7	21.4
79.4	62.8	20.1
88.5	86.5	20.3
94.0	93.4	21.6
95.8	116.6	19.9
97.9	135.0	21.4
99.0	138.7	21.4
99.7	137.4	21.3

REFERENCES

1. Hansen, M. and Anderko, K., *Constitution of Binary Alloys,* McGraw-Hill, New York, 1958.
2. Linde, J. O., *Ann. Phys. Folge,* 5(14), 353, 1932.
3. Matthiessen, A., *Philos. Trans. R. Soc. London A,* 150, 161, 1860.

Ag-Cd (Silver-Cadmium)

Hansen and Anderko[1] show that about 40 a/o Cd will dissolve in Ag in the α-phase and about 3 a/o Ag will dissolve in Cd at room temperature in the η-phase.

Köster and Rothenbacher[2] studied the resistivity of a $Ag_{76}Cd_{24}$ sample and found that the resistivity was 7.20 $\mu\Omega$cm for a deformed sample (probably 90% deformed), 6.17 $\mu\Omega$cm for a probably recrystallized sample, and 5.67 $\mu\Omega$cm for a sample in an ordered state obtained by slow cooling. The effect of deformation on the resistivity of CdAg alloys with up to 3.3 a/o Ag was studied.[3,4] Samples were deformed in torsion at liquid N_2 temperatures. The isochronous annealing spectrum is explained on the basis of line and point defects. Of intermetallic phases, the ξ-β transformation was studied,[5] whereas Reference 6 investigated $\varrho(T)$ of CdAg from room temperature to 615°C. Tong and Wayman[7] measured the resistivity of Ag-Cd to determine martensitic transformation.

Dukin and Aleksandrov[8] gave a critical account of pseudopotential calculations and the resistivity increase due to 1% impurity. They gave $[\Delta\varrho/\Delta c]_{exp}$ = 0.36 $\mu\Omega$cm/(a/o)Ag in CdAg. Linde[9] gives for $\Delta\varrho/\Delta c$ for dilute alloys a value of 0.38 $\mu\Omega$cm/(a/o)Cd for AgCd.

REFERENCES

1. Hansen, M. and Anderko, K., *Constitution of Binary Alloys,* McGraw-Hill, New York, 1958.
2. Köster, W. and Rothenbacher, P., *Z. Metallkd.,* 58, 93, 1967.
3. Sprusil, B., Vostry, P., Maisnar, J., and Selecka, M., *Czech. J. Phys. Sect. B,* 28, 813, 1978.
4. Vostry, P., *Phys. Status Solidi A,* 48, K39, 1978.
5. Kittl, J. E., Serebrinsky, H., and Gomex, M. P., *Acta Metall.,* 15, 1703, 1967.
6. Takano, K., *J. Phys. Soc. Jpn.,* 26, 362, 1969.
7. Tong, H. C. and Wayman, C. M., *Scr. Metall.,* 7, 215, 1973.
8. Dukin, V. V. and Aleksandrov, B. N., *Pseudopotential Calculations of the Residual Resistivities of Dilute Solid Alloys Based on Normal Metals, Physics of Low Temperatures,* Akademia Nauk Uk SSR Academy of Science, USSR, Kharkov, 1978, 1.
9. Linde, J. O., *Helv. Phys. Acta ,* 41, 1013, 1968.

Ag-Ce (Silver-Cerium)

Hansen and Anderko[1] report that three intermetallic compounds exist: $CeAg_3$, $CeAg_2$, and CeAg. The resistivity of AgCe is given in Reference 2.

REFERENCES

1. Hansen, M. and Anderko, K., *Constitution of Binary Alloys,* McGraw-Hill, New York, 1958.
2. Kadomatsu, H, Kursiu, M., and Fujiwara, H., *Phys. Lett. A,* 70, 472 1979.

Ag-Cu (Silver-Copper)

The Cu-Ag equilibrium phase diagram is a simple eutectic whose terminal primary solid solution phases, α-Cu and α-Ag, exhibit very limited solubility. The solubility of Ag in α-Cu at 200°C is about 0.06% Ag and the solubility of Cu in α-Ag at 200°C is 0.35% Cu.

The composition dependence of the resistivity of Cu-Ag alloy system at 25°C (see Figure 1) shows that the resistivity is relatively independent of composition for the two-phase alloys spanning the Ag-5% Cu to Ag-95% Cu composition range. In this range, the resistivity is approximately 1.9 $\mu\Omega$cm.

The room temperature resistivity of the α-Cu primary solid solution is a linear function of the silver concentration with a slope of 0.2 $\mu\Omega$cm/(a/o)Ag (see Figure 2). The concentration coefficient of the resistivity at 273°K of the Ag primary phase is dϱ/dc = 0.1 μΩcm/(a/o)Cu. The deviation from Matthiessen's rule for the α-Cu primary phase is dependent on temperature and the Ag content (see Figure 3) as shown by Black.[9] The peak at low temperatures (∼60°K) in the DMR is characteristic of many dilute nonmagnetic alloys.

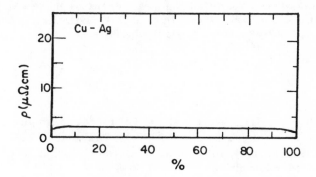

FIGURE 1.   Resistivity (μΩcm) of Ag-Cu alloy system (abscissa gives w/o Cu).[5]

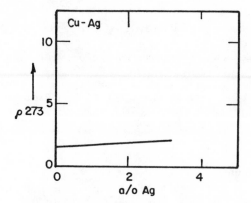

FIGURE 2.   Resistivity at 0°C of the α-Cu primary phase as a function of the Ag concentration.[7]

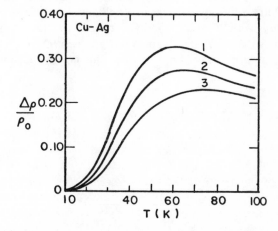

FIGURE 3.   The DMR for α-Cu alloys at various temperatures where (1) $ϱ_o$ = 0.03 μΩcm, (2) $ϱ_o$ = 0.6 μΩcm, and (3) $ϱ_o$ = 0.12 μΩcm.[9]

REFERENCES

1. Kurnakov, N. S., Pushin, N., and Senkowski, N., *Z. Anorg. Chem.*, 68, 123, 1910.
2. Matthiessen, A., *Philos. Trans. R. Soc. London*, 150, 161, 1860.
3. Matthiessen, A. and Vogt, C., *Ann. Phys.*, 2, 19, 1864.
4. Matthiessen, A. and Vogt, C., *Philos. Trans. R. Soc. London*, 154, 167, 1865.
5. *International Critical Tables*, Vol. 6, Washburn, E. W., Ed., McGraw-Hill, New York, 1929, 202.
6. MacDonald, D. K. C. and Pearson, W. B., *Acta Metall.*, 3, 392, 1955.
7. Gerritsen, A. N., *Handbook of Physics*, Springer-Verlag, Berlin, 1956.
8. Linde, J. O., *Ann. Phys.*, 14, 353, 1932.
9. Black, J. E., *Can. J. Phys.*, 52, 345, 1974.

Ag-Ga (Silver-Gallium)

Ag can dissolve up to 11.8 a/o Ga at 200°C.[1] Linde[2] determined $\Delta\varrho/c$ for AgGa alloys with up to 2.14 a/o Ga. $\Delta\varrho/c$ = 2.30, 2.31, 2.25, and 2.27 $\mu\Omega$cm/(a/o)Ga for samples with 0.49, 1.13, 1.74, and 2.14 a/o Ga, respectively, at 18°C; more recently he gave for $\Delta\varrho/\Delta c$ for dilute alloys a value of 1.40 $\mu\Omega$cm/(a/o)Ga.[4]

Dukin and Aleksandrov[3] gave a critical account of pseudopotential calculations and the resistivity increase due to 1% impurity. They gave $[\Delta\varrho/\Delta c]_{exp}$ = 3.3 $\mu\Omega$cm/(a/o)Ag in GaAg.

REFERENCES

1. Hansen, M. and Anderko, K., *Constitution of Binary Alloys*, McGraw-Hill, New York, 1958.
2. Linde, J. O., *Ann. Phys. Folge*, 5(4), 353, 1932.
3. Dukin, V. V. and Aleksandrov, B. N., *Pseudopotential Calculations of the Residual Resistivities of Dilute Solid Alloys Based on Normal Metals, Physics of Low Temperatures*, Akademia Nauk Uk SSR Academy of Science, USSR, Kharkov, 1978, 1.
4. Linde, J. O., *Helv. Phys. Acta*, 41, 1013, 1968.

Ag-Ge (Silver-Germanium)

Small amounts of Ge will dissolve in Ag. Ag is essentially insoluble in Ge.[1] The phase diagram shows one eutectic reaction at 651°C, with $Ag_{90.4}Ge_{9.6}$ + Ge ↔ liquid (25.9 a/o Ge).

Linde[2] found $\Delta\varrho/c$ = 5.46, 5.38, and 5.32 $\mu\Omega$cm/(a/o)Ge for alloys with 0.61, 1.20, and 1.88 a/o Ge at 18°C. This nearly composition-independent value of $\Delta\varrho/c$ implies that AgGe formed solid solutions. Ag was of Kahlbaum purity. Soifer et al.[3] measured the resistivity of alloys with 10 to 32.3 a/o Ge from 20 to 1150°C. Table 1 gives part of the results. Samples were prepared from Ag of grade Sr 999 and Ge of semiconductor purity. Samples are not always single phase. This means that the resistivity is not uniquely defined.

Table 1
SPECIFIC RESISTIVITY OF Ag-Ge ALLOYS

10 a/o Ge		15.4 a/o Ge		17.9 a/o Ge		25 a/o Ge		32.3 a/o Ge	
Temp. (°C)	ϱ ($\mu\Omega$cm)	Temp. (°C)	ϱ ($\mu\Omega$cm)	Temp. (°C)	ϱ ($\mu\Omega$cm)	Temp. (°C)	ϱ ($\mu\Omega$cm)	Temp. (°C)	ϱ ($\mu\Omega$cm)
20	37	20	67	20	87	20	156	—	—
118	38	178	65	100	85	138	150	325	188
172	37	273	64	200	78	248	144	425	173
248	37	350	60	325	75	368	135	500	165
350	37	445	49	473	68	445	116		
408	29	545	53	538	71	545	109		
445	28	625	56	570	69	620	107		
535	28	645	60	605	80				
580	34								
648	41								

REFERENCES

1. Hansen, M. and Anderko, K., *Constitution of Binary Alloys,* McGraw-Hill, New York, 1958.
2. Linde, J. O., *Ann. Phys. Folge,* 5(14), 353, 1932.
3. Soifer, L. M., Izmailov, V. A., and Kashin, V. I., *Teplofiz. Vys. Temp.,* 12, 669, 1974.

Ag-Hg (Silver-Mercury)

The phase diagram of Ag-Hg shows extensive solubility of Hg in Ag (more than 30 a/o Hg dissolves in Ag at room temperature), two intermetallic phases (ε and γ), and practically no solubility of Ag in solid Hg.

Resistivity and thermopower measurements on AgHg from 20 to 300°K show that data above 160°K follow the Nordheim-Gorter relationship.[2,3] Linde[4] found that $\Delta\varrho/c$ = 0.820, 0.785, 0.761, 0.788, and 0.785 $\mu\Omega$cm/(a/o)Hg for alloys with 0.623, 0.72, 1.04, 1.51, and 2.16 a/o Hg, respectively, at 18°C. The nearly composition-independent value of $\Delta\varrho/c$ shows that the samples were in the solid solution range.

Dukin and Aleksandrov[5] gave a critical account of pseudopotential calculations and the resistivity increase due to 1% impurity. They gave $[\Delta\varrho/\Delta c]_{exp}$ = 2.39 $\mu\Omega$cm/(a/o)Ag in HgAg.

REFERENCES

1. Hansen, M. and Anderko, K., *Constitution of Binary Alloys,* McGraw-Hill, New York, 1958.
2. Craig, R. and Crisp, R. S., *Thermoelectricity in Metallic Conductors,* Plenum Press, New York, 1978, 57.
3. Craig, R., Crisp, R. S., Blatt, F. J., and Schroder, P. A., *Thermoelectricity in Metallic Conductors,* Plenum Press, New York, 1977, 51.
4. Linde, J. O., *Ann. Phys. Folge,* 5(14), 353, 1932.
5. Dukin, V. V. and Aleksandrov, B. N., *Pseudopotential Calculations of the Residual Resistivities of Dilute Solid Alloys Based on Normal Metals, Physics of Low Temperatures,* Akademia Nauk Uk SSR Academy of Science, USSR, Kharkov, 1978, 1.

Ag-In (Silver-Indium)

About 20 a/o Li will dissolve in Ag. This is nearly independent of temperature.[1] Ag is essentially insoluble in solid In.

Linde[2] studied only alloys with up to 1.495 a/o In. He found for $\Delta\varrho/c$ values of

1.76, 1.75, and 1.72 $\mu\Omega$cm/(a/o)In for alloys with 0.47, 0.88, and 1.495 a/o In, respectively, at 18°C. The effect of stress on ϱ was determined by Berry and Orehotsky[3] (anelastic piezoresistance), whereas Uemura and Satow[4] studied the order-disorder transformation Ag_3In. $\varrho(T)$ showed a λ-type anomaly with a maximum at 214°C.

Dukin and Aleksandrov[5] gave a critical account of pseudopotential calculations and the resistivity increase due to 1% impurity. They gave $[\Delta\varrho/\Delta c]_{exp} = 1.3\ \mu\Omega$cm/(a/o)Ag in InAg.

REFERENCES

1. Hansen, M. and Anderko, K., *Constitution of Binary Alloys,* McGraw-Hill, New York, 1958.
2. Linde, J. O., *Ann. Phys. Folge,* 5(14), 353, 1932.
3. Berry, B. S. and Orehotsky, J. L., *Philos Mag.,* 9, 467, 1964.
4. Uemura, O. and Satow, T., *Trans. Jpn. Inst. Met.,* 14, 199, 1973.
5. Dukin, V. V. and Aleksandrov, B. N., *Pseudopotential Calculations of the Residual Resistivities of Dilute Solid Alloys Based on Normal Metals, Physics of Low Temperatures,* Akademia Nauk Uk SSR Academy of Science, USSR, Kharkov, 1978, 1.

Ag-Li (Silver-Lithium)

Krill and Lapierre[1] measured the resistivity of LiAg alloys with up to 6 a/o Ag from liquid He temperatures to about 65°K. They investigated deviations from Matthiessen's rule. They found that $d\varrho_o/dc \simeq 1.6\ \mu\Omega$cm/(a/o)Ag.

Dukin and Aleksandrov[2] gave a critical account of pseudopotential calculations and the resistivity increase due to 1% impurity. They gave $[\Delta\varrho/\Delta c]_{exp} = 1.26\ \mu\Omega$cm/(a/o)Ag in LiAg.

REFERENCES

1. Krill, G. and Lapierre, M. F., *Solid State Commun.,* 9, 835, 1971.
2. Dukin, V. V. and Aleksandrov, B. N., *Pseudopotential Calculations of the Residual Resistivities of Dilute Solid Alloys Based on Normal Metals, Physics of Low Temperatures,* Akademia Nauk Uk SSR Academy of Science, USSR, Kharkov, 1978, 1.

Ag-Mg (Silver-Magnesium)

Up to 25 a/o Mg can dissolve in Ag between 200 and 760°C.[1] Intermediate phases will form around the composition MgAg and Mg_3Ag. The maximum solubility of Ag in solid Mg is 4 a/o Ag at the eutectic temperature of 471°C. Resistivity measurements[2-8] were used to study internal oxidation and dispersion hardening processes.

Dukin and Aleksandrov[9] gave a critical account of pseudopotential calculations and the resistivity increase due to 1% impurity. They gave $[\Delta\varrho/\Delta]_{exp} = 0.625\ \mu\Omega$cm/(a/o)Ag in MgAg. Linde[10] gives for $\Delta\varrho/\Delta c$ for dilute alloys a value of 0.5 $\mu\Omega$cm/(a/o)Mg for AgMg.

Gangulee and Beever[8] prepared AgMg alloys with up to 26.4 a/o Mg from 99.99% pure Ag and 99.99% pure Mg. The solidified ingots were homogenized at 823°K for 10 days, swaged, drawn, and annealed at 773°K for 24 hr, followed by a quench or a slow cool (15 days). $\varrho(c)$ of these alloys is shown in Figure 1. The resistivity of alloys of the β-structure at −195°C is given in Table 1.

Smirnov and Kurnakov[7] determined the resistivity and the temperature coefficient of Ag-Mg alloys at 25.25°C. Table 2 gives their results.

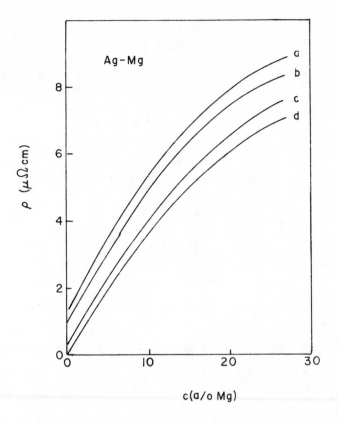

FIGURE 1.   Resistivity isotherms for <u>Ag</u>Mg alloys (°K). a = 273; b = 195; c = 78; d = 4.

Table 1

Mg	44.2	49.5	50.1	(a/o)
$\varrho(-195°C)$	3.446	1.089	0.815	($\mu\Omega$cm)

Table 2
Ag-Mg

Mg (w/o)	ϱ ($\mu\Omega$cm)	α^a (10^{-3}/°C)	Mg (w/o)	ϱ ($\mu\Omega$cm)	α^a (10^{-3}/°C)
0.00	1.637	—	23.1	7.74	2.05
0.04	1.78	3.81	28.00	11.74	1.79
0.1	2.14	3.21	32.21	13.53	2.02
0.17	2.23	—	34.73	17.99	1.78
0.24	2.52	—	35.1	15.01	1.81
0.82	3.41	—	38.28	16.93	2.46
1.55	4.65	—	39.57	16.53	2.90
2.91	6.28	—	39.68	16.23	3.09
6.75	6.85	1.93	43.67	22.05	1.59
7.55	6.86	—	44.39	19.7	2.32
8.89	6.83	1.61	62.67	14.31	2.10
9.60	7.05	1.71	72.31	11.25	—
10.31	7.25	—	79.34	9.91	—
11.31	7.28	—	83.88	9.42	—
13.14	7.67	1.65	89.29	8.54	—
17.11	6.77	1.99	89.46	8.19	2.56
17.87	5.88	2.34	91.75	7.17	—
18.65	4.87	3.10	93.45	7.27	—
19.82	5.73	2.70	96.2	6.87	—
20.26	5.82	—	98.08	5.96	—
21.6	6.27	2.31	100.0	4.81	3.98
23.03	7.82	—	—	—	—

^{a} 25 to 100°C.

REFERENCES

1. Hansen, M. and Anderko, K., *Constitution of Binary Alloys,* McGraw-Hill, New York, 1958.
2. Charrin, L., Combe, A., and Moya, G., *J. Thermal Anal.,* 14, 89, 1978.
3. Besterci, M. and Prochaka, V., *Kovove Mater.,* 12, 258, 1974.
4. Roy, S. K., Coble, R. L., Gangulee, A., and Beever, M. B., *Trans. Metall. Soc. AIME,* 242, 272, 1968.
5. Robinson, P. M. and Beever, M. B., *Acta Metall.,* 13, 647, 1965.
6. Mannchen, W., *Z. Metallkd.,* 23, 193, 1931.
7. Smirnow, W. J. and Kurnakov, N. S., *Z. Anorg. Allg. Chem.,* 72, 31, 1911.
8. Gangulee, A. and Beever, M. B., *Trans. Metall. Soc. AIME,* 242, 272, 1968.
9. Dukin, V. V. and Aleksandrov, B. N., *Pseudopotential Calculations of the Residual Resistivities of Dilute Solid Alloys Based on Normal Metals, Physics of Low Temperatures,* Akademia Nauk Uk SSR Academy of Science, USSR, Kharkov, 1978, 1.
10. Linde, J. O., *Helv. Phys. Acta,* 41, 1013, 1968.
11. Smirnov, B. and Kurnakov, A., *Z. Anorg. Allg. Chem.,* 72, 31, 1911.

Ag-Mn (Silver-Manganese)

About 26 a/o Mn will dissolve under equilibrium conditions in Ag at 300°C, practically no Ag will dissolve in Mn at room temperature, and about 5.5 a/o Ag will dissolve at 471°C, where the eutectic reaction ε + (Mg) $\rightleftharpoons$ liquid (87.5 a/o Mg) takes place. ε is an intermediate phase near Mg_3Ag. The β' intermediate phase around MgAg has a very broad composition range.

Most of the resistivity measurements deal with <u>Ag</u>Mn alloys to study exchange interaction and the Kondo effect.[2-23] References 11 and 22 give ϱ of alloys with higher Mn concentration, whereas Reference 12 gives values of the Lorenz number.

Jha and Jericho[7] measured the resistivity of <u>Ag</u>Mn alloys with up to 1.1 a/o Mn at temperatures from 0.3 to 20°K. These samples were prepared from 6N Cominco Ag and 4N Mn. The alloys were prepared by a succession of dilution of a master alloy. They were melted inductively in a graphite crucible inside a He-filled (1 atm) quartz tube. Electron beam microprobe analysis and resistivity measurements showed that the samples were homogeneous. After machining, the samples were annealed at 700°C for 24 hr in high vacuum. The resistivity is shown in Figures 1 to 3. Dilute alloys show a resistance minimum and, at lower temperatures, the ln T temperature dependence typical for the Kondo effect. The minimum and maximum in the $\varrho(T)$ curves are similar to that found by Malm and Woods for AgMn alloys[10] and for CuMn,[13] CdMn, and ZnMn. Hedgcock and Rizzuto[15] also used 6N pure Ag. $\varrho(T)$ becomes a linear function of T for higher Mn concentration below the maximum in $\varrho(T)$ (see Figure 2). The position of the maximum in ϱ, T_m is in first approximation a linear function of c(Mn).

Otter[16] measured the resistivity of alloys with up to 11.14 a/o Mn from −200 to +500°C. Ag was 99.99⁺% pure and Mn was 99.9⁺% pure (see Figure 4). He found that the resistivity followed essentially a function of the form A + BT, but the resistivities do not follow Matthiessen's rule. This was also observed by Linde.[17-21] $d\varrho/dT$ for alloys with high manganese content at high temperature is less than half that of pure Cu.

Gerritsen and Linde[22] measured the resistivity of AgMn alloys with up to 17.8 a/o Mn from 1 to 273°K. Their samples were prepared by induction heating of pure Ag (RRR ≈ 140) and Mg (probable impurities: 0.004% Si, 0.003% Fe, 0.01% Co) in evacuated and sealed silica tubes. After rolling, samples were annealed at 850°C, drawn into wires, and reannealed for 3 hr at 450°C. The resistivity of the samples is given in Figure 5. Figure 6 shows $r_T = r_T$ (ideal) + r_o, with $r_T = R_T/R_{273}$. R_T is the resistance at the wire temperature, T. This graph shows clearly the resistance minimum. Linde[24] gave for $\Delta\varrho/\Delta c$ for dilute alloys 1.60 $\mu\Omega$cm/(a/o)Mn.

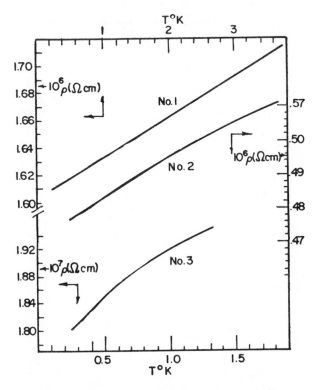

FIGURE 1.   Resistivity of AgMn. (1) 1.12 a/o Mn; (2) 0.332 a/o Mn; (3) 0.558 a/o Mn.

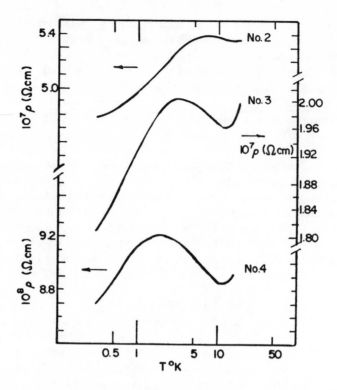

FIGURE 2.   Resistivity of A̲gMn. (2) 0.332 a/o Mn; (3) 0.125 a/o Mn; (4) 0.0558 a/o Mn.

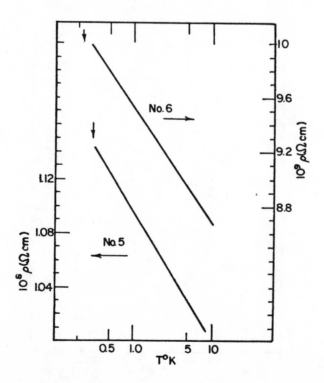

FIGURE 3.   Resistivity of A̲gMn. (5) 0.007 a/o Mn; (6) 0.005 a/o Mn.

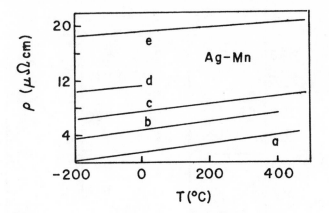

FIGURE 4. Resistivity of <u>Ag</u>Mn. (a) pure Ag; (b) 1.93 a/o Mn; (c) 3.85 a/o Mn; (d) 757 a/o Mn; (e) 11.14 a/o Mn.

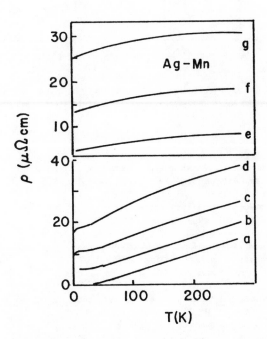

FIGURE 5. Resistivity of <u>Ag</u>Mn alloys with up to 17.8 a/o Mn. (a) 0; (b) 3.2; (c) 7.3; (d) 14.4; (e) 4.75; (f) 10.3; (g) 17.8.

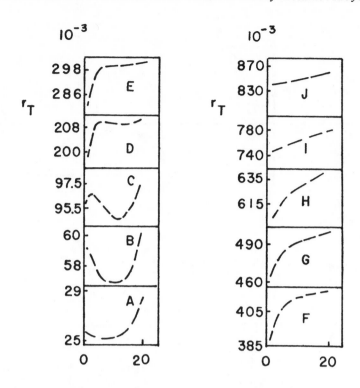

FIGURE 6. r_T of AgMn alloys (a/o Mn). (A) 0.02; (B) 0.05; (C) 0.11; (D) 0.24; (E) 0.40; (F) 0.73; (G) 1.44; (H) 4.35; (I) 10.3; (J) 17.8

REFERENCES

1. Hansen, M. and Anderko, K., *Constitution of Binary Alloys,* McGraw-Hill, New York, 1958.
2. Karagyozyan, A. G. and Artsruni, A. A., *IV. Akad., Nauk Arm. SSR Fiz.,* 12, 473, 1977.
3. Ramos, E. D., *J. Low Temp. Phys.,* 20, 547, 1975.
4. Tindall, D. A., *Phys. Rev. B,* 9, 3113, 1974.
5. Smith, D. A. and Smith, G. B., *Phys. Rev. B,* 4, 191, 1971.
6. Jha, D., *Indian J. Phys.,* 45, 74, 1971.
7. Jha, D. and Jericho, M. H., *Phys. Rev. B,* 3, 147, 1971.
8. Gangulee, A. and Beever, M. D., *Philos. Mag.,* 20, 519, 1969.
9. Chari, M. S. R., *Nuclear Physics and Solid State Physics Symposium Digest,* Powai, India, 1968, 2831.
10. Malm, H. L. and Woods, S. B., *Can. J. Phys.,* 44, 2293, 1966.
11. Köster, W. and Rothenbacher, P., *Z. Metallkd.,* 58, 248, 1967.
12. Chari, M. S. R., *Nature (London),* 193, 968, 1962.
13. Fujiwara, K., Hirabayashi, M., Watanaba, D., and S. Ogawa, *J. Phys. Soc. Jpn.,* 13, 167, 1958.
14. Clarebrough, L. M. and Nicholas, J. F., *Aust. J. Sci. Res. A,* 3, 284, 1950.
15. Hedgcock, F. and Rizzuto, C., *Phys. Rev.,* 163, 517, 1967.
16. Otter, F. A., *J. Appl. Phys.,* 27, 198, 1956.
17. Linde, J. O., *Ann. Phys.,* 10, 52, 1931.
18. Linde, J. O., *Ann. Phys.,* 14, 353, 1932.
19. Linde, J. O., *Ann. Phys.,* 15, 219, 1932.
20. Linde, J. O., *Ann. Phys.,* 30, 151, 1937.
21. Linde, J. O., *Appl. Sci. Res. Sec. B,* 4, 73, 1954.
22. Gerritsen, A. N. and Linde, J. O., *Physica,* 17, 573, 1951.
23. Schmitt, R. W. and Jacobs, I. S., *J. Phys. Chem. Solids,* 3, 324, 1957.
24. Linde, J. O., *Helv. Phys. Acta,* 41, 1013, 1968.

Ag-Pb (Silver-Lead)

Hansen and Anderko[1] state that up to 2.8 a/o Pb will dissolve in Ag at 600°C and 0.35 a/o Pb will dissolve in Ag at 250°C. The solubility of Ag in Pb is very restricted; only 0.19 a/o Ag will dissolve in Pb at the eutectic temperature of 304°C, and 0.04 a/o Ag will dissolve in Pb at 100°C. The phase diagrams show one eutectic reaction and no evidence of intermediate phases.

Cohen et al.[2] measured ϱ of PbAg between −196 to + 240°C. Alloys were prepared from 99.999% Ag and 99.9999% Pb. $\Delta\varrho/\Delta c$(a/o)Ag = 4.7 $\mu\Omega$cm/(a/o)Ag, independent of temperature.

Dukin and Aleksandrov[3] gave a critical account of pseudopotential calculations and the resistivity increase due to 1% impurity. They gave $[\Delta\varrho/\Delta c]_{exp}$ = 3.4 $\mu\Omega$cm/(a/o)Ag in PbAg. Linde[5] gave 5.1 $\mu\Omega$cm/(a/o)Pb for AgPb. Table 1 gives room temperature resistivity data.[4]

Table 1
ROOM
TEMPERATURE
RESISTIVITY
DATA[4]

Pb (w/o)	ϱ ($\mu\Omega$cm)	T (°C)
0.00	1.54	0
0.96	2.294	23.4
14.5	3.225	23.8
32.4	9.85	13.9
49.0	13.17	16.5
65.8	14.41	15.6
78.3	17.15	26.1
95.1	19.11	24.3
97.9	19.47	25.3

REFERENCES

1. Hansen, M. and Anderko, K., *Constitution of Binary Alloys,* McGraw-Hill, New York, 1958.
2. Cohen, B. M., Turnbull, D., and Warburton, W. K., *Phys. Rev. B,* 16, 2491, 1977.
3. Dukin, V. V. and Aleksandrov, B. N., *Pseudopotential Calculations of the Residual Resistivities of Dilute Solid Alloys Based on Normal Metals, Physics of Low Temperatures,* Akademia Nauk Uk SSR Academy of Science, USSR, Kharkov, 1978, 1.
4. *International Critical Tables,* Vol. 6, McGraw-Hill, New York, 1929, 156.
5. Linde, J. O., *Helv. Phys. Acta,* 41, 1013, 1968.

Ag-Pd (Silver-Palladium)

Ag-Pd forms a complete series of solid solutions.[1] There is no evidence of any solid state transformation. Magnetic measurements first suggested that the density of states curve was nearly flat from $Ag_{50}Pd_{50}$ to Ag and rises rapidly with decreasing Ag concentration for alloys with less than 50 a/o Ag. This model was confirmed in specific heat measurements at LHe temperatures.

This led to extensive investigations of electrical transport properties. It showed that the resistivity of Ag-Pd alloys[2-31] does not follow Nordheim's rule. One has to consider that the density of d-states is composition dependent. This modifies the scattering process and is described in detail, e.g., in the work by Coles and Taylor[26] and Dugdale and Guenault.[25]

Figure 1 gives $\varrho(T)$ for several alloys.[21] $\varrho(T)$ is nearly linear in temperature for alloys with up to 30 a/o Pd. The residual resistivity is in this range also a nearly linear function of composition. On the Pd-rich side, Matthiessen's rule breaks down at much lower Ag concentration.

Murani[10] measured the resistivity of Pd-Ag alloys for temperatures up to 25°K. The Pd-rich region with up to 50 a/o Ag shows a resistance minima, in good qualitative agreement with the results by Edwards et al.[14] The next alloy with 62 a/o Ag shows already the positive T^2 dependence. This behavior is explained on the basis of localized d-states, statistical clusters of Pd, localized enhancement effects, or spin fluctuations (see Figure 2).

A high-temperature maximum and minimum of $\varrho(\varrho_{max}$ at about 350°K, ϱ_{min} at about 750°K) (see Figure 16 in the chapter entitled "Introduction") was found for a $Pd_{60}Ag_{40}$ alloy by Ahmed and Greig.[9] They attributed it to a temperature dependent decrease of the impurity resistivity due to a reduction in s-d scattering. Arajs et al.[5] studied similar alloys (Pd, 4N pure; Ag, 5N pure). The arc-melted ingots were sealed in evacuated silica tubes and homogenized at about 1300°K for 60 hr. Wires were drawn, again resealed, annealed at 1200°K for 1 hr, and then rapidly cooled. They could not confirm the maximum and minimum in $\varrho(T)$ found by Ahmed and Greig at elevated temperatures. The data by Arajs et al.[5] are in good agreement with data in most of the earlier works. However Ricker and Pflüger[21] also found a maximum (T ≈ 300°K) and minimum (T ≈ 500°K) in a $Pd_{60}Ag_{90}$ alloy. These samples were annealed in Ar atmospheres for 30 min and quenched in air. The discrepancies in these measurements are really remarkable, considering the purity of Ag and Pd used in these experiments.

Linde[30] gives for $\Delta\varrho/\Delta c$ for dilute alloys a value of 0.44 $\mu\Omega$cm/(a/o)Pd for $\underline{Ag}Pd$. Table 1 gives older room temperature data.[31]

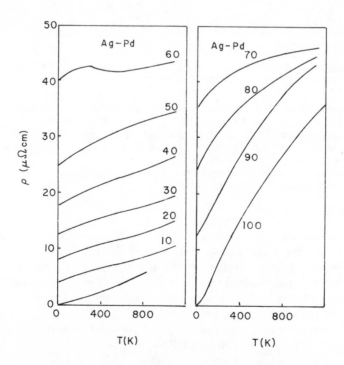

FIGURE 1.   Resistivity of Ag-Pd alloys. Number gives a/o Pd.

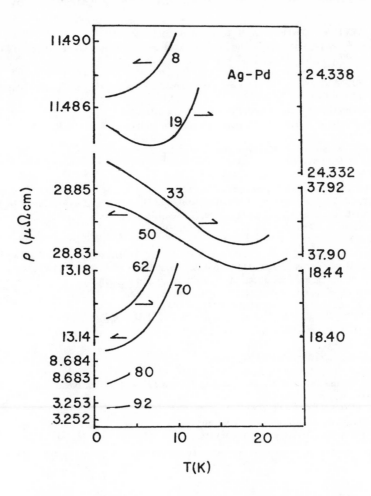

FIGURE 2. Resistivity of Ag-Pd alloys at low temperature. Number gives a/o Ag.[10]

Table 1

Pd (w/o)	ϱ ($\mu\Omega$cm)	α (10^{-3}/°C)
0	1.57	4.1
10	6.05	0.91
20	10.43	0.47
30	15.4	0.36
40	21.75	0.36
50	32.8	0.26
60	42.0	0.03 — 0.07
70	38.77	0.25
80	30.68	0.66
90	20.6	1.17
100	10.56	3.28

REFERENCES

1. Hansen, M. and Anderko, K., *Constitution of Binary Alloys,* McGraw-Hill, New York, 1958.
2. Chupina, L. I., Zenovev, V. E., Polyakova, V. P., Lukyanova, G. P., Savitskii, E. M., and Geld, P. V., *Fiz. Met. Metalloved.,* 48, 476, 1979.

3. Chiu, J. C. H. and Devine, R. A. B, *Can. J. Phys.,* 55, 1218, 1977.
4. Rao, K. V., Rapp, O., Johannesson, Ch., Astrom, H. U., Budnick, J. I., Burch, T. J., and Niculescu, V., *Physica B and C,* 86-88, 2, 831, 1976.
5. Arajs, S., Rao, K. V., Yao, Y. D., and Teoh, W., *Phys. Rev. E,* 15, 2429, 1977.
6. Babic, E. and Marchnic, Z., *Fiz. Proc. 5th Yugoslav Symp. Physics of Condensed Matter Suppl.,* 8, Sarajevo, 222, 1976.
7. Fort, D. and Harris, I. R., *J. Less Common Met.,* 41, 313, 1975.
8. Platov, Yu. M., Pletnev, M. N., Popov, V. I., and Sadykhov, S. I. O., *Fiz. Met. Metalloved.,* 39, 1290, 1975.
9. Ahmed, H. M. and Greig, D., *Phys. Rev. Lett.,* 32, 833, 1974.
10. Murani, A. P., *Phys. Rev. Lett.,* 33, 91, 1974.
11. Foiles, C. L., *AIP Conf. Proc.,* 24, 439, 1974.
12. Szalrenski, A. W. and Barenowski, B., *Phys. Status Solidi A,* 9, 435, 1972.
13. Dekhtyar, I. Ya, Madatova, E. G., and Fedchenko, R. G., *Ukr. Fiz. Zh. (Russ. Ed.),* 16, 915, 1971.
14. Edwards, L. R., Chen, C. W., and Legvold, S., *Solid State Commun.,* 8, 1403, 1970.
15. Kjollerstrom, B., *Solid State Commun.,* 7, 705, 1969.
16. Barnard, R. D., *J. Phys. Chem.,* 2, 2114, 1969.
17. Kagan, G. E., Gottsov, V. A., and Aks, V. Yu, *Ukr. Fiz. Zh. (Russ. Ed.),* 14, 1750, 1969.
18. Chen, C. W., Edwards, L. R., and Legvold, S., *Phys. Status Solidi,* 26, 611, 1968.
19. Westerlund, R. W. and Nicholson, M. E., *Acta Metall.,* 14, 569, 1966.
20. Seemann, H. H. and Rennollet, G., *Z. Phys.,* 196, 486, 1966.
21. Ricker, T. and Pflüger, E., *Z. Metallkd.,* 57, 39, 1966.
22. Schroeder, P. A., Wolf, R., and Woollman, J. A., *Phys. Rev. Sec. A,* 138, 105, 1965.
23. Simon, P. R. F., *Proc. 9th Int. Conf. on Low Temp. Phys.,* Plenum Press, New York, 1965, Part B, 1045.
24. Rado, K. K., *Acta Metall.,* 10, 900, 1962.
25. Dugdale, J. S. and Guenault, A. M., *Philos. Mag.,* 13, 503, 1966.
26. Coles, B. R. and Taylor, J. C., *Proc. R. Soc. London Ser. A,* 267, 139, 1962.
27. Verboven, E., *Philos. Mag.,* 5, 753, 1960.
28. Knook, B., *Physica,* 24, 174, 1958.
29. Otter, F. A., Jr., *J. Appl. Phys.,* 27, 197, 1956.
30. Linde, J. O., *Helv. Phys. Acta.,* 41, 1013, 1968.
31. *International Critical Tables,* Vol. 6, Washburn, E. W., Ed., McGraw-Hill, New York, 1929.

Ag-Pt (Silver-Platinum)

The phase diagram of Ag-Pt shows a very limited degree of solubility of Pt in Ag and several intermediate phases.[1] At 1185°C, 22.5 a/o Ag will dissolve in Pt. Several percent Ag will dissolve in Pt near room temperature, but data are uncertain.

Otter[2] measured the resistivity of Ag-Pt alloys with up to 2.25 a/o Pt from −200 to +500°C. $\delta\varrho/\delta T$ is essential constant and about the same as Cu. Linde[3] gave $\Delta\varrho/c$ = 1.51 $\mu\Omega$cm/(a/o)Pt at −100°C and $\Delta\varrho/c$ = 1.57 $\mu\Omega$cm/(a/o)Pt at 18°C[3] or 1.7 $\mu\Omega$cm/(a/o)Pt.[4] A summary of older data on the room temperature resistivity is given in Table 1.[5]

Table 1

Pt (w/o)	ϱ ($\mu\Omega$cm)
0.0	2.17
10.39	9.18
20.59	18.14
31.46	29.14
37.89	31.10

REFERENCES

1. Hansen, M. and Anderko, K., *Constitution of Binary Alloys,* McGraw-Hill, New York, 1958.
2. Otter, F. A., Jr., *J. Appl. Phys.,* 27, 197, 1956.
3. Linde, J. O., *Helv. Phys. Acta,* 41, 1013, 1968.
4. Johanson, C. H. and Linde, J. O., *Ann. Phys. Folge,* 6(4), 458, 1930.
5. *International Critical Tables,* Vol. 6, McGraw-Hill, New York, 1929, 156.

Ag-Sb (Silver-Antimony)

The phase diagram shows[1] that several percent Sb can dissolve in Ag. Jericho[2] measured ϱ of $\underline{Ag}Sb$ alloys with up to 4.85 a/o in the LHe temperature range. The resistance was constant within 1% for most single-crystal samples except for the alloy with 2.00% Sb, where ϱ = 14.63 $\mu\Omega$cm for T <1°K, then ϱ dropped to 14.45 $\mu\Omega$cm at 1.5°K and stayed constant.

a/o Sn	4.85	5.52	2.00
ϱ(4°K) ($\mu\Omega$cm)	30.82	36.13	14.45

Linde[3] gives for $\Delta\varrho/\Delta c$ for dilute alloys 7.2 $\mu\Omega$cm/(a/o)Sb for $\underline{Ag}$Sb.

REFERENCES

1. Hansen, M and Anderko, K., *Constitution of Binary Alloys,* McGraw-Hill, New York, 1958.
2. Jericho, M. H., *Philos. Trans. R. Soc. London Ser. A,* 257, 385, 1965.
3. Linde, J. O., *Helv. Phys. Acta,* 41, 1013, 1968.

Ag-Sn (Silver-Tin)

Silver can dissolve several percent of Sn at room temperature.[1] Jericho[2] determined ϱ of single-crystal alloys with up to 8% Sn in the liquid He temperature range. Figure 1 shows his experimental results. Linde[3] found that $\Delta\varrho/c$ = 4.3 $\mu\Omega$cm/(a/o)Sn for an alloy with 0.98 a/o Sn for T = −180, −100, and +18°C.

Dukin and Aleksandrov[4] gave a critical account of pseudopotential calculations and the resistivity increase due to 1% impurity. They gave $[\Delta\varrho/\Delta c]_{exp}$ = 1.94 $\mu\Omega$cm/(a/o)Ag in $\underline{Sn}$Ag.

Table 1 gives older resistivity data near room temperature.[5]

Table 1
RESISTIVITY DATA
NEAR ROOM
TEMPERATURE[5]

Sn (w/o)	ϱ ($\mu\Omega$cm)	T (°C)
0.00	1.54	0.0
0.65	4.31	20.7
1.41	6.43	20.6
4	16.92	54.09
51.8	10.73	20.6
68.3	12.23	19.8
76.4	12.58	23.3
86.6	13.35	20.1
90.7	13.32	20.1
92.8	13.31	20.3
95.1	13.43	20.3
96.4	13.43	20.3
99.0	13.54	21.9

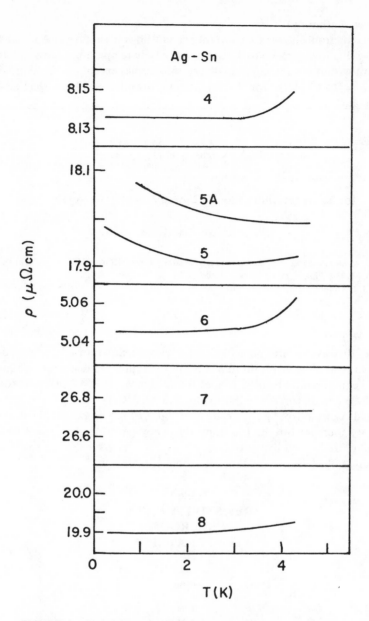

FIGURE 1.   Resistivity temperature diagram of Ag-Sn. (4) 2 a/o Sn (sample annealed at 820°C for 60 hr); (5) 5 a/o Sn (sample annealed at 800°C for 78 hr); (6) 0.8 a/o Sn (sample annealed at 850°C for 70 hr); (7) 8 a/o Sn (sample annealed at 750°C for 90 hr); (8) 3 a/o Sn (sample annealed at 800°C for 130 hr).

REFERENCES

1. Hansen, M. and Anderko, K., *Constitution of Binary Alloys,* McGraw-Hill, New York, 1958.
2. Jericho, M. H., *Philos. Trans. R. Soc. London Ser. A,* 257, 385, 1965.
3. Linde, J. O., *Ann. Phys. Folge,* 5, 14, 353, 1932.
4. Dukin, V. V. and Aleksandrov, B. N., *Pseudopotential Calculations of the Residual Resistivities of Dilute Solid Alloys Based on Normal Metals, Physics of Low Temperatures,* Akademia Nauk Uk SSR Academy of Science, USSR, Kharkov, 1978, 1.
5. *International Critical Tables,* Vol. 6, McGraw-Hill, New York, 1929, 156.

Ag-Tl (Silver-Telluride)

Dukin and Aleksandrov[1] gave a critical account of pseudopotential calulations and the resistivity increase due to 1% impurity. They gave $[\Delta\varrho/\Delta c]_{exp} = 3.7\ \mu\Omega cm/(a/o)Ag$ for TlAg. Linde[2] gives for $\Delta\varrho/\Delta c$ for dilute alloys a value of $2.26\ \mu\Omega cm/(a/o)Tl$ for AgTl.

REFERENCES

1. **Dukin, V. V. and Aleksandrov, B. N.,** *Pseudopotential Calculations of the Residual Resistivities of Dilute Solid Alloys Based on Normal Metals, Physics of Low Temperatures,* Akademia Nauk Uk SSR Academy of Science, USSR, Kharkov, 1978, 1.
2. **Linde, J. O.,** *Helv. Phys. Acta,* 41, 1013, 1968.

Ag-Yb (Silver-Ytterbium)

Murani[1] measured the resistivity of some AgYb alloys which show that Yb ions in Ag cause magnetic scattering of conduction electrons. Figure 1 shows the characteristic minimum in $\varrho(T)$ at low temperatures. The alloys were prepared by melting 6N purity Ag and 2N8 purity Yb and then homogenizing the sample at 800°C. Sample a and b in Figure 1 have a nominal composition of 0.3 a/o Yb. The total Fe + Mn impurity content is less than 0.1 ppm. Stage 1 represents data after an anneal of 3 days at 800°C for sample a and 1 day at 800°C for sample b. A further anneal at 800°C for 3 days for both samples lead to stage 2, and stage 3 corresponds to third anneal at 800°C for 3 days. These anneals lead to a removal of excess Yb.

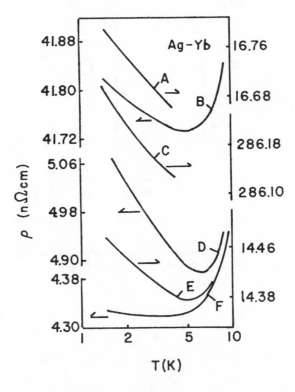

FIGURE 1. Resistivity of AgYb alloys. A = (b), stage 3; B = (b), Stage 2; C = (b), Stage 1; D = (a), Stage 2; E = (a), Stage 1; F = Ag.

REFERENCES

1. **Murani, A. P.,** *Solid State Commun.,* 14, 199, 1974.

Ag-Zn (Silver-Zinc)

The phase diagram of Ag-Zn is similar to the one of Cu-Zn with an α-phase extending from pure Ag to 40.2 a/o Zn at 258°C, several intermediate phases, and a Zn-rich phase which can dissolve less than 1 a/o Ag at room temperature.[1] Resistivity measurements[2-10] were frequently used to study phase transformations. Reference 10 gives for $\Delta\varrho/c$ of <u>Zn</u>Ag single crystal alloys: $\Delta\varrho/c$ = 1.1 $\mu\Omega$cm/(a/o)Ag. Reference 8 gives 0.4 ± 0.02 $\mu\Omega$cm/(a/o)Ag. The temperature dependence of $\varrho(T)-\varrho_o$ for pure Zn and Zn - 2(a/o) Ag is given in Figure 1. Figure 2 gives ϱ of Ag - 48 a/o Zn. Linde[11] gave for $\Delta\varrho/\Delta c$ for dilute alloys a value of 0.63 $\mu\Omega$cm/(a/o)Zn for <u>Ag</u>Zn.

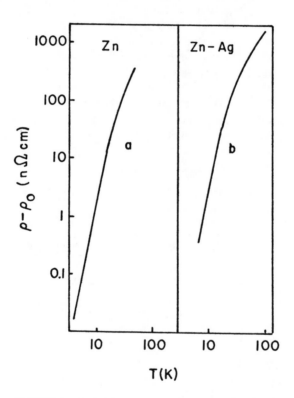

FIGURE 1.   Resistivity of Zn and $Zn_{98}Ag_2$. (a) Pure Zn: ϱ_o = 40.42 nΩcm; (b) 2 a/o Ag: ϱ_o = 1905.6nΩcm.

69

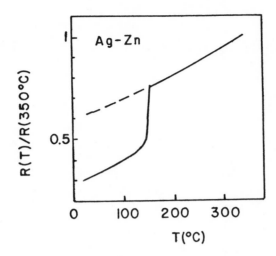

FIGURE 2. Resistivity of $As_{52}Zn_{48}$. Dashed lines: quenched sample, 48 a/o Zn.

REFERENCES

1. Hansen, M. and Anderko, K., *Constitution of Binary Alloys*, McGraw-Hill, New York, 1958.
2. Arias, D. and Kittl, J. E., *Metallography* 11(4), 429, 1978.
3. Kubo, H., Wayman, C. M., and Shimizu, K., *Metall. Trans. A*, 8(3), 493, 1977.
4. Platov, Yu. M., Pletnev, M. N., Popov, V. I., and Sadykhov, S. I., *Fiz. Met. Metalloved.*, 39(6), 1290, 1975.
5. Salvadori, P., Bubuc, E., Krsnik, R., and Rizzuto, C., *J. Phys. F*, 3, L195, 1973.
6. Ayers, J. D, 5th Spring Meet. Metallurgical Soc. of AIME (abstracts only received) Philadelphia, New York, II&155, Metallurgical Soc. AIME, May 29 to June 1, 1973.
7. Bystrov, L. N., Ivanov, L. I., and Platov, Yu. M., *Phys. Status Solidi A*, 7(2), 617, 1971.
8. Farrell, D. E., Tripp, J. H., and Harding, T. J., *Phys. Rev. B*, 1(12), 4533, 1970.
9. Noguchi, S., *J. Phys. Soc. Jpn.*, 17(12), 1844, 1962.
10. Way, H. E., *Phys. Rev.*, 50, 1181, 1936.
11. Linde, J. O., *Helv. Phys. Acta*, 41, 1013, 1968.

ALUMINUM (Al)

Al-Au (Aluminum-Gold)

There is practically no solubility of Au in Al[1] but up to 6 a/o Al will dissolve in Au at 300°C. The compounds $AuAl_2$, $AuAl$, Au_2Al, and possibly Au_5Al_2 form in equilibrium.

Panova et al.[2] measured the resistivity of AlAu alloys with up to 0.13 a/o Au from 1.2 to 300°K. Samples were prepared from 99.999% pure elements, drawn to wires, and annealed at about 500°C. It is proposed that a sharp increase in impurity resistance in AlAu alloys is due to an anomalously perturbed lattice vibration of the so-called quasilocal frequencies in the phonon spectrum.

Köster and Hank[3] studied the Hall constant, thermopower, and the resistivity of ordered (anneal at 150°C for 60 hr), disordered (quench from 450°C), and deformed (90% cold work) AuAl alloys with up to 12 a/o Al. $\varrho(T)$ is for all states a nearly linear function of composition and the same for ordered and disordered alloy ($\varrho = 4.03$ $\mu\Omega$cm for $Au_{99}Al_1$ and 20.3 or 20.7 $\mu\Omega$cm for $Au_{88}Al_{12}$ in the disordered or ordered state, respectively). This gives for $\Delta\varrho/\Delta c = 1.86$ $\mu\Omega$cm/(a/o)Al. Linde[5] obtained 1.87 $\mu\Omega$/(a/o)Al.

Jan and Pearson[4] prepared a $AuAl_2$ compound which has a cubic structure of the fluorite type by melting components in sealed evacuated tubes and passing several molten zones through them. That gave essential single-phase single crystals, except for a few small globular inclusions of a second phase in $AuGa_2$. Figure 1 gives the electrical resistivity of these compounds.

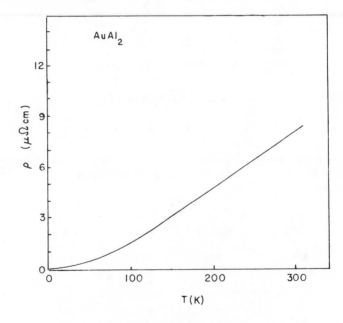

FIGURE 1.   Resistivity of $AuAl_2$.

REFERENCES

1. Hansen, M. and Anderko, K., *Constitution of Binary Alloys,* 2nd ed., McGraw-Hill, New York, 1958; Mondolfo, L. F., *Aluminum Alloys Structure and Properties,* Butterworths, London, 1976.
2. Panova, G. Kh., Zhernov, A. P., Kutaitsev, V. I., *Sov. Phys. JETP,* 29, 59, 1969.
3. Köster, W. and Hank, J., *Z. Metallkd.,* 56, 846, 1947.
4. Jan, J. P. and Pearson, W. P., *Philos. Mag.,* 8, 279, 1963.
5. Linde, J. O., *Helv. Phys. Acta,* 41, 1013, 1968.

Al-Be (Aluminum-Beryllium)

Dukin and Aleksandrov[1] gave a critical account of pseudopotential calculations and the resistivity increase due to 1% impurity. They gave $[\Delta\varrho/\Delta c]_{exp} = 0.4\ \mu\Omega cm/(a/o)Be$ in AlBe.

REFERENCES

1. Dukin, V. V. and Aleksandrov, B. N., *Pseudopotential Calculations of the Residual Resistivities of Dilute Solid Alloys Based on Normal Metals, Physics of Low Temperatures,* Akademia Nauk Uk SSR Academy of Science, USSR, Kharkov, 1978, 1.

Al-Bi (Aluminum-Bismuth)

The *International Critical Tables* gives the following room temperature data for Al-Bi.[1]

Bi (w/o)	ϱ(Annealed) ($\mu\Omega$cm)	ϱ(Tempered) ($\mu\Omega$cm)
0	2.49	2.60
6.52	2.92	3.12
15.56	3.29	3.51
28.7	3.60	3.83
60.8	7.81	8.55
78.4	19.28	19.46
89.4	45.67	45.88
97.05	73.0	91.75
98.5	106.4	120.5
99.4	117.6	142.9
100.0	107.5	117.6
100.0	107.5	164.0

REFERENCES

1. *International Critical Tables,* Vol. 6, McGraw-Hill, New York, 1929, 156.

Al-Ca (Aluminum-Calcium)

The *International Critical Tables* gives the following resistivity values for Al-Ca at room temperature.[1]

Ca (w/o)	ϱ ($\mu\Omega$cm)	Ca (w/o)	ϱ ($\mu\Omega$cm)
3.07	15.8	31.3	118.0
6.0	20.2	32.8	112.0
9.1	27.1	37.7	110.0
12.5	38.4	50.0	141.0
19.4	49.5	60.0	171.0
23.9	99.5	75.0	189.0
25	113.4	78.8	112.0
28.5	117.0		

REFERENCES

1. *International Critical Tables,* Vol. 6, McGraw-Hill, New York, 1929, 156.

Al-Cd (Aluminum-Cadmium)

The two metals are almost immiscible in the liquid state.[1] The solid solubility of Cd in Al is reported to be less than 0.25 a/o at 550°C and less than 0.05 a/o at 200°C. There is practically no solubility of Al in solid Cd.

Gefen et al.[2] studied the effect of hydrostatic pressure on the cubic orthorhombic phase transformation in Au - 47.5 (a/o) Cd. Samples were prepared by melting Au and Cd of 99.999% purity in quartz tubing. $(\Delta\varrho/\Delta T)_p = 0.02$ $\mu\Omega$cm/°C for both phases, and $(\Delta\varrho/\Delta P)_T = -0.04$ $\mu\Omega$cm/K bar, again for both phases. The transformation shows a hysteresis.

Dukin and Aleksandrov[3] gave a critical account of pseudopotential calculations and the resistivity increase due to 1% impurity. They gave $[\Delta\varrho/\Delta c]_{exp} = 0.6$ $\mu\Omega$cm/(a/o)Cd in AlCd.

REFERENCES

1. Hansen, M. and Anderko, K., *Constitution of Binary Alloys*, 2nd ed., McGraw-Hill, New York, 1958; Mondolfo, L. F., *Aluminum Alloys Structure and Properties*, Butterworths, London, 1976.
2. Gefen, Y., Halwany, A., and Rosen, M., *Philos. Mag.*, 28(1), 1, 1973.
3. Dukin, V. V. and Aleksandrov, B. N., *Pseudopotential Calculations of the Residual Resistivities of Dilute Solid Alloys Based on Normal Metals*, Physics of Low Temperatures, Akademia Nauk Uk SSR Academy of Science, USSR, Kharkov, 1978, 1.

Al-Cr (Aluminum-Chromium)

Al and Cr form several intermediate phases.[1] The solubility of Cr in Al is very limited, but more than 28 a/o Al should dissolve in Cr. The resistivity in the Al-Cr alloy system deals essentially only with Al-rich and Cr-rich alloys.[2-23]

The resistivity of AlCr alloys was measured by Kedves et al.[20,22] over an extended temperature range. The alloys were prepared from 99.999% Al and 99.9 + % pure Cr. The temperature dependence of the impurity resistance is given in Figure 1. It can be approximated in the temperature range 78 to 450°K by the equation $\Delta\varrho(T) = \Delta\varrho(0)[1 - (T/\Theta_1)^2]$. $\Theta_1 = 1370 \pm 50$°K for AlCr. The 0°K impurity resistivity is 8.3 $\pm$ 0.1 $\mu\Omega$cm/(a/o)Cr. The value for AlCr is about 15% higher than found by Babic et al.[23] Similar results were found earlier by Caplin and Rizzuto[17] and Boato et al.[21] Caplin and Rizzuto[17] measured $\varrho(T)$ of dilute AlCr alloys and found also that $\varrho(T) = \varrho_0 (1 - T/\Theta)^2$. $\Theta = 1200 \pm 400$°K, and $\varrho_0 = 8.4$ $\mu\Omega$cm/(a/o)Cr. Data were explained on the basis of localized spin fluctuations.

Caplin and Rizzuto[17] studied the validity of Matthiessen's rule on Al-rich alloys. Samples were prepared by quenching from the melt, annealing a few hours just below the melting point, and finally quenching in water.

The resistivity of Cr-rich alloys is characterized by a resistance anomaly due to antiferromagnetic ordering. This leads to a maximum in $\varrho(T)$. The onset of antiferromagnetism is associated frequently with the minimum in $\varrho(T)$ at T just above this maximum or with the minimum in $d\varrho/dT$. Sousa et al.[2] studied electrical transport coefficients of a $Cr_{99.94}Al_{0.06}$ alloy. Figure 2 gives $\varrho(T)$. Figure 3 gives $d\varrho/dT$. They associate the minimum in $d\varrho/\delta T$ at 310°K with the onset of antiferromagnetism.

Arajs et al.[18] measured the resistivity of CrAl alloys over a larger composition range than used by Sousa et al.[2] These alloys had 0.3 to 6.2 a/o Al. Samples were prepared by arc-melting. Figure 4 gives experimental results. Surprisingly, T_N first decreases, then increases with increasing Al concentration (see Table 1). Data are in good agreement with measurements by Köster et al.[9]

Babic et al.[12] measured the low-temperature resistivity of four AlCr alloys with up to 3.6 a/o Cr. The impurity resistivity at low temperatures follows a T^2 dependence.

Figure 5 shows that a second minima exists below the minima found by Arajs et al.[18] This is similar to the findings on Co-Cr alloys.

Chakrabarti and Beck[8] measured the Hall coefficient and the resistivity of Cr-Al alloys with up to 33.3 a/o Al. Samples were prepared by arc-melting electrolytic chromium of approximately 99.95% purity and Al with a purity better than 99.99%. Figures 6 and 7 give $\varrho(T)$ curves. Alloys with about 15 to 25 a/o Al may be narrow gap semiconductors, since they have negative $d\varrho/dT$ and high resistivity values.

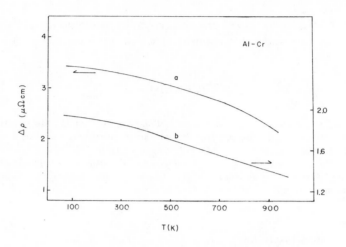

FIGURE 1.   Temperature dependence of the impurity resistivity of <u>Al</u>Cr Alloys with (a) 0.42 a/o Cr and (b) 0.23 a/o Cr.

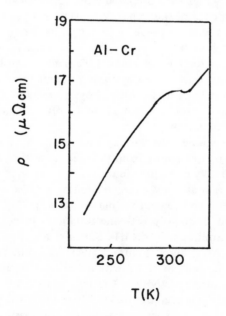

FIGURE 2.   Resistivity of <u>Al</u>Cr with 0.06 a/o.

75

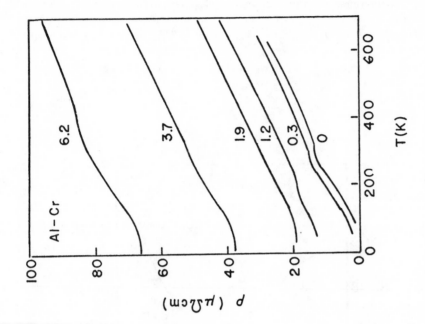

FIGURE 4.   Resistivity of A̲lCr alloys with up to 6.7 a/o Cr. (Number gives a/o Cr.)

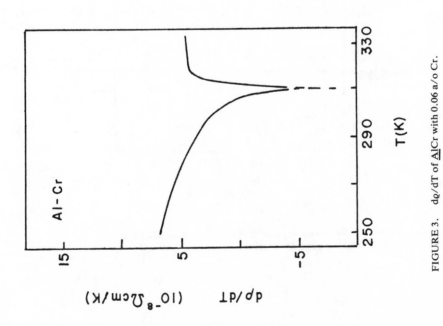

FIGURE 3.   dρ/dT of A̲lCr with 0.06 a/o Cr.

Table 1

Al concentration (a/o)	$T_N(°K)$	$\varrho(4.2°K)\,\mu\Omega cm$
0	312 ± 1	0.08
0.3	300 ± 5	2.32
1.2	202 ± 5	12.27
1.9	310 ± 10	19.14
3.7	338 ± 10	37.34
6.2	440 ± 10	64.32

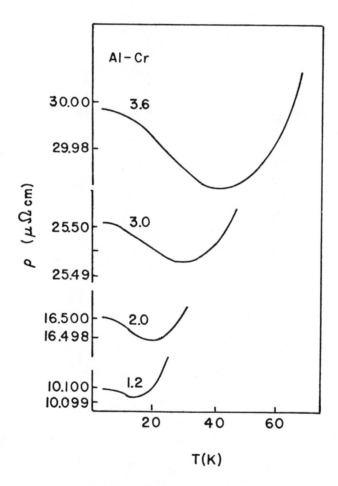

FIGURE 5.   Resistivity of A̲lCr alloys with up to 3.6 a/o Cr below 70°K. (Number gives a/o Cr.)

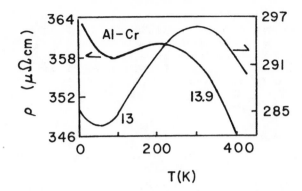

FIGURE 6.　Resistivity of Al-Cr alloys with 13 and 13.9 a/o Al.

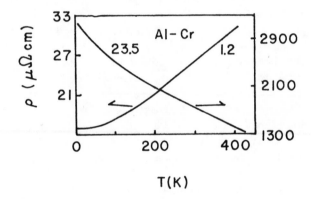

FIGURE 7.　Resistivity of Al-Cr alloys with 1.2 and 23.5 a/o Al.

REFERENCES

1. Hansen, M. and Anderko, K., *Constitution of Binary Alloys,* 2nd ed., McGraw-Hill, New York, 1958; Mondolfo, L. F., *Aluminum Alloys Structures and Properties,* Butterworths, London, 1976.
2. Sousa, J. B., Pinto, R. S., Amado, M. M., Moreira, J. M., Braga, M. E., Ausloos, M., and Balberg, I., *Solid State Commun.,* 31, 209, 1979.
3. Elk, K., Richler, J., and Christoph, V., *Phys. Status Solidi B,* 91, K39, 1979.
4. Beekmans, N. M., Brouha, M., Van Daal, H. J., Lee, M. J. G., Perz, J. M., and Fawcett, E., *Trans. Met.* 15, 429, 1978.
5. Garg, N. K., Singh, N., and Ramachandran, T. R., *J. Mater. Sci.,* 13, 1136, 1978.
6. Sousa, J. B., Amado, M. M., Pinto, R. P., Moreira, J. M., Braga, M. E., Ausloos, M., *J. Phys. Colloq.,* 39, C-6, Pt. 2, 1978.
7. Kedves, F. J., Hordos, M., and Schuszter, F., *Phys. Status Solidi A,* 38, K123, 1977.
8. Chakrabarti, D. J. and Beck, P. A., *J. Phys. Chem. Solids,* 32, 1609, 1971.
9. Köster, W., Wechtel, E., and Grube, K., *Z. Metallkd.,* 54, 393, 1963.
10. Hamzic, A., Babic, E., and Leontic, B., *Mater. Sci. Eng.,* 23, 271, 1976.
11. Babic, E., Krsnik, R., Leontic, B., Vucic, Z., and Zoric, J., *1st Eur. Conf. on Condensed Matter Summaries,* European Phys. Soc., Geneva, 1971, 36.
12. Babic, E., Krsnik, R., and Rizzuto, C., *Solid State Commun.,* 13, 1027, 1973.
13. Kedves, F. J., *Phys. Status Solidi A,* 14, 561, 1972.
14. Babic, E., Ford, P. J., Rizzuto, C., and Salamoni, E., *Solid State Commun.,* 11, 519, 1972.
15. Ichikawa, B. and Ohashi, T., *Trans. Jap. Inst. Met.* 12, 179, 1971.
16. Toth, J., Report KFKI-71-75, Hungarian Academy of Science, Budapest, 1971, 8.
17. Caplin, A. D. and Rizzuto, C., *Phys. Rev. Lett.,* 21, 746, 1968.

18. Arajs, S., Reeves, N. L., and Anderson, E. E., *J. Appl. Phys.*, 42, 1691, 1971.
19. Caplin, A. D. and Rizzuto, C., *Aust. J. Phys.*, 24, 309, 1971.
20. Kedves, K. J. and Gergely, L., *Phys. Status Solidi A*, 38, K31, 1967.
21. Boato, G., Bugo, M., and Rizzuto, C., *Nuovo Cimento*, 45, 226, 1966.
22. Kedves, F. J., Hordos, M., and Gergely, L., *Solid State Commun.*, 11, 1067, 1972.
23. Babic, E., Krsnik, R., Leontic, B., Ocko, M., Vucik, Z., and Zorik, I., *Solid State Commun.*, 10, 691, 1972.

Al-Cu (Aluminum-Copper)

The phase diagram of Al-Cu[1,2] shows that up to 2.53 a/o Cu can dissolve in Al at 547°C, but only 0.18 a/o Cu can dissolve at 267°C, and 19.6 a/o Al can dissolve in Cu at 565°C. Reference 1 lists ten intermediate phases. Resistivity measurements[3-82] are mostly used to study transformation processes, e.g., the formation of Gunier-Preston zones.

Zubchenko et al.[3] studied the concentration dependence of the residual resistivity of Cu-rich alloys with 6 to 17 a/o Al, disordered by plastic deformation at room temperature or ordered by annealing. These alloys show in the CuAl single-phase solid solution range a complex behavior.[4,8,19,20] Zubchenko et al.[3] prepared alloys from elements of 99.99% purity in an arc furnace. Samples were several times remelted and annealed at 900°C for 125 hr. They were then forged and drawn and residual resistivities were determined. Short-range order (SRO) was studied with electrical resistivity measurements[9-18] on alloys with up to 18 a/o Al. The conclusions reached in these studies were that the SRO structure in α-CuAl is heterogeneous. Two distinct processes seem to occur which cannot be explained with a statistical model of homogeneous SRO. The model of disperse order seems to be best suited for interpreting the results.[18] At high Al concentration ($\sim$18 a/o Al), the disperse order seems to change to a microdomain structure.

Bradley and Stringer[21] measured Hall coefficients and resistivities of Al-rich alloys with up to 1.5 a/o Cu. Samples were prepared from 5N elements. They were homogenized, cold rolled, solution treated at 500°C, and quenched immediately before measurements. Room temperature measurements (20°C) gave the following values.

Alloy	$\varrho(\mu\Omega cm)$
Al + 0.2 a/o Cu	2.84
0.4	3.02
0.5	3.15
0.6	3.17
0.8	3.36
1	3.55
1.2	3.65
1.5	3.74

Papastaikoudis et al.[22] studied the resistivity of dilute AlCu alloys between 4.2 and 360°K. The phonon part of the resistivity at low temperatures follows a power law of the form ϱ(thermal) α T^n, with $3 < n < 3.6$. The residual resistivity per atomic percent Cu was 0.7 $\mu\Omega cm/(a/o)$Cu. This is lower than the calculated value of 0.810 $\mu\Omega cm/(a/o)$Cu.[23] The temperature at which the maximum deviations from Matthiessen's rule occurs shifts to higher values with concentration as $c^{1/5}$. Deviations have negative slopes at high temperatures.

The resistivity of the Θ phase (Al$_2$Cu) is of the order of 5 to 6 $\mu\Omega cm$ at 250°K and reaches 15.8 $\mu\Omega cm$ at 800°K.[2] Linde[83] gives for $\Delta\varrho/\Delta c$ for dilute alloys 1.25 $\mu\Omega cm/(a/o)$Al for CuAl.

The *International Critical Tables*[84] gives the following room temperature resistivity data.

Cu (w/o)	ϱ(annealed) ($\mu\Omega$cm)
0	2.49
3.80	3.23
8.66	3.59
18.5	3.94
28.2	4.31
39.2	5.08
48.4	5.75
51.3	6.06
54	6.58
60.5	7.09
66.9	7.69
70.3	8.27
74.4	10.20
77.1	11.6
78	11.3
78.6	28.1
79.8	28.9
82.6	21.2
86.1	15.8
89	13.7
90.5	11.8
93.1	10.85
95.3	9.95
98.8	5.85
100	1.54

REFERENCES

1. Hansen, M. and Anderko, K., *Constitution of Binary Alloys,* 2nd ed., McGraw-Hill, New York, 1958.
2. Mondolfo, L. F., *Aluminum Alloys Structure and Properties,* Butterworths, London, 1976.
3. Zubchenko, V. S., Kulish, N. P., and Petrenko, P. V., *Phys. Met. Metallogr.,* 47(3), 31, 1979.
4. Panin, V. Ye, Fadin, V. P., and Kutznetsov, L. D., *Fiz. Met. Metalloved.,* 19, 316, 1965.
5. Panin, V. Ye and Zenkova, E. K., *Dokl. Akad. Nauk SSSR,* 129, 1024, 1959.
6. Borie, B. and Sparks, C., *J. Acta Cryst.,* 17, 827, 1964.
7. Williams, R. O., *Met. Trans.,* 5, 1843, 1974.
8. Ivenerova, V. I., Katsnelson, A. A., and Revkevitch, G. P., *Izv. Akad. Nauk SSSR Neogr. Mater.,* 2, 823, 1966; *Fiz. Met. Metalloved.,* 26, 1064, 1968.
9. Wechsler, M. S. and Kernohan, R. H., *J. Phys. Chem. Solids,* 7, 301, 1958.
10. Wechsler, M. S. and Kernohan, R. H., *Acta Metall.,* 7, 599, 1959.
11. Wechsler, M. S. and Kernohan, R. H., *J. Phys. Chem. Solids,* 18, 175, 1961.
12. Radelaar, S., *J. Phys. Chem Solids,* 27, 1375, 1966.
13. Radelaar, S., Thesis, Delft.
14. van den Beukel, A., Coremans, P. C. J., and Vriejhoef, M. M. A., *Phys. Status Solidi,* 19, 177, 1967.
15. Trieb, L., Siebinger, K., and Aubauer, H. P., *Scr. Metall.,* 7, 245, 1973.
16. Veight, G., Trieb, L., Püschl, W., and Aubauer, H. P., *Phys. Status Solidi A,* 27, 59, 1975.
17. Veigth, G., Trieb, L., Püschl, W., and Aubauer, H. P., *Scr. Metall.,* 9, 737, 1975.
18. Trieb, L. and Veith, G., *Acta Metall.,* 26, 185, 1978.
19. Köster, W. and Rave, H. P., *Z. Metallkd.,* 55, 750, 1964.
20. Sidora, T. S., Paning, V. Ye, and Bol'shanina, M. A., *Fiz. Met. Metalloved.,* 14, 750, 1962.
21. Bradley, M. J. and Stringer, J. J., *Phys. F,* 4, 839, 1974.
22. Papastaikoudis, C., Papathanasopoulos, K., Rocofylon, E., and Tefles, W., *Phys. Rev.,* 14, 3399, 1976.
23. Caplin, A. D. and Rizzuto, C., *J. Phys. C,* 3, L117, 1970.
24. Esmail, E., Grabec, I., and Krasevec, V., *J. Phys. D,* 12(2), 265, 1979.
25. Gorecki, T., Krol, S., and Tokarski, M., *Trans. Jpn. Inst. Met.,* 20(1), 24, 1979.
26. Eguchi, T., Kinoshita, C., and Tomokiyo, Y., *Trans. Jpn. Inst. Met.,* 19(4), 198, 1978.
27. Sutra, G., *C.R. Acad. Sci. Paris,* 236, 2391, 1953.

28. Kubo, S., Yamauchi, G., and Arita, K., *Jpn. J. Appl. Phys.*, 16(3), 447, 1977.
29. Tomokiyo, Y., Kaku, K., and Eguchi, T., *Trans. Japan Inst. Met.*, 15(1), 39, 1974.
30. Kinoshita, C., Tomokiyo, Y., Matsuda, H., and Eguchi, T., *Trans. Jpn. Inst. Met.*, 14(2), 91, 1973.
31. Tomokiyo, Y., Kaku, K., and Eguchi, T., *J. Jpn. Inst. Met.*, 36(4), 329, 1972.
32. Matsura, K. and Koda, S., *J. Phys. Soc. Jpn.*, 15, 2106, 1960.
33. Sinha, T. H. D. and Prasad, R. S., *Indian J. Pure Appl. Phys.*, 9(4), 246, 1971.
34. Gandig, W. and Warlimont, H., *Z. Metallkd.*, 60(5), 488, 1969.
35. Scattergood, R. O. and Beever, M. B., *Philos. Mag.*, 22,(177), 501, 1970.
36. Köster, W. and Rothenbacher, P., *Z. Metallkd.*, 58(2), 93, 1967.
37. Schule, W., *Phys. Lett.*, 14(2), 81, 1965.
38. Panin, V. E., Zenkova, E. K., and Fadin, V. P., *Fiz. Met. Metalloved.*, 13(1), 86, 1962.
39. Kernohan, R. H. and Wechsler, M. S., *J. Phys. Chem. Solids*, 18(23), 178, 1961.
40. Hibbard, W. R., Jr., *Acta Metall.*, 7(8), 565, 1959.
41. Pecijare, O. and Janssen, S., *C.R. Acad. Sci.*, 245(15), 1228, 1957.
42. Darmois, G. and Janssen, S., *C.R. Acad. Sci.*, 243(20), 1496, 1956.
43. Boutillier, A., *C. R.*, 208, 361, 1939.
44. Schilling, J. S. and Crone, J., *J. Phys. F*, 6(11), 2097, 1976.
45. Netchaev, J. S., *J. Nucl. Mater.*, 69(1,2), 741, 1978.
46. Netchaev, J. S., *J. Nucl. Mater.*, 69(1-2), 810, 1978.
47. Osamura, K., Furuichi, S., Kanbayashi, K., Takamuku, S., and Murakami, Y., *Mem. Fac. Eng. Kyoto Univ.*, 40, 115, 1978.
48. Papustaikoudis, C., *Z. Naturforsch. Teil A*, 32, 327, 1977.
49. Papastaikoudis, C., *Z. Naturforsch. Teil A*, 32, 327, 1977.
50. Osamura, K., Furuichi, S., Kanbayashi, K., Takamuku, S., and Murakami, Y., *J. Jpn. Inst. Met.*, 40, 1228, 1976.
51. Onodera, Y., and Hirano, K., *J. Mater. Sci.*, 11, 809, 1971.
52. Naumov, N. M., Zueva, N. M., and Krastilevskii, A. A., *Defektoskopiya*, 12(3), 116, 1976.
53. Papastaikoudis, C., Papathanasopoulos, K., Rocofylou, E., and Tseifes, W., *Phys. Rev. B*, 14, 3394, 1976.
54. Osamura, K., Furuichi, S., Kanbayashi, K., Takamuku, S., and Murakami, Y., *J. Jpn. Inst. Met.*, 40, 1228, 1976.
55. Pabi, S. K., *Indian J. Technol.*, 14, 345, 1976.
56. Hyun Kee Cho, *J. Korean Inst. Met.*, 14, 368, 1976.
57. Bradley, J. M., *J. Phys. F*, 4, 839, 1974.
58. Yoshida, H. and Kodaka, H., *Ann. Rep. Res. React. Inst. Kyoto Univ.*, 7, 103, 1974.
59. Wahi, R. P. and von Heimendahl, M., *Z. Metallkd.*, 65, 442, 1974.
60. Satyauarayan, R. G., Jayapalan, A. K., and Anantharaman, T. R., *Curr. Sci. India*, 42, 6, 1973.
61. Hori, H. and Hirano, K., *J. Jpn. Inst. Met.*, 37, 142, 1973.
62. Horouchi, T., Liu, K., and Marakami, Y., *J. Jpn. Inst. Met.*, 37, 1120, 1973.
63. Thompson, G. E. and Noble, B., *Met. Sci. J.*, 7, 32, 1973.
64. Satyanarayana K. G., Jayapalan, K., and Anatharaman, T. R., *Curr. Sci. India*, 42, 6, 1973.
65. Mikrukov, V. E. and Karagefyan, A. G., *Inzh. Fiz. Zh.*, 4(12), 90, 1961.
66. Romanova, A. V. and Persion, Z. V., *Ukr. Fiz. Zh.*, 17(10), 1747, 1972.
67. Asano, K. and Hirano, K., *Trans. Jpn. Inst. Met.*, 13, 112, 1972.
68. Lysenko, N. Ya., Paskal, Yu. I., and Shikhmanter, L. D., *Izv. Vu. Fiz. (USSR)*, 1, 145, 1971.
69. Asano, K. and Hirano, K., *J. Jpn. Inst. Met.*, 35, 364, 1971.
70. Yamada, M., Shimiau, T., and Tanaka, K., *J. Jpn. Inst. Met.*, 35, 476, 1971.
71. Kawano, O., Hirouchi, T., Yoshida, H., and Murakami, Y., *J. Jpn. Inst. Met.*, 35, 1001, 1971.
72. Matsuo, S., Hagenoya, I., and Hirata, T., *Trans. Natl. Res. Inst. Met. Jpn.*, 13, 223, 1971.
73. Matsuo, S., *Trans. Natl. Res. Inst. Met. Jpn.*, 13, 57, 1971.
74. Leroy, M., *Scr. Met.*, 5, 759, 1971.
75. Miraille, J. P., Leroy, M., Briehet, C., and Lacombe, P., *Scr. Met.*, 5, 1061, 1971.
76. Bernhardt, W., *Neue Hutte*, 16, 740, 1971.
77. Matsura, K., *Trans. J. Inst. Met.*, 2, 125, 1963.
78. Miraille, J. P., Leroy, M., Brichet, C., and Lacombe, P., *Scr. Metall.*, 5, 1061, 1971.
79. Spektor, E. N. and Khayuten, S. G., *Fiz. Met. Metalloved.*, 32(5), 1091, 1971.
80. Noble, B., McLauchlen, I. R., and Thompson, G., *Acta Metall.*, 18, 339, 1970.
81. Köster, W. and Hornbogen, E., *Z. Metallkd.*, 59, 787, 1968.
82. Turnbull, D., Rosenbaum, H. S., and Treaflis, H. N., *Acta Metall.*, 8, 277, 1968.
83. Linde, J. O., *Helv. Phys. Acta*, 41, 1013, 1968.
84. *International Critical Tables*, 6, McGraw-Hill, New York, 1929, 156.

Al-Fe (Aluminum-Iron)

The α-Fe crystal structure can dissolve large amounts of Al ($\sim$ 50 a/o) before exceeding the solubility limit, while Fe is insoluble in Al. The F_3Al_1 and Fe_1Al_1 compositions in this dominant α-FeAl primary phase will order below 550°C. Three compositionally constructed intermediate phases (ζ-FeAl$_2$, η-Fe$_2$Al$_5$, and Θ-FeAl$_3$) are stable at 25°C.

The room temperature resistivity of the α-FeAl alloys[1-2] as a function of Al concentration (see Figure 1) is relatively independent of the thermal treatment except for compositions near Fe$_3$Al$_1$ where ordering effects on the measured resistivity values are apparent.[3]

At 4.6°K the resistivity is linearly related to the Al concentration[4-6] in the dilute FeAl alloys (see Figure 2) with a coefficient of $d\varrho/dc = 5.1$ $\mu\Omega$cm/(a/o)Al. The DMR at 296°K is small and relatively concentration insensitive in these dilute FeAl alloys (Figure 3).[5]

The resistivity of the more concentrated FeAl primary phase alloys between 36 and 52 a/o Al has been measured at 4.2, 77, and 298°K.[7] The resistivity was found to decrease appreciably with increasing concentration and was relatively temperature independent below 77°K.

Resistivity studies have also been made on the very dilute AlFe alloys.[8-9] The resistivity of Fe$_3$Al and especially the order-disorder transition at 813.6°K is consistent with the scaling arguments.[11] Tabulated room temperature data for annealed and tempered alloys are given in Table 1.

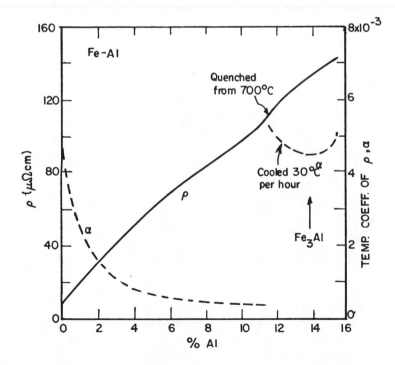

FIGURE 1. Resistivity of FeAl alloys as quenched and as cooled at 30°C/hr from 700°C (also temperature coefficient).[10]

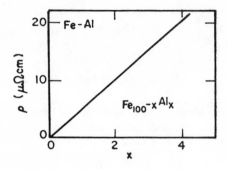

FIGURE 2.   Resistivity at 4.2°K as a function of composition in F̲eAl alloys.[4]

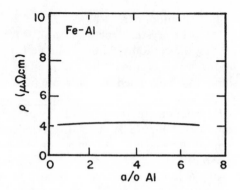

FIGURE 3.   Apparent deviations from Matthiessen's rule at room temperature for some binary iron alloys vs. solute concentration.[5]

Table 1

Fe (w/o)	ϱ(Annealed) (μΩcm)	ϱ(Tempered) (μΩcm)
0	2.44	2.60
12.8	6.49	6.76
24.3	13.3	14.2
33.6	43.3	44.8
40.5[a]	140.8	166.6

[a] Al_2Fe.

REFERENCES

1. Sykes, C. and Evans, H., *Proc. R. Soc. London Ser. A*, 145, 529, 1934.
2. Sykes, C. and Evans, H., *J. Iron Steel Inst.*, 131, 225, 1935.
3. Bennett, W. D., *J. Iron St. Inst.*, 171, 372, 1952.
4. Dorleijn, J. W. F., *Philips Res. Rep.*, 31, 287, 1976.
5. Arajs, S., Schwerer, F. C., and Fisher, R. M., *Phys. Status Solidi*, 33, 731, 1969.
6. Campbell, I. A., Fert, A., and Pomeroy, A. R., *Philos. Mag.*, 15, 977, 1967.
7. Riviere, J. P., *Phys. Status Solidi*, 45, 157, 1978.
8. Hedgcock, F. T., Muir, W. B., and Wallingford, E., *Can. J. Pys.*, 38, 376, 1960.
9. Caplin. A. D. and Rizzuto, C., *J. Phys. C*, 3, L117, 1970.
10. Bozorth, R. M., *Ferromagnetism*, Van Nostrand, Princeton, N.J., 1951.
11. Chakraborty, D. P. and Parks, R. D., *Phys. Rev. B*, 6195, 1978.
12. *International Critical Tables*, Vol. 6, McGraw-Hill, New York, 1929, 156.

Al-Ga (Aluminum-Gallium)

The phase diagrams given by Hansen and Anderko[1] show practically no solubility of Al and Ga in each other. A eutectic reaction takes place at about 27°C.

Sato et al.[2] prepared A̲lGa alloys from 99.999% pure Al or Al with RRR ≈ 1300 and Ga which was 99.999% pure. Components were molten in a vacuum. After rolling, samples were annealed for 3 hr at 400°C in vacuum. The residual resistivities are given in Table 1. The data are in good agreement with results by Carter[3] and Zoller,[4] who obtained $\Delta \varrho/c = 0.21$ or 0.22 μΩcm/(a/o)Ga. Reference 5 obtained a much lower value: $\varrho/c = 0.065$ μΩcm/(a/o)Ga. Klopin et al.[6] measured the resistivity of Al-Ga alloys with up to 0.035 (a/o)Ga between 2 and 40°K. They obtained higher $\Delta\varrho/c$ values.

Dukin and Aleksandrov[9] gave a critical account of pseudopotential calculations and the resistivity increase due to 1% impurity. They gave $[\Delta\varrho/\Delta c]_{exp} = 0.25 \; \mu\Omega cm/$ (a/o)Ga in $\underline{Al}Ga$ and $[\Delta\varrho/\Delta c]_{exp} = 0.27 \; \mu\Omega cm/$(a/o)Al in $\underline{Ga}Al$. Resistivity measurements have been used to study the effect of radiation.[7-8]

Table 1

Ga (a/o)	$\varrho_o(\mu\Omega cm)$
0.049	0.011
0.11	0.022
0.20	0.040
0.84	0.17
3.5	0.56
0.18	0.034
0.50	0.092

REFERENCES

1. Hansen, M. and Anderko, K., *Constitution of Binary Alloys*, McGraw-Hill, New York, 1958; Mondolfo, L. F., *Aluminum Alloys Structure and Properties*, Butterworths, London, 1976.
2. Sato, H., Babouchi, T., and Youemitsu, K., *Phys. Status Solidi B*, 89, 571, 1978.
3. Carter, R. L., Dissertation, University of Michigan, Ann Arbor, 1971.
4. Zoller, H., *Metallurgica (Manchester)*, 11, 378, 1958; *Phys. Rev.*, 186, 697, 1969.
5. Papastaikoudis, C., Rocoffylon, E., Tselfes, W., and Chountas, K., *Z. Phys. B*, 25, 131, 1976.
6. Klopin, M. N., Panova, G. Kh., and Samoilov, B. N., *Sov. Phys. JETP*, 45(2), 287, 1977.
7. Dimitrov, O., Dimitrov, C., Rosner, R., and Boning, K., *Radiat. Eff.*, 30, 135, 1976.
8. Kontoleon, N., Papathanassopoulos, K., and Chountas, K., *Radiat. Eff.*, 27, 251, 1976.
9. Dukin, V. V. and Aleksandrov, B. N., *Pseudopotential Calculations of the Residual Resistivities of Dilute Solid Alloys Based on Normal Metals, Physics of Low Temperatures*, Akademia Nauk Uk SSR Academy of Science, USSR, Kharkov, 1978, 1.

Al-Ge (Aluminum-Germanium)

Al is practically insoluble in Ge, but about 0.28 Ge will dissolve in Al at 30.3°C, the eutectic temperature.[1] Most of the resistivity measurements[2-12] deal with resistivity changes in supersaturated $\underline{Al}Ge$ alloys. Equilibrium systems at high pressures, where the solubility range increases, have been given.[3] Sato et al.[10] prepared Al alloys with Ge impurities from 99.999% pure Al or Al with RRR $\approx$ 1300 and 99.999% pure Ge. After melting, ingots were homogenized, cold rolled, and finally annealed for 3 hr at 400° in vacuum. Table 1 gives residual resistivities ϱ_o. They obtain $\Delta\varrho/\Delta c$(a/o)Ge = 0.73 $\mu\Omega cm/$(a/o)Ge, in good agreement with data by Böning et al.[11] [$\Delta\varrho/c = 0.72$ $\mu\Omega cm/$(a/o)Ge] and Fickett.[12] Kontoleon et al.[5] studied deviations from Matthiessen's rule in $\underline{Al}Ge$ alloys with up to 0.5 (a/o)Ge in the temperature range from 4.2 to 320°K. $\varrho \propto T^{2 \; <n<3.6}$ for 10°K < T < 60°K. $\Delta\varrho_o/c = 0.74 \; \mu\Omega cm/$(a/o)Ge was in good agreement with References 10 and 11. Soifer et al.[9] determined the resistivity of alloys with 10 to 32.3 a/o Ge in Al between 850 and 1050°C.

Banova et al.[3] prepared $\underline{Al}Ge$ wires. The resistivity is given as a function of composition in Figure 1. The change in slope indicates the change from a one-phase to a two-phase sample.

Table 1

a/o Ge	0.010	0.043	0.092	0.13	0.15	0.46	0.74
$\varrho_o(\mu\Omega cm)$	0.011	0.041	0.052	0.090	0.12	0.32	0.49

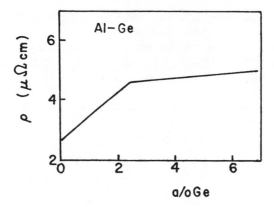

FIGURE 1.   Concentration dependence of the resistivity of A̲lGe alloys. Samples were quenched from 400°C.

REFERENCES

1. Hansen, M. and Anderko, K., *Constitution of Binary Alloys,* 2nd ed., McGraw-Hill, New York, 1958; Mondolfo, L. F., *Aluminum Alloys Structure and Properties,* Butterworths, London, 1976.
2. Larson, B. C. and Haubold, H. G., *J. Nucl. Mater., Int. Conf. on the Properties of Atomic Defects in Metals,* Vol. 69(1-2), Vol. 70(1-2), 758, 1976.
3. Banova, S. M., Korsunskaya, I. A., Kuznetsov, G. M. and Sergeev, V. A., *Fiz. Met. Metalloved,* 46, 521, 1978.
4. Kontoleon, N., Papathanassopoulos, K., and Chountas, K., *Phys. Lett. A,* 53, 413, 1975.
5. Kontoleon, N., Papathanassopoulos, K., Chountas, K., and Papastaikoudis, C., *J. Phys. F,* 4, 2109, 1974.
6. Beller, M., Furnrohr, H., and Gerold, V., *Phys. Stat. Solidi A,* 17, 435, 1973.
7. Beller, M., *Z. Metallkd.,* 64, 387, 1973.
8. Beller, M., *Z. Metallkd.,* 63, 663, 1972.
9. Soifer, L. M., Izmailov, V. A., and Kashin, V. I., *High Temp.,* 12, 581, 1974.
10. Sato, H., Babauchi, T., and Yonemitusu, K., *Phys. Stat. Solidi B,* 89, 571, 1978.
11. Böning, K., Pfänder, K., Rosner, P., and Schlüter, M., *J. Phys. Fl,* 1176, 1975.
12. Fickett, F. R., *Cryogenics,* 11, 349, 1971.

Al-In (Aluminum-Indium)

The solubility of Al in In and vice versa is very small, even at the melting temperature of Al or In.[1] No intermediate phase or compounds form. Resistivity measurements in these alloys[2-5] deal mostly with recovery processes in quenched alloys, leading to very small ($\sim 10^{-9}\Omega\text{cm}^2$) resistivity changes.

REFERENCES

1. Hansen, M. and Anderko, K., *Constitution of Binary Alloys,* 2nd ed., McGraw-Hill, New York, 1958; Mondolfo, L. F., *Aluminum Alloys Structure and Properties,* Butterworths, London, 1976.
2. Epperson, J. E., Fuernrohr, P., and Gerold, V., *Philos. Mag.,* 29, 1189, 1974.
3. Roebuck, B. and Entwistle, K. M., *Philos. Mag.,* 25, 153, 1972.
4. Plumbridge, W. J., *Scr. Metall.,* 4, 85, 1970.
5. Plumbridge, W. J., *Philos. Mag.,* 20, 707, 1969.

Al-Li (Aluminum-Lithium)

Hansen and Anderko[1] report no noticeable solubility of Al in Li, and about 3 a/o Li may dissolve in Al at 200°C. LiAl is stable over a composition range from $\sim$45 to 56 a/o Li at room temperature. The compound Li_2Al is stable to 522°C. Resistivity measurements[2-8] deal mostly with the elucidation of precipitation processes, vacancies, and recovery.

Ceresara et al.[4] measured the resistivity of AlLi alloys with up to 6.72 a/o Li. The alloys were prepared from 99.995% purity Al. The cast and cold rolled samples were drawn to wires and then annealed for 5 min at 400°C. This lead to a recrystallized alloy. These alloys are all in the primary solid solution range. Their resistivity at 78°K is a linear function of composition, with $d\varrho/dc = 0.80$ $\mu\Omega$cm/(a/o)Li. The binding energy of Li to vacancies was calculated to be 0.25 ± 0.03 eV.

Dukin and Aleksandrov[9] gave a critical account of pseudopotential calculations and the resistivity increase due to 1% impurity. They gave $[\Delta\varrho/\Delta c]_{exp} = 7.1$ $\mu\Omega$cm/(a/o)Al in LiAl and $[\Delta\varrho/\Delta c]_{exp} = 0.9$ $\mu\Omega$cm/(a/o)Li in AlLi.

REFERENCES

1. Hansen, M. and Anderko, K., *Constitution of Binary Alloys,* 2nd ed., McGraw-Hill, New York, 1958; Mondolfo, L. F., *Aluminum Alloys Structure and Properties,* Butterworths, London, 1976.
2. Kamel, R., Ali, A. R., and Farid, Z., *Phys. Status Solidi A,* 45, 47, 1978.
3. Kamel, R., Ali, A. R., and Farid, Z., *Phys. Status Solidi A,* 46, 697, 1978.
4. Ceresara, S., Giarda, A., and Sanchez, A., *Philos. Mag.,* 35, 97, 1977.
5. Kawala, S., *J. Sci. Hiroshima Univ. Ser. A,* 40, 43, 1971.
6. Epperson, J. E., *Philos. Mag.,* 29, 1189, 1974.
7. Fridlyander, I. N., Sandler, V. S., and Nikolskaya, T. I., *Metalloved. Term. Obrab. Met.,* 14(3), 41, 1972.
8. Noble, B. and Thompson, G. E., *Met. Sci. J.,* 5, 114, 1971.
9. Dukin, V. V. and Aleksandrov. B. N., *Pseudopotential Calculations of the Residual Resistivities of Dilute Solid Alloys Based on Normal Metals, Physics of Low Temperatures,* Akademia Nauk Uk SSR Academy of Science, USSR, Kharkov, 1978, 1.

Al-Mg (Aluminum-Magnesium)

A total of 2.1 a/o Mg will dissolve in Al and 1.3 a/o Al will dissolve in Mg at 100°C.[1] Two intermediate phases, β (stable near 38 a/o Mg) and γ (stable between 53 and 60 a/o Mg), have been found. There is some evidence for an ε-phase near 43 a/o Mg.

Resistivity measurements[2-17] deal mostly with aging processes. Reference 3 gives $\varrho(T,c)$ of splat-cooled (metastable) alloys.

The validity of Matthiessen's rule was investigated[7,9,16,17] on AlMg samples. Caplin and Rizzuto[16] studied samples which were prepared by quenching from the melt, annealing just below the melting point, and quenching in water. The residual resistivity was 0.50 nΩcm for an alloy with 0.003 a/o Mg. Klopkin et al.[17] measured the resistivity of alloys with up to 0.155 a/o Mg between 2 to 40°K. They obtained a value of $\Delta\varrho/\Delta c = 0.43$ $\mu\Omega$cm/(a/o)Mg. Fujita and Ohtsuka[4] prepared AlMg samples from nominally 99.999% pure Al. They were annealed at 550°C for 30 min and slowly cooled to room temperature. The total low-temperature resistivity can be represented by $\varrho = \varrho_o + \alpha T^3$, with $\varrho_o/c = 0.46$ $\mu\Omega$cm/(a/o)Mg.

Rapp and Fogelholm[10] measured the resistivity of Al-Mg alloys with up to 0.5 a/o Mg between 0 and 100°C. Figure 1 gives results. $\Delta\varrho/c = 0.80 \pm 0.07$ $\mu\Omega$cm/(a/o)Mg, and $d\varrho/dT = 11.34 \pm 0.05$ $\mu\Omega$cm/°K. The results were analyzed on the basis of spin fluctuations and electron phonon interactions.

Dukin and Aleksandrov[18] gave a critical account of pseudopotential calculations and the resistivity increase due to 1% impurity. They gave $[\Delta\varrho/\Delta c]_{exp} = 2.1$ $\mu\Omega$cm/(a/o)Al in MgAl.

Klaffky et al.[9] prepared Al-Mg alloys from 99.99% pure Al and 99.99% pure Mg. They obtained the following residual resistivity values.

		ϱ_0 ($\mu\Omega$cm)
Al + 5% Mg	Swaged	2.045
	Swaged and an- nealed	1.869 — 1.862
Al + 7% Mg	Swaged	2.744 — 2.521
	Swaged and an- nealed	2.424 — 2.359
Al + 3% Mg	Swaged	1.376
	Swaged and an- nealed	1.273
Al + 10% Mg	Swaged and ma- chined	4.250

Table 1 gives room temperature resistivities.[19]

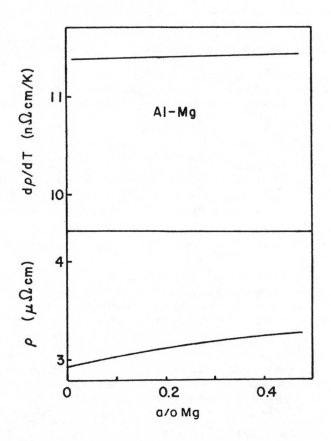

FIGURE 1.   Resistivity and its temperature coefficient of A̲lMg at 45°C.

Table 1
ROOM TEMPERATURE
RESISTIVITIES

Mg (w/o)	ϱ(annealed) ($\mu\Omega$cm)	ϱ(tempered) ($\mu\Omega$cm)
0	2.49	2.60
3.76	6.02	6.45
6.92	6.99	7.09
14.46	10.41	10.42
18.06	12.20	12.36
21.79	14.71	14.97
29.73	26.60	28.40
31.23	27.79	31.15
35.86	40.66	37.18
40.0	49.26	45.46
47.75	38.03	37.31
52.28	45.06	43.48
53.89	38.48	36.36
54.95	32.57	31.46
57.69	22.07	23.09
60.29	30.77	29.86
64.74	24.52	27.40
73.02	19.92	22.84
85.26	16.47	17.30
92.85	14.22	14.01
100.0	40.64	42.54

REFERENCES

1. Hansen, M. and Anderko, K., *Constitution of Binary Alloys,* 2nd ed., McGraw-Hill, New York, 1958; Mondolfo, L. F., *Aluminum Alloys Structure and Properties,* Butterworths, London, 1976.
2. Takahashi, T. and Kojima, Y., *J. Jpn. Inst. Met.,* 42, 396, 1978.
3. Yakunin, A. A., Osipov, I. I., Tkach, V. I., and Lysenko, A. B., *Fiz. Met. Metalloved.,* 43, 140, 1977.
4. Fujita, T. and Ohtsuka, T., *J. Low Temp. Phys.,* 29, 333, 1977.
5. Tuli, R. and Grover, P. S., *Ind. J. Pure Appl. Phys.,* 14, 938, 1976.
6. Govila, R. K., Sharma, C. P., and Rajan, T. V., *Ind. J. Technol.,* 14, 398, 1976.
7. Papathanassopoulos, K. and Rokotyllou, E., *Solid State Commun.,* 19, 665, 1976.
8. Fukuchi, M. and Watanabe, K., *J. Jpn. Inst. Met.,* 39, 493, 1975.
9. Klaffky, R. W., Mohan, N. S., and Damon, D. H., *J. Phys. Chem. Solids,* 36(10), 1147, 1975.
10. Rapp, O. and Fogelholm, R., *Solid State Commun.,* 15, 1291, 1974.
11. Muromachi, S., *J. Jpn. Inst. Met.,* 38, 130, 1974.
12. Chatterjee, D. K. and Entwistle, K. M., *J. Inst. Met.,* 101, 53, 1973.
13. Mayer, H., Boning, K., Dimitrov, C., and Dimitrov, D., *Phys. Status Solidi A,* 15, K91, 1973.
14. Clark, A. F. and Tryon, P. V., *Cryogenics,* 12, 451, 1972.
15. Baba, Y., *J. Jpn. Inst. Met.,* 36, 335, 1972.
16. Caplin. A. D. and Rizzuto, C., *Aust. J. Phys.,* 24, 309, 1971.
17. Klopkin, M. M., Panova, G. Kh., and Samoilov, B. N., *Sov. Phys. JETP,* 45, 287, 1977.
18. Dukin, V. V. and Aleksandrov, B. N., *Pseudopotential Calculations of the Residual Resistivities of Dilute Solid Alloys Based on Normal Metals, Physics of Low Temperatures,* Akademia Nauk Uk SSR Academy of Science, USSR, Kharkov, 1978, 1.
19. *International Critical Tables,* Vol. 6, McGraw-Hill, New York, 1929, 156.

Al-Mn (Aluminum-Manganese)

Only small amounts of Mn will dissolve in Al.[1] A few percent of Al will dissolve in α-Mn, but more than 20 a/o Al will dissolve in β-, γ-, and δ-Mn solid solutions. Several compounds exist. Resistivity measurements[2-22] are mostly concerned with the aging processes except for some investigations of very dilute alloys.

The electrical resistivity of dilute AlMn alloys was measured by Kovacs-Csetenyi et al.[8] These alloys can be regarded as nearly magnetic with local susceptibility enhancement. The localized spin fluctuations are responsible for the quadratic temperature dependence of ϱ at low temperatures. The samples in this study were prepared from 5N pure Al and 4N pure Mn. After cold drawing, the samples were annealed for 1 hr at 620°C. The low-temperature impurity resistivity $\Delta\varrho(T)$ can be fitted to the equation $\Delta\varrho(T) = \Delta\varrho(0) [1 - (T/\Theta_1)^2]$ with $\Theta_1 = 670 \pm 30°$K. At higher temperatures, $\Delta\varrho(T) \sim 1 - T/\Theta_2$ with $\Theta_2 \simeq 1600°$K. Caplin and Rizzuto[18] found at low temperatures a similar behavior, with $\Theta_1 = 530 \pm 30°$K. The linear temperature dependence of $\varrho(T)$ above 150°K was found not only by these authors, but also by Zuckermann.[9] He used the theory of localized spin fluctuation to obtain theoretical expressions for $\varrho(T)$. Theory and experiments agree for $\varrho(T)$, but agreement for $N_d(E)$, the density of states for localized electrons at the Fermi energy level, is poor.

Babic et al.[4,7,19-21] measured the resistivity of AlMn alloys with 0.56, 1.05, and 1.7 a/o Mn between 4 and 200°K. Figures 1 and 2 give their results. The impurity resistivity follows again a temperature dependence of $\Delta\varrho(T) = \varrho_o[1 - (T/\Theta_1)^2]$ at low temperatures, and $\varrho(T) = \varrho_o^1(1 - T/\Theta_2)$ at high temperatures, with $\Theta_1 = 530 \pm 30°$K and $\Theta_2 = 1600°$K. The transition from one stage to the other takes place at about 85°K. The data are explained on the basis of localized spin fluctuations on the virtual bound states formed by the atoms. $\varrho_o = 8 \ \mu\Omega$cm/(a/o)Mn.

Data at higher temperatures have been obtained by Rapp and Fogelholm.[22] They measured the resistivity of Al-Mn alloys with up to 0.5 a/o Mn between 0 and 100°C. Figure 3 gives the experimental results. $\Delta\varrho/\Delta c = 7.6 \pm 0.07 \ \mu\Omega$cm/(a/o)Mn. Data are analyzed on the basis of spin fluctuations and electron phonon interactions.

Dunlop and Grüner[3] studied the susceptibility and resistivity of $Al_{11}Mn_4$ samples in the low-temperature modification. Samples were annealed for 72 hr at 880°C. Less than 5% of a second phase may be present. Figure 4 shows $\varrho(T)$ curves of two samples. The different shapes of the curves are associated with deviations from stochiometry. The low temperature up turn is associated with the Kondo effect produced by off-stochiometric Mn atoms. Table 1 gives room temperature resistivities.[23]

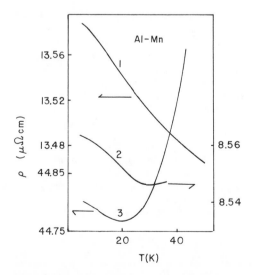

FIGURE 1.   Resistivity of AlMn alloys. (1) 1.7 a/o Mn; (2) 1.05 a/o Mn; (3) 0.58 a/o Mn.

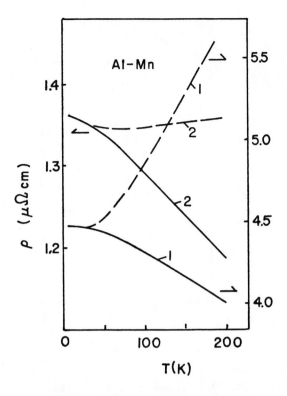

FIGURE 2. Resistivity of Al-Mn alloys. (1) 0.58 a/o
Mn and (2) 1.7 a/o Mn. Dashed line, total resistivity;
full lines, impurity resistivity.

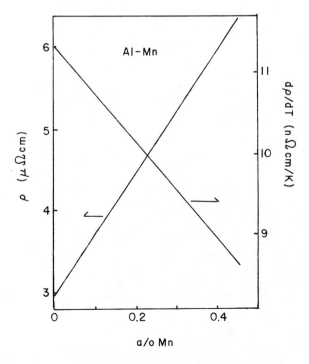

FIGURE 3. Resistivity and dϱ/dT of <u>Al</u>Mn alloys at 45°C.

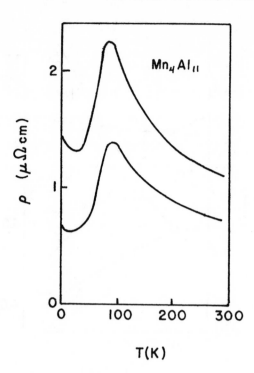

FIGURE 4.   Resistivity of two Mn_4Al_{11} alloys.

Table 1

Mn (w/o)	ϱ(Annealed) ($\mu\Omega$cm)	ϱ(Tempered) ($\mu\Omega$cm)
0	2.49	2.60
6.3	14.27	14.6
12.6	20.1	20.7
21.4	35.7	34.6
32.8	84.0	96.6
40.4	500	294
46.9	2000	714

REFERENCES

1. Hansen, M. and Anderko, K., *Constitution of Binary Alloys,* 2nd ed., McGraw-Hill, New York, 1958; Mondolfo, L. F., *Aluminum Alloys Structure and Properties,* Butterworths, London, 1976.
2. Singh, V., Lahiri, D. P., and Ramachandran, T. R., *Scr. Metall.,* 12, 213, 1978.
3. Dunlop, J. B. and Gruner, G., *Solid State Commun.,* 18, 827, 1976.
4. Hamzik, A., Babic, E., and Leontic, B., *Mater. Sci. Eng.,* 23, 271, 1976.
5. Lahiri, D. P., Jena, A. K., and Ramachandran, T. R., *J. Mat. Sci.,* 10, 1458, 1975.
6. Nagahama, K. and Miki, I., *Trans. Jpn. Inst. Met.,* 15, 185, 1974.
7. Babic, E., Krsnik, R., Rizzuto, C., *Solid State Commun.,* 13, 1027, 1973.
8. Kovacs-Csetenyi, E., Kedves, F. J., Gergely, L., and Gruner G., *J. Phys. F,* 2, 499, 1972.
9. Zuckermann, M. J., *J. Phys. F,* 2, L25, 1972.
10. Rizzuto, C., *Phys. Rev. Lett.,* 27, 805, 1971.
11. Baba, Y., *Trans. Jpn. Inst. Met.,* 13, 76, 1972.
12. Hirata, T., *Trans. Natl. Res. Inst. Met.,* 14, 1, 1972.
13. Kovacs, I., Lendvai, J., and Nagy, E., *Acta Metall.,* 20, 975, 1972.
14. Ito, T., Furuya, T., Matsura, K., and Watanabe, K., *Trans. Jpn. Inst. Met.,* 12, 379, 1971.

15. Dimitrov, C. and Dimitrov, O., *Phys. Status Solidi B,* 34, 545, 1969.
16. Collings, E. W., Hedgcock, F. T., Muir, W. B., and Muto, Y., *Philos. Mag.,* 10, 159, 1964.
17. Fridlyander, I. N., Sandler, V. S., and Nikolskaya, T. I., *Fiz. Met. Metalloved.,* 32, 767, 1971.
18. Caplin, A. D. and Rizzuto, C., *Phys. Rev. Lett.,* 8, 243, 1968.
19. Babic, E., Ford, P. J., Rizzuto, C., and Salamoni, E., *Solid State Commun.,* 11, 519, 1972.
20. Babic, E., Krsnik, R., Leontic, B., Vacic, Z., Zovic, I., and Rizzuto, C., *Phys. Rev. Lett.,* 27, 805, 1971.
21. Babic, E., Krsnik, R., and Rizzuto, C., *Solid State Commun.,* 13, 1027, 1973.
22. Rapp, O. and Fogelholm, R., *Solid State Commun.,* 15, 1291, 1974.
23. *International Critical Tables,* Vol. 6, McGraw-Hill, New York, 1929, 156.

Al-Na (Aluminum-Sodium)

Dukin and Aleksandrov[1] gave a critical account of pseudopotential calculations and the resistivity increase due to 1% impurity. They gave $[\Delta\varrho/\Delta c]_{exp} = 2.0 \ \mu\Omega cm/(a/o)$Na in AlNa.

REFERENCES

1. Dukin, V. V. and Aleksandrov, B. N., *Pseudopotential Calculations of the Residual Resistivities of Dilute Solid Alloys Based on Normal Metals, Physics of Low Temperatures,* Akademia Nauk Uk SSR Academy of Science, USSR, Kharkov, 1978, 1.

Al-Ni (Aluminum-Nickel)

The Al-Ni phase diagram shows that up to 10% Al can be dissolved in solid Ni at 25°C, while no appreciable amount of Ni can be dissolved in aluminum. Four intermediate phases exist in the concentrated alloy compositions at room temperature.

The concentration dependence of the residual resistivity of the primary NiAl phase yields a coefficient of 2.1 $\mu\Omega cm/(a/o)$ Al at 4.2°K.

The resistivity of the β' - NiAl intermediate phase has been investigated as a function of both concentration and temperature. The composition dependence at 298°K has a minimum near the equiatomic stoichiometry (see Figure 1) and the temperature dependence shows a rapid increase in resistivity for temperatures above 15°K (Figure 2). Similar results were obtained by Jacobi et al.,[8] who measured ϱ of alloys with 40 to 52 a/o Al at 77 and 298°K.

The resistivity of the α'-Ni_3Al intermediate phase has also been measured as a function of composition and temperature.[5] The residual resistivity and the room temperature resistivity as a function of composition is shown in Figure 3. Room temperature resistivity data over a wide composition range is given in Table 1.[7]

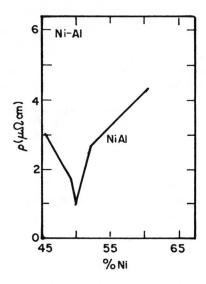

FIGURE 1.   Resistivity of NiAl, as a function of composition.[7]

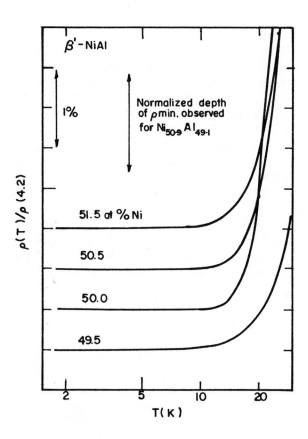

FIGURE 2.   Temperature dependence of resistivity ratio of NiAl.[2]

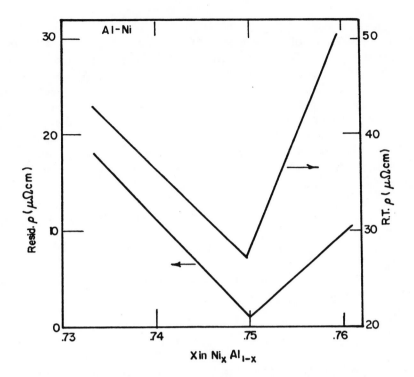

FIGURE 3.   Residual (lower curve) and room temperature resistivity (upper curve) in $Ni_x Al_{1-x}$.[6]

Table 1

Ni (w/o)	ϱ(Annealed) ($\mu\Omega$cm)	ϱ(Tempered) ($\mu\Omega$cm)
0	2.49	2.60
11.4	4.14	4.0
22.4	7.81	7.04
30.2	9.9	9.18
42.4	28.8	15.15
44.6	34.8	23.5
83.5	62.9	80.6
86.5	59.5	64.5
88.5	75.8	72.5
92.8	61.4	67.1
96.6	24.9	31.25
98.4	17.7	17.8
100.0	11.96	13.14

REFERENCES

1. Dorleijn, J. W. F. and Miedema, A. R., *J. Phys. F*, 5, 487, 1975.
2. Yoshitomi, T., Ochiai, Y., and Brittian, J. O., *Solid State Commun.*, 20, 741, 1976.
3. Butler, S. R., Hanlon, J. E., and Wasilewski, R. J., *J. Phys. Chem. Solids*, 30, 1929, 1969.
4. Guseva, L. N., *Dokl. Akad. Nauk SSSR*, 77, 415, 1951.
5. Fluitman, H. J., Boom, R., DeChatel, P. F., Schinkel, C. J., Tilanus, J. L. L., and DeVries, B. R., *J. Phys. F*, 3, 109, 1973.
6. Yamaguchi, T., Kiewit, D. A., Aoki, T., and Brittian, J. O., *J. Appl. Phys.*, 39, 231, 1968.
7. *International Critical Tables*, Vol. 6, McGraw-Hill, New York, 1929, 156.
8. Jacobi, H., Vassor, B., and Engel, H. J., *J. Phys. Chem. Solids*, 30, 1261, 1969.

Al-Si (Aluminum-Silicon)

A total of 1.59 a/o Si will dissolve at 577°C in Al, and 0.29 a/o Si will dissolve at 350°C. Al is according to lattice parameter measurements insoluble in Si.[1] The phase diagram shows one eutectic reaction at 577°C: $Al_{98.5}Si_{1.5} + Si_{99.5}Al_{0.5} \rightleftharpoons l$ ($Al_{88.9}Si_{12.1}$).

Resistivity measurements[2-7] were used to evaluate a two-band model,[2] the state of Si atoms,[3] atomic segregation for alloys with 8 to 9% Si over the temperature range of 500 to 1000°C,[5] and clustering in quenched dilute alloys.[7] Bradley and Stringer[2] measured the Hall coefficient and resistivity in AlSi alloys. They prepared samples from 5N pure elements. After homogenizing and cold rolling, the samples were solution treated at 500°C and water quenched. Measurements followed immediately. The following resistivity values were observed at 293°K.

	Al + 0.2 a/o Si	0.4 a/o Si	0.6 a/o Si	0.8 a/o Si	1 a/o Si
ϱ(293°K) in $\mu\Omega$cm	2.65	2.81	3.08	3.19	3.33

Dukin and Aleksandrov[8] gave a critical account of pseudopotential calculations and the resistivity increase due to 1% impurity. They gave $[\Delta\varrho/\Delta c]_{exp} = 0.6 \ \mu\Omega$cm/(a/o)Si in AlSi.

REFERENCES

1. Hansen, M. and Anderko, K., *Constitution of Binary Alloys,* 2nd ed., McGraw-Hill, New York, 1958; Mondolfo, L. F., *Aluminum Alloys Structure and Properties,* Butterworths, London, 1976.
2. Bradley, J. M. and Stringer, J., *J. Phys. F,* 4, 839, 1974.
3. Kittaka, T., *J. Sci. Hiroshima Univ. A-U,* 32, 173, 1968.
4. Guillet, L. and Ballay, M., *C. R.,* 190, 473, 1930.
5. Krushenko, G. G., Shpakov, V. I., Nikitin, V. I. and Torshilova, S. I., *Izv. Akad. Nauk SSSR Met.,* 4, 203, 1977.
6. Mii, H., Senoo, H., and Fujishiro, I., *Jpn. J. Appl. Phys.,* 15, 777, 1976.
7. Sayed, H. El. and Kovacs, I., *Phys. Status Solidi A,* 24, 123, 1974.
8. Dukin, V. V. and Aleksandrov, B. N., *Pseudopotential Calculations of the Residual Resistivities of Dilute Solid Alloys Based on Normal Metals,* Physics of Low Temperatures, Akademia Nauk Uk SSR Academy of Science, USSR, Kharkov, 1978, 1.

Al-Sn (Aluminum-Tin)

Dukin and Aleksandrov[1] gave a critical account of pseudopotential calculations and the resistivity increase due to 1% impurity. They gave $[\Delta\varrho/\Delta c]_{exp} = 0.9 \ \mu\Omega$cm/(a/o)Sn in AlSn and $[\Delta\varrho/\Delta c]_{exp} = 0.25 \ \mu\Omega$cm/(a/o)Al in SnAl.

REFERENCES

1. Dukin, V. V. and Aleksandrov, B. N., *Pseudopotential Calculations of the Residual Resistivities of Dilute Solid Alloys Based on Normal Metals,* Physics of Low Temperatures, Akademia Nauk Uk SSR Academy of Science, USSR, Kharkov, 1978, 1.

Al-Ti (Aluminum-Titanium)

More than 35 a/o Al will dissolve in α- and β- Ti, but Ti is nearly insoluble in Al. Two intermediate phases, TiAl and $TiAl_3$, the latter with a narrow solubility range, have been observed.[1] Resistivity measurements[2-12] deal mostly with fcc Al-rich alloys.

Babic et al.[2] conducted measurements of the resistivity of rapidly quenched Al-Ti alloys with up to 0.5 a/o Ti. They gave values for the residual resistivity ϱ_o and $\varrho(T) - \varrho(o)$ (see Table 1).

Rayetskiy[9] studied $\varrho(T)$ of T̲iAl alloys with up to 5 w/o Al at higher temperature. Technical Ti was used to prepare samples. Figure 1 shows $\varrho(293°K)$ and $\varrho(78°K)$ as a function of composition. A series of commercial alloys, including one T̲iAl alloy with 3.9 a/o Al, was studied by Gusev et al.[12].

Kornilov et al.[10] measured $\varrho(T)$ of Ti-Al alloys with up to 17.5% Al from room temperature to 1200°C. Samples were melted twice, annealed for 100 hr at 900°C, 200 hr at 800°C, 100 hr at 700°C, and furnace cooled. Starting materials were TG-00 and 99.9% pure Al. Figure 2 gives the experimental results. The drop in ϱ above 900°C is due to the α-β phase transformation. $\varrho(T)$ values were obtained for the case of phase transformation by waiting for equilibrium. Figure 3 gives the temperature for the onset and completion of the phase transformation obtained from the $\varrho(T)$ curves.

Zelenkov and Osokin[5] studied TiAl₃ with and without third element impurities. TiAl₃ was prepared from iodide grade Ti and Grade AV000 aluminum. The alloys were melted in an arc furnace under Ar, annealed for 50 hr at 1100°C in vacuum, and then cooled at a rate of 3 to 5°C/min. Figure 4 shows $\Delta\varrho/\varrho(20°C)$ as a function of temperature. $\Delta\varrho/\varrho(20°C)$ is not a straight line. According to these authors, $\varrho(T)$ increases at temperatures above 800°C as a result of a decrease in the degree of long-range order. Figure 5 shows $\varrho(T)$ of a TiAl₃ alloy which was after arc-melting and annealing quenched in oil.[7]

Hake et al.[4] measured $\varrho(T)$ for hcp T̲iAl alloys. Samples were prepared from "iodide process" and were typically of 99.92 w/o purity. The Ti-Al sample with 1.06 a/o Al showed a resistance minimum which the authors associated with localized moments. The following resistivity values were found for the alloy: $\varrho(273°K) = 55.9$ $\mu\Omega$cm, $\varrho(4.2°K) = 3.22$ $\mu\Omega$cm, $T_{min} = 12.5°K$. $[\varrho(4.2°K) - \varrho(12.5°K)]/\varrho(4.2°K) = 0.0028$.

Table 1
IMPURITY CONCENTRATION

(a/o)Ti	ϱ_o(nΩcm)	$\varrho_o(10°K) -$ ϱ_o(nΩcm)	$\varrho(14°K) -$ ϱ_o(nΩcm)	$\varrho(20°K) -$ ϱ_o(nΩcm)
0	22	0.22	0.65	2.15
0.05	280	0.33	1.00	—
0.2	1190	0.40	1.05	3.40
0.5	2900	0.42	1.20	3.75

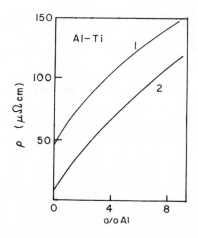

FIGURE 1. Resitivity as a function of Al concentration (°K). (1) 293; (2) 78.

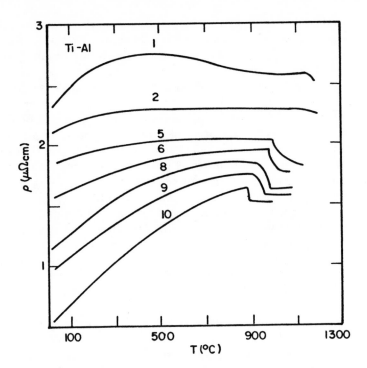

FIGURE 2.   Resistivity of <u>Al</u>Ti alloys: drop in ϱ at $\sim$900°C indicates phase transformation.

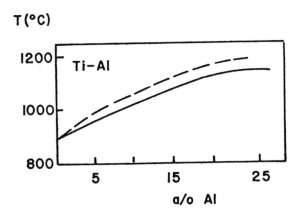

FIGURE 3. $\alpha \leftrightarrow \beta$ transformation temperatures in Ti-Al. Full line, $\beta \rightarrow \alpha$: dashed line, $\alpha \rightarrow \beta$.

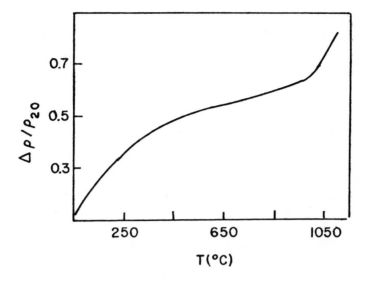

FIGURE 4. $\Delta\varrho/\varrho_{20}$ of Ti$_3$Al.

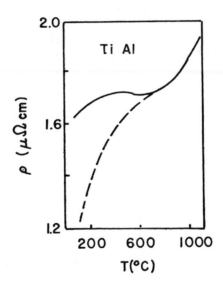

FIGURE 5.   Temperature dependence of the
resistivity: Ti$_3$Al. Full line: quenched sample;
dashed line: annealed sample.

REFERENCES

1. Hansen, M. and Anderko, K., *Constitution of Binary Alloys*, 2nd ed., McGraw-Hill, New York, 1958. Mondolfo, L. F., *Aluminum Alloys Structures and Properties*, Butterworths, London, 1976.
2. Babic, E., Krsnik, R., and Ocko, M., *J. Phys. F*, 6, 73, 1976.
3. Kornilov, I. I., Nartova, T. T., and Chernyshova, S. P., *Izv. Akad, Nauk SSSR Met.*, 6, 192, 1976.
4. Hake, R. R., Leslie, D. H., and Berlincourt, T. G., *Phys. Rev.*, 127, 170, 1962.
5. Zelenkov, I. A. and Osokin, E. N., *Poroshk. Metall.*, 15, 44, 1976.
6. Chernyshova, S. P., Nartova, T. T., Shapovaloa, O. M., and Naidam, V. M., *Met. Sci. Heat Treat.*, 17, 620, 1975.
7. Zelenkov, I. A., Osipenko, I. A., and Osokin, E. N., *Poroshk. Metall.*, 15, 101, 1976.
8. Namboodhiri, T. K. G., Herman, H., and McMahon, C. J., Jr., *Metall. Trans.*, 4, 1323, 1973.

9. Rayetskiy, V. M., *Phys. Met. Metallogr.*, 23(4), 749, 1967.
10. Kornilov, I. I., Mikheyev, V. S., and Konstantinov, K. M., *Phys. Met. Metallogr.*, 16(1), 56, 1963.
11. Zelenkov, I. A., Osipenko, I. A., and Osokin, E. N., *Sov. Powder Metall. Met. Ceram.*, 15, 327, 1976.
12. Gusev, Ye V., Lashko, N. F., and Khatsinskaya, I. M., *Phys. Met. Metallogr.*, 16, 56, 1963.

Al-V (Aluminum-Vanadium)

More than 40 a/o Al will dissolve in V and 0.2 a/o V will dissolve at 500°C. Several intermetallic phases exist.[1]

Most of the resistivity measurements[2-8] deal with the resistivity of Al-rich alloys. Caplin and Rizzuto[3] prepared samples by quenching from the melt, annealing a few hours close to the melting temperature, and quenching in water. The residual resistivities were 0.56 and 2.57 $\mu\Omega$cm for alloys with 0.083 and 0.38 a/o V, respectively. This gives for $\Delta\varrho/\Delta c = 6.8\ \mu\Omega$cm/(a/o)V.

Babic et al.[4] studied the low-temperature resistivity of rapidly quenched <u>Al</u>V alloys in the temperature range from 1.5 to 20°K. Both T^3 and T^5 temperature dependencies were observed. The latter term is more important at higher temperatures. Results are given in Table 1.

The alloy $Al_{10}V$ has been studied in detail,[2,6-8] since this well-ordered crystal has one Al atom whose environment is unusual and gives rise to a low-lying vibrational mode. Caplin et al.[8] measured ϱ of $Al_{10}V$. They showed that the resistivity reveals a "soft mode" with a characteristic temperature. This is associated with excess Al atoms which may occupy "holes" in the $Al_{10}V$ unit cell. Figure 1 gives $\varrho(T)-\varrho_0$ of $Al_{10}V$.

Table 1

(a/o)V	ϱ_o(nΩcm)	$d\varrho/dT\|_{273°K}$ ($\mu\Omega$cm/°K)
0.005	55	11.2
0.015	94	11.4
0.06	450	11.6
0.1	660	11.4
0.3	2360	11.3
0.55	4100	11.6
0.7	5300	11.3
0.85	6380	11.5

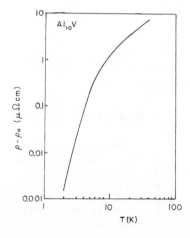

FIGURE 1.   Temperature-dependent resistivity of $Al_{10}V$.

REFERENCES

1. Hansen, M. and Anderko, K., *Constitution of Binary Alloys,* 2nd ed., McGraw-Hill, New York, 1958; Mondolfo, L. F., *Aluminum Alloys Structure and Properties,* Butterworths, London, 1976.
2. Caplin, A. D. and Nicholson, L. K., *J. Phys. F.,* 8, 51, 1978.
3. Caplin, A. D. and Rizzuto, C., *Aust. J. Phys.,* 24, 309, 1971.
4. Babic, E., Kronik, R., and Ocko, M., *J. Phys. F.,* 6, 73, 1976.
5. Babic, E., Kronik, R., and Ocko, M., *Proc. Int. Conf. Low Temp. Phys.,* 14(13), 105, 1975.
6. Cooper, J. R., *Phys. Rev. B,* 9, 2778, 1974.
7. Caplin, A. D. and Grüner, G., *4th Int. Conf. on Solid Compounds of Trans. Elements,* Switzerland University, Geneva, 1973, 205.
8. Caplin, A. D., Duulop, J. B., and Grüner, G., *Phys. Rev. Lett.,* 30, 1138, 1973.

Al-Zn (Aluminum-Zinc)

The solubility of Al in Zn is small, 2.4 a/o Al in Zn at 380°C and about 0.15 a/o Al in Zn at room temperature.[1] Only 1.7 a/o Zn will dissolve in Al at 100°C, but up to 66.5 a/o Zn will dissolve at 382°C.

The resistivity measurements of Al-Zn alloys deal mostly with the formation of Gunier-Preston zones;[8-62] only a few studies are concerned with equilibrium systems.[2-7] A survey of resistivities of Al-Zn alloys over the whole composition and temperature range has been made by Yrkov et al.[2] They prepared Al-Zn alloys from chemical grade elements. They were cast in the form of cylinders and annealed for 120 hr at 250 ± 5°C and furnace cooled. The temperature dependence of ϱ for all alloys is shown in Figures 1 and 2. $\varrho(T)$ of alloys with 10, 20, and 90 w/o Zn follows essentially straight lines. Alloys with 30 to 80 w/o Zn have sharp kinks in the $\varrho(T)$ curves. These kinks indicate a change from a one-phase to a two-phase sample. Isotherms of the resistivity, together with the phase diagram, are shown in Figure 3. Naturally, one has to keep in mind that the resistivity of two-phase systems depends on the spacial distributions of the two phases. ϱ is not uniquely defined. The miscibility gap in Figure 3 can be determined from resistivity measurements.

Klopin et al.[3] measured the resistivity of Al-Zn alloys between 2 and 40°K and obtained for the residual resistivity change with impurity concentration $\Delta\varrho/\Delta c = 0.21$ $\mu\Omega$cm/(a/o)Zn. The maximum Zn concentration was 0.42 a/o.

Osamura et al.[4] prepared Al-Zn alloys with up to 10 a/o Zn, using 99.998% pure Al. After cold rolling, samples were solution treated at 450°C for 1 hr, then cooled to 300°C and held for 1 hr. Finally the samples were quenched into water at 0°C and immersed into LN_2. Table 1 gives ϱ for T = 4.2 and 77°K. The data are in good agreement with older measurements.

The homogeneous alloys at low concentration are discussed in respect to deviations in Matthiessen's rule by Papastaikoudis et al.[28] Dukin and Aleksandrov[58] gave a critical account of pseudopotential calculations and the resistivity increase due to 1% impurity. They gave $[\Delta\varrho/\Delta c]_{exp} = 0.49$ $\mu\Omega$cm/(a/o)Al in ZnAl.

Table 2 gives results from earlier measurements at room temperature on the complete composition range.[62]

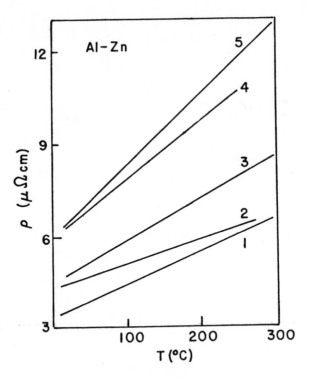

FIGURE 1.   Resistivity of Al-Zn alloys. (1) Al; (2) 10 w/o Zn; (3) 20 w/o Zn; (4) 90 w/o Zn; (5) Zn.

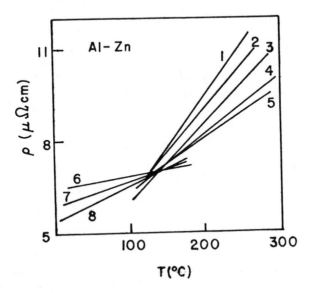

FIGURE 2.   Resistivity of Al-Zn alloys. (1) 90 w/o Zn; (2) 80 w/o Zn; (3) 60 w/o Zn; (4) 50 w/o Zn; (5) 30 w/o Zn; (6) 30 w/o Zn; (7) 50 w/o Zn; (8) 60 w/o Zn.

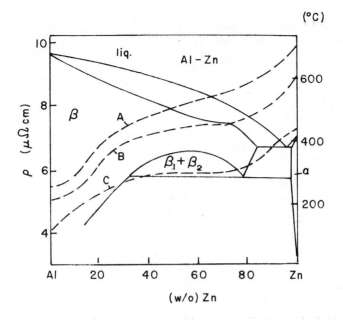

FIGURE 3. Phase diagram of Al-Zn (full lines) and isotherms (dashed lines) (°C). (A) 250; (B) 200; (C) 100.

Table 1

(a/o) Zn	$\varrho(4.2°K)$ ($\mu\Omega$cm)	$\varrho(77°K)$ ($\mu\Omega$cm)
0	0.00231	0.237
0.072	0.0180	0.265
0.125	—	0.722
0.224	0.0500	0.300
1.0	0.229	0.434
2.0	0.459	0.670
3.4	0.680	0.948
4.9	1.210	1.365
5.9	1.387	1.589
6.8	1.580	1.877
8.5	1.981	2.222
10.0	2.190	2.691

Table 2

Zn (w/o)	ϱ(Room temperature) ($\mu\Omega$cm)
0.00	2.77
4.89	3.34
8.02	3.67
10.00	4.01
11.98	4.43
15.08	4.63
17.80	4.83
19.44	5.05
25.00	5.04
30.82	4.88
35.20	5.09
41.08	5.06
48.95	5.14
57.17	5.56
63.77	5.47
70.30	5.56
75.30	5.38
77.95	5.53
80.00	5.5
82.30	5.65
86.10	5.7
96.05	5.88

REFERENCES

1. Hansen, M. and Anderko, K., *Constitution of Binary Alloys,* 2nd ed., McGraw Hill, New York, 1958; Mondolfo, L. F., *Aluminum Alloys Structure and Properties,* Butterworths, London, 1976.
2. Yrkov, V. A., Dutysheva, O., and Okolykhina, L. B., *Phys. Met. Metallogr.,* 20(4), 33, 1966.

3. Klopin, M. N., Panova, G. K. H., and Samoilov, B. n., *Sov. Phys. JETP,* 45(2), 287, 1977.
4. Osamura, K., Hiraoka, Y., and Murakami, Y., *Philos. Mag.,* 28, 321, 1973.
5. Smurgeveski, J. E., Herman, H., and Pollack, S. R., *Acta Metall.,* 17, 883, 1969.
6. Minault, J., Delafond, J., Junqua, A., Naudon, A., and Grilhe, J., *Philos. Mag. B,* 38, 255, 1978.
7. Rivaud, G., Guillot, J., and Grilhe, J., *Scr. Metall.,* 6, 411, 1972.
8. Ohta, M., Yamada, M., Kanalani, T., Hida, M., and Sakakibara, A., *J. Jpn. Inst. Met.,* 42, 946, 1978.
9. Ohta, M., Sakakibara, A., Yamada, M., and Kanadani, T., *Phys. Status Solidi A,* 48, K141, 1978.
10. Bossac, H., Fabian, H. G., and Loffler, H., *Phys. Status Solidi A,* 48, 369, 1978.
11. Sato, T., Kojima, Y., and Takahashi, T., *J. Jpn. Inst. Met.,* 42, 702, 1978.
12. Ohta, M., Kanadani, T., Sakakibara, A., and Yamada, M., *J. Jpn. Inst. Met.,* 42, 954, 1978.
13. Hillel, A. J. and Edwards, J. T., *Philos. Mag.,* 35, 1231, 1977.
14. Tomlinson, P. G. and Carbotte, J. P., *Can. J. Phys.,* 55, 747, 1977.
15. Perrin, J. and Rossiter, P. L., *Philos. Mag.,* 36, 109, 1977.
16. Lasek, J., *Czech. J. Phys. Sect. B.,* 27(9), 1059, 1977.
17. Satyanarayana, K. G. and Hirano, K., *Trans. Jpn. Inst. Met.,* 18, 403, 1977.
18. Papastaikoudis, C., *Z. Nabtuarf. A,* 32, 327, 1977.
19. Kital, T., Tokuhara, J., and Matsuda, H., *Bull. Kyushu Inst. Technol. Sci. Technol.,* 35, 117, 1977.
20. Lendvai, J., Ungar, T., and Kovacs, I., *Philos. Mag.,* 35, 1119, 1977.
21. Guyot, P. and Simon, J. P., *Scr. Metall.,* 11, 751, 1977.
22. Loffler, H. and Fabian, H. G., *Phys. Status Solidi A,* 33, 345, 1976.
23. Lasek, J., *Czech. J. Phys. B.,* 26, 184, 1976.
24. Ohta, M., Tanadaui, T., and Maeda, H., *J. Jpn. Inst. Met.,* 40, 1199, 1976.
25. Kitaⁱ, T., Tokuhara, J., and Matsuda, H., *Bull. Kyushu Inst. Technol. Sci. Technol.,* 33, 167, 1976.
26. Ohta, M., Kanadani, T., and Maeda, H., *J. Jpn. Inst. Met.,* 40, 1199, 1976.
27. Hiraoka, Y., Osamura, K., and Murakami, Y., *J. Jpn. Inst. Met.,* 40, 1134, 1976.
28. Papastaikoudis, C., Papathanasopoulos, K., and Rocofyllou, E., *J. Phys. F,* 5, 231, 19 .
29. Ciach, R., Dutkiewicz, J., Kroggel, R., Loffler, H., and Wendrock, G., *Krist. Tech.,* 10, 123, 1975.
30. Prasad, G., Sing, H. P., and Rajan, T. V., *Md. J. Technol.,* 13, 274, 1975.
31. Ceresara, S., *Philos. Mag.,* 29, 1245, 1974.
32. Prasad, G., Singh, H. P., and Rajan, T. V., *Ind. J. Technol.,* 12, 502, 1974.
33. Ciach, R., Kroggel, R., Loffler, H., and Wendrock, G., *Arch. Hutn.,* 19, 183, 1974.
34. Hori, M. and Hirano, K., *J. Jpn. Inst. Met.,* 37, 135, 1973.
35. Osamura, K., Hiraoka, Y., and Murakami, Y., *Philos. Mag.,* 28, 809, 1973.
36. Osamura, K., Hiraoka, Y., and Murakami, Y., *Philos. Mag.,* 28, 321, 1973.
37. Anantharaman, T. R. and Satyanarayana, K. G., *Scr. Metall.,* 7, 189, 1973.
38. Murakami, Y., Kawano, O., and Yoshida, H., *Annu. Rep. Res. Reactor Inst. Kyoto Univ.,* 3, 139, 1970.
39. Murakami, M., Kawano, O., and Murakami, Y., *J. Inst. Met.,* 99, 60, 1971.
40. Umakoshi, Y., Yamaguchi, M., and Mima, G., *Trans. Jpn. Inst. Met.,* 12, 7, 1971.
41. Ohta, M. and Hashimoto, F., *J. Jpn. Inst. Met.,* 36, 321, 1972.
42. Raman, K. S., *Scr. Metall.,* 5, 791, 1971.
43. Henman, H. and Cohen, J. B., *Nature (London),* 191, 63, 1961.
44. Hovi, V. and Erling, Y., *Ann. Acad. Sci. Fenn. A VI,* 20, 11, 1959.
45. Bartsch, G., *Acta Metall.,* 12, 270, 1964.
46. Hashimoto, F. and Ohta, M., *J. Phys. Soc. Jpn.,* 19, 150, 1964.
47. Schula, W., *Philos. Mag.,* 10, 913, 1964.
48. Lasek, J., *Phys. Status Solidi,* 5, K117, 1964.
49. Krishna Rao, K., Katz, L. E., and Herman, H., *Mater. Sci. Eng.,* 1, 263, 1967.
50. Turnball, D., Rosenbaum, H. S., and Treaflis, H. N., *Acta Metall.,* 8, 277, 1968.
51. Dartyge, E., Dartyge, J. M., Lambert, M., and Guiner, A., *J. Phys.,* 30, 82, 1969.
52. Dwarakadasa, E. S., Ramen, K. S., and Vasu, K. T., *Scr. Metall.,* 3, 327, 1969.
53. Raman, K. S., Das, E. S. D., and Vasu, K. I., *Scr. Metall.,* 4, 197, 1970.
54. Zelyavskii, V. B. and Kharikov, E. L., *Ukr. Fiz, Zh.,* 17, 1873, 1972.
55. Murakami, M., Kikuchi, M., Kawano, O., and Murakami, Y., *J. Jpn. Inst. Met.,* 36, 1164, 1972.
56. Dwarakadasa, E. S., *Scr. Metall.,* 6, 187, 1972.
57. Renaud, G., Riviere, J. P., and Grilhe, J., *Scr. Metall.,* 6, 65, 1972.
58. Dukin, V. V. and Aleksandrov, B. N., *Pseudopotential Calculations of the Residual Resistivities of Dilute Solid Alloys Based on Normal Metals, Physics of Low Temperatures,* Akademia Nauk Uk SSR Academy of Science, USSR, Kharkov, 1978, 1.
59. Hirano, K. and Hori, H., *J. Jpn. Inst. Met.,* 36, 97, 1972.
60. Ohta, M. and Hashimoto, F., *J. Phys. Soc. Jpn.,* 19, 1987, 1964.

61. Ceresara, S. and Fiorini, P., *Mater. Sci. Eng.,* 10, 205, 1972.
62. *International Critical Tables,* Vol. 6, McGraw-Hill, New York, 1929, 156.

Al-Zr (Aluminum-Zirconium)

More than 11 a/o Al can dissolve in α-Ti at 940°C and about 26 a/o Al can dissolve in β-Ti at 1350°C.[1] Hansen and Anderko[1] list nine intermetallic compounds. Zr is practically not soluble in Al.

Schulson and Turner[2] prepared a Zr_3Al sample by nonconsumable arc-melting elements of 99.999% purity. After rolling at 1375°K in the two-phase field, they obtained finally a fully ordered polycrystalline matrix. The total concentration of second-phase particles (α-Zr and Zr_2Al) was less than 1%. Chief impurities were O($\approx$120 ppp), Fe($\approx$100 ppm), Cr($\approx$12 ppm) C($\leqslant$110 ppm), Hf ($\approx$45 ppm), and Cu($\approx$17 ppm). The experimental results can be described with the equation $\varrho = 45.8 + 0.128\,T - 3.96 \times 10^{-5}\,T^2\ \mu\Omega cm$. T is given in °C. The sample is metallic since $\delta\varrho/\delta T > 0$. Recovery and recrystallization were studied by Caresara et al.,[3] and the effect of superheating was studied by Varich et al.[4]

REFERENCES

1. Hansen, M. and Anderko, K., *Constitution of Binary Alloys,* 2nd ed., McGraw-Hill, New York, 1958.
2. Schulson, E. M. and Turner, R. B., *Phys. Status Solidi A,* 50, 83, 1978.
3. Caresara, S., Conserva, M., and Fiorini, P., *Mater. Sci. Eng.,* 9, 19, 1972.
4. Varich, N. I., Lyukevich, R. B., and Kolmortteva, L. F., *Fiz. Met. Metalloved.,* 27, 361, 1969.

ARSENIC (As)

As-Cu (Arsenic-Copper)

The Cu-rich side of the Cu-As phase diagram shows both a modest solubility limit of 6% as in the α-Cu primary solid solution and the presence of an intermediate phase centered at a Cu-25% As composition. This intermediate phase is suspected to be receptive to the order-disorder transformation. The As-rich end of the diagram is not established.

The electrical resistivity of the Cu-As alloys at 25°C for compositions up to approximately Cu-40 a/o As have been determined by a number of investigators[1-6] and is shown in Figure 1. The resistivity of the Cu-25 a/o As intermediate phase is approximately 60 $\mu\Omega$cm at 25°C. The resistivity of the α-Cu primary phase at 17°C is linearly related to the As concentration (see Figure 2) over almost the entire composition range of the primary phase with $d\varrho/dc(290°K) = 6.7$ $\mu\Omega$cm/(a/o)As. This value of the concentration coefficient of resistivity does not appear to change significantly with temperature (see Figure 3).

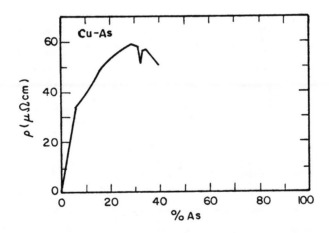

FIGURE 1.   Resistivity ($\mu\Omega$cm) of the Cu-As alloys as a function of composition (Abscissa in weight percent).[7]

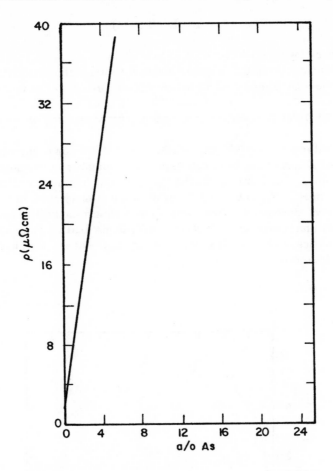

FIGURE 2.   Composition dependence of the residual resistivity of the α-Cu primary solid solution alloys at 290°K.[6]

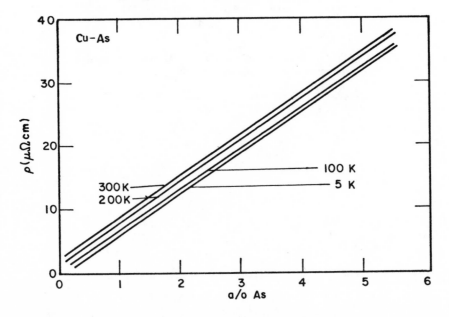

FIGURE 3.   Composition dependence of the resistivity of CuAs alloys at selected temperatures.[6]

REFERENCES

1. Pushin, N. and Dishler, K., *J. Russ. P. Chem.,* 44, 125, 1912.
2. Mathiessen, A. and Holzmann, P., *Philos. Trans. R. Soc. London,* 150, 85, 1860.
3. Matthiessen, A. and Vogt, C., *Ann. Phys.,* 2, 19, 1864.
4. Matthiessen, A. and Vogt, C., *Philos. Trans. R. Soc. London,* 154, 167, 1865.
5. Linde, J. O., *Ann. Phys.,* 15, 219, 1932.
6. Crisp, R. S., Henry, W. G., Schroeder, P. A., and Wilson, R. W., *J. Chem. Eng. Data,* 11, 556, 1966.
7. *International Critical Tables,* Vol. 6, McGraw-Hill, New York, 1929, 203.

As-Fe (Arsenic-Iron)

The Fe-As phase diagram is not established in the As-rich end. The Fe-rich end indicates a moderate solutibility of As in α-Fe and the existence of two intermetallic compounds, Fe_2As and FeAs. The residual resistivity of α-Fe As is linearly related to the As concentration (see Figure 1).

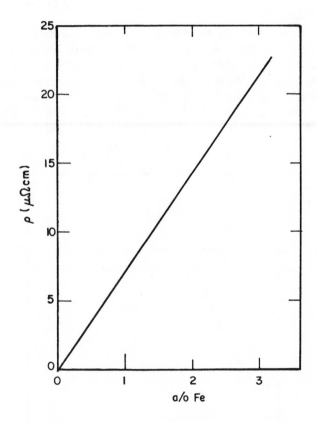

FIGURE 1.   Resistivity of FeAs alloys at 4.2°K vs. composition.[1]

REFERENCES

1. Biolluz, A., Thesis, Strasbourg, 1978.

As-Mn (Arsenic-Manganese)

Hansen and Anderko[1] suggest that a few percent of As will dissolve in Mn. The intermediate phases, Mn_2As, Mn_4As_3, and MnAs, are given in their phase diagram.

Bärner[2] grew MnAs whiskers at 950°C in As vapor on a polycrystalline substrate. He found $\varrho(24°C) = 6.2 \pm 0.5 \times 10^{-4}$ Ωcm. The sample should be paramagnetic at this temperature. Fisher and Pearson[3] obtained $\varrho(50°C) = 6.7 \times 10^{-4}$ Ωcm. The resistivity increases at the transition temperature ($T_n\uparrow = 23°C$, $\Delta T_u = 3 - 6°C$) by a factor of 2.5.[2]

REFERENCES

1. Hansen, M. and Anderko, K., *Constitution of Binary Alloys,* 2nd ed., McGraw-Hill, New York, 1958.
2. Bärner, K., *Phys. Lett. A,* 35, 333, 1971.
3. Fisher, G. and Pearson, W. B., *Can. J. Phys.,* 36, 1010, 1957.

As-Ni (Arsenic-Nickel)

The limit of solid solubility of As in Ni has not been accurately established at 29°C. The limit appears to be about 4 a/o As at 900°C. Three intermetallic compounds, Ni_5As_2, Ni_3As_2, and NiAs, apparently exist in the Ni-rich end of the phase diagram, where all the alloy compositions except for these stoichiometric compounds exist as two-phase structures. The As end of the phase diagram has not been determined.

The residual resistivity of several dilute NiAs alloys has been determined, and their concentration dependence (see Figure 1) yields a coefficient of $d\varrho/dc = 4$ $\mu\Omega$cm/(a/o)As.

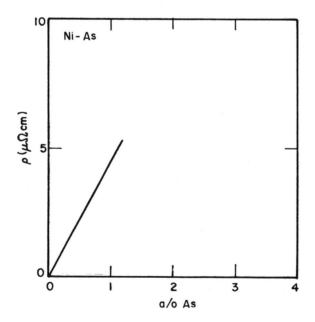

FIGURE 1.   Residual resistivity of NiAs alloys vs. concentration.[1]

REFERENCES

1. Hugel, J., *J. Phys. F,* 3, 1723, 1973.

As-Pb (Arsenic-Lead)

Dukin and Aleksandrov[1] gave a critical account of pseudopotential calculations and the resistivity increase due to 1% impurity. They gave $[\Delta\varrho/\Delta c]_{exp} = 1.15$ $\mu\Omega cm/(a/o)As$ in PbAs.

REFERENCES

1. Dukin, V. V. and Aleksandrov, B. N., *Pseudopotential Calculations of the Residual Resistivities of Dilute Solid Alloys Based on Normal Metals, Physics of Low Temperatures*, Akademia Nauk Uk SSR Academy of Science, USSR, Kharkov, 1978, 1.

As-Sn (Arsenic-Tin)

Dukin and Aleksandrov[1] gave a critical account of pseudopotential calculations and the resistivity increase due to 1% impurity. They gave $[\Delta\varrho/\Delta c]_{exp} = 2.0$ $\mu\Omega cm/(a/o)As$ in SnAs.

REFERENCES

1. Dukin, V. V. and Aleksandrov, B. N., *Pseudopotential Calculations of the Residual Resistivities of Dilute Solid Alloys Based on Normal Metals, Physics of Low Temperatures*, Akademia Nauk Uk SSR Academy of Science, USSR, Kharkov, 1978, 1.

As-Tl (Arsenic-Thallium)

Dukin and Aleksandrov[1] gave a critical account of pseudopotential calculations and resistivity increase due to 1% impurity. They gave $[\Delta\varrho/\Delta c]_{exp} = 4.6$ $\mu\Omega cm/(a/o)As$ in TlAs.

REFERENCES

1. Dukin, V. V. and Aleksandrov, B. N., *Pseudopotential Calculations of the Residual Resistivities of Dilute Solid Alloys Based on Normal Metals, Physics of Low Temperatures*, Akademia Nauk Uk SSR Academy of Science, USSR, Kharkov, 1978, 1.

As-U (Arsenic-Uranium)

Henkie and Bazan[1] prepared U_3As_4 single crystals by the chemical transport method. These materials crystallize in the bcc lattice of the Th_3P_4 type. Below 198°K, U_3As_4 shows magnetic ordering, with a very large magnetic anisotropy. The easy axis of magnetization is the <111> direction. The electrical resistivity of the compounds at 300°K is

Orientation	Resistivity measured paralled to			
	<100>	<110>	111	Polycrystalline
Resistivity	499	491	499	393 ($\mu\Omega cm$)
	485	—	—	—

$\varrho(T)/\varrho$ (300°K) is given in Figure 1. Multiple cooling can modify $\varrho(T)/\varrho(300°K)$.

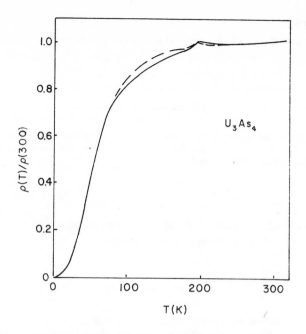

FIGURE 1.   Dashed line, current parallel (100); full line, current parallel (110).

REFERENCES

1. Henkie, Z. and Bazan, C., *Phys. Status Solidi A*, 5, 259, 1971.

As-V (Arsenic-Vanadium)

Henkie and Bazan[1] prepared V_3As_4 single crystals by the chemical transport method. These materials crystallize in the bcc lattice of the Th_3P_4 type. Below 198°K, V_3As_4 shows magnetic ordering, with a very large magnetic anisotropy. The easy axis of magnetization is the <111> direction. The electrical resistivity of the compounds at 300°K is

	Measured parallel to			
	<100>	<110>	111	Polycrystalline
V_3As_4	499	491	499	393 $\mu\Omega$cm
	485	—	—	—

These authors measured $\varrho(T)/\varrho$ (300°K). Multiple cooling can modify $\varrho(T)/\varrho(300°K)$.

REFERENCES

1. Henkie, Z. and Bazan, C., *Phys. Status Solidi A*, 5, 259, 1971.

GOLD (Au)

Au-Bi (Gold-Bismuth)

The elements are essentially insoluble in each other.[1] They form the compound Au_2Bi which decomposes at 373°C. Table 1[2] summarizes older data.

Table 1

Au (w/o)	ϱ (μΩcm)	T (°C)
1.1	152.5	24.0
2.3	154.3	21.6
4.5	141.3	19.9
10.6	123.2	21.9
19.23	108.4	22.6
32.2	84.6	13.7
48.66	52.2	14.3

REERENCES

1. Hansen, M. and Anderko, K., *Constitution of Binary Alloys*, 2nd ed., McGraw-Hill, New York, 1958.
2. *International Critical Tables*, Vol. 6, McGraw-Hill, New York, 1929, 156.

Au-Cd (Gold-Cadmium)

The existence of seven intermediate phases has been reported.[1] The Au-rich primary solid solution extends to about 17 a/o Cd at room temperature and to a maximum of 32.5 a/o Cd at 625°C. The maximum solubility for Au in Cd is 7.9 a/o Au at 309°C. The resistivity of AuCd alloys has been measured.[2] Their data are given in Table 1. Linde[10] gives for $\Delta\varrho/\Delta c$ for dilute alloys 0.63 μΩcm/(a/o)Cd. Data from the *International Critical Tables* are given in Table 2. Most of the resistivity measurement in the Au-Cd system are associated with the study of defects,[3-8] especially with the investigation of the martensitic transformation of $Au_{52.5}Cd_{47.5}$ and $Au_{50.3}Cd_{49.5}$ (Figures 1 to 3).

Miura et al.[3] prepared Au-Cd samples by melting 99.99% purity Au and 99.999% purity Cd in quartz sealed tubes at about 1373°K. The single crystals were grown by the Tammann method. They were homogenized at 923°K for 15 hr and quenched. Figure 1 shows the transformation behavior of annealed and slowly cooled Au-47.5 a/o Cd and Au.-49 a/o Cd and of a Au-47.5 a/o Cd sample quenched from 703 to 353°K. A clear interpretation of the mechanism responsible for the absolute value of the resistance change is presently not available.

Gefen et al.[4] investigated the pressure dependence of the martensitic $\beta \leftrightarrow \beta'$ transition (β = CsCl structure, β'-orthorhombic). Samples were prepared by melting 99.999% purity Au and 99.999% Cd in evacuated quartz tubing 30 min. Finally, the quartz tube was tilted and the melt was cast. The resistivity change with temperature was about rates of 1°C/min. Figure 3 shows results for zero pressure of a sample with 47.5 a/o Cd. The results agree well with the data obtained by Wechsler and Read.[7] They prepared $Au_{100-x}Cd_x$ compounds with x = 47.5 or 49.5 a/o Cd by a modified Bridgeman technique. The samples were quenched from 420 to 70°C. Both samples are in the stable CsCl structure. Further cooling leads then to transformation into the orthorhombic phase. The difference in ϱ above the transition temperature could be associated with different degrees of order. However, these compounds are slightly ordered even at high temperatures. Wechsler and Read associate the difference in the CsCl structure compounds with different amounts of excess vacancies.

Dukin and Aleksandrov[9] gave a critical account of pseudopotential calculations and the resistivity increase due to 1% impurities. They gave $[\Delta\varrho/\Delta c]_{exp} = 0.64$ $\mu\Omega$cm/(a/o)Au in C̲d̲Au.

Table 1

(a/o)Cd	Disordered alloys $\varrho(\mu\Omega\text{cm})$	Cold worked alloys $\varrho(\mu\Omega\text{cm})$	Ordered $\varrho(\mu\Omega\text{cm})$
1	2.80	2.85	2.80
2	3.30	3.40	3.31
4	4.33	4.49	4.36
8	6.18	6.55	6.28
14	8.65	9.34	8.80
20	10.90	11.83	9.37

Table 2

Cd (w/o)	ϱ(Tempered, 25°C) ($\mu\Omega$cm)	Temperature coefficient α (Tempered, 25°C) (10^{-3}/°C)
0	2.20	4.45
5.05	4.04	2.62
14.44	6.81	2.41
30.36	12.36	1.35
35.44	15.4	1.31
40.77	10.08	2.16
43.23	10.03	2.34
44.05	10.23	2.34
46.34	9.2	2.67
48.4	7.71	6.24
49.76	7.39	6.36
50.27	7.47	5.67
51.36	8.72	3.06
55.89	14.31	1.89
60.32	17.7	1.93
70.21	14.73	2.46
72.93	14.51	2.73
74.07	13.9	3.50
74.89	13.27	4.17
75.87	15.03	2.71
78.56	33.4	1.64
81.89	28.15	2.35
84.93	26.45	2.35
88.01	18.65	2.97
89.87	16.34	2.60
91.08	15.17	2.50
92.17	14.3	2.20
92.95	12.02	2.84
94.05	10.9	3.11
94.95	9.49	3.31
98.1	9.205	3.39
99.08	8.95	3.01
100	7.26	4.32

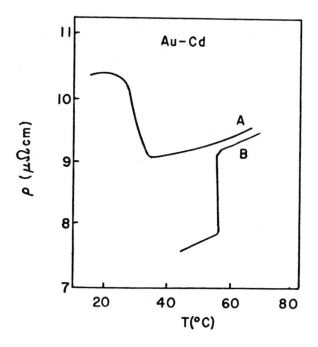

FIGURE 1. Resistivity of $Au_{52.5}Cd_{47.5}$. (A) Quench from 410
to 70°C; (B) slowly cooled 47.5 a/o Cd.

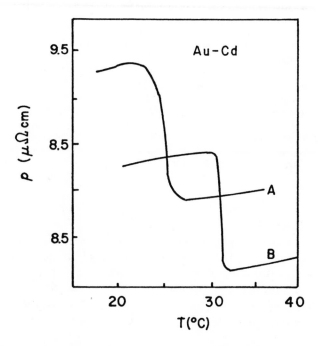

FIGURE 2. Resistivity of $Au_{50.5}Cd_{49.5}$. (A) Quench from 410
to 70°C; (B) slowly cooled.

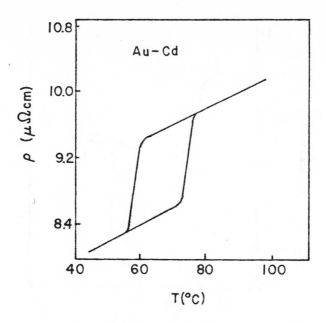

FIGURE 3.   Resistivity of $Au_{52.5}Cd_{47.5}$.

REFERENCES

1. Hansen, M. and Anderko, K., *Constitution of Binary Alloys*, 2nd ed., McGraw-Hill, New York, 1958.
2. Köster, W. and Hauk, J., *Z. Metallkd.*, 56, 846, 1965.
3. Miura, S., Hori, F., and Nakanishi, N., *Scr. Metall.*, 12, 265, 1978.
4. Gefen, Y., Halwany, A., and Rosen, M., *Philos. Mag.*, 28, 1, 1973.
5. Livingston, H., *J. Appl. Phys.*, 43, 4944, 1972.
6. sturm, W. J. and Wechsler, M. S., *J. Appl. Phys.*, 28, 1509, 1957.
7. Wechsler, M. S. and Read, T. A., *J. Appl. Phys.*, 27, 194, 1956.
8. Nakanishi, N. and Wayman, C. M., *Trans. Jpn. Inst. Met.*, 4, 179, 1963.
9. Dukin, V. V. and Aleksandrov, B. N., *Pseudopotential Calculations of the Residual Resistivities of Dilute Solid Alloys Based on Normal Metals, Physics of Low Temperatures,* Akademia Nauk Uk SSR Academy of Science, USSR, Kharkov, 1978, 1.
10. Linde, J. O., *Helv. Phys. Acta,* 41, 1013, 198.
11. *International Critical Tables,* Vol. 6, McGraw-Hill, New York, 1929, 156.

Au-Co (Gold-Cobalt)

Below 400°C Au and Co are not mutually soluble in each other, and their alloys are a simple two-phase mixture of Au and Co for all compositions. The electrical resistance of dilute AuCo alloys have been measured at low temperatures and a resistance minimum is frequently observed (see Figure 1). Stewart and Huebner[3] studied the validity of Matthiessen's rule and found $\Delta\varrho/c = 1.25$ $\mu\Omega$cm/(a/o)Co in $\underline{Au}$Co. Linde[2] gives for $\Delta\varrho/\Delta c$ for dilute alloys a value of 5.8 $\mu\Omega$cm/(a/o)Co for $\underline{Au}$Co.

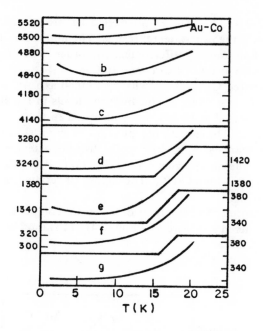

FIGURE 1.   Resistivity minimum in AuCo alloys.
Resistivity given in arbitrary units. (a) 0.44% Co, (b)
0.32% Co, (c) 0.25% Co, (d) 0.16% Co, (e) 0.055%
Co, (f) Au.

REFERENCES

1. Gerritsen, A. N., *Physica*, 25, 489, 1959.
2. Linde, J. O., *Helv. Phys. Acta*, 41, 1013, 1968.
3. Stewart, R. G. and Huebner, R. P., *Phys. Rev. B*, 1, 3323, 1970.

Au-Cr (Gold-Chromium)

The phase diagram of Au-Cr indicates[1] that about 24 a/o Cr can dissolve in Au at 1072°C and about 6 a/o Au can dissolve in Cr at the sample temperature. Only one intermediate phase is shown which is stable between 1033 and 1152°C.[1] However, Toth et al.[10] reported the existence of a Au$_4$Cr phase.

The resistivity of Au-Cr alloys[2-16] is interesting because of magnetic effects. Cr in Au leads to a resistance minimum;[4,9,11] Au in Cr modifies the Neel temperature.[2] AuCr alloys with 3.3 and 10.6 a/o Cr have been regarded as examples of spin-glasses.[3,7] It has been proposed that alloys with a few percent Cr may be suitable materials for resistance standards.[14,15]

Inouo et al.[8] prepared alloys from 99.99% Au and 99.9% Cr in a graphite crucible in vacuum at 1400°C. The samples were homogenized at about 1000°C for several hours in evacuated and sealed quartz tubes. After working, the samples were reannealed at about 1000°C for 1 hr and quenched into water to obtain the disordered state. $\varrho(T)$ for the alloys is shown in Figure 1. The alloys show a broad peak in $\varrho(T)$ above the Neel temperature. Changes in ordering may take place at higher temperatures. They may be responsible for part of the decrease in ϱ with increasing temperature.

Shiozaki et al.[4] prepared AuCr alloys by induction melting appropriate amounts of 99.99% pure Au and 99.9% pure Cr in an Ar atmosphere; 99.99% pure Cr was used for an alloy with 25.0 a/o Cr. The ingots were drawn to wires after several recovery anneals. Figure 2 gives the experimental results. The minimum in ϱ is found only for alloys which show long-range antiferromagnetic order. The depth of the resistivity minimum increases with increasing Cr concentration.

Eroglu et al.[2] prepared C̲r̲Au alloys with up to 0.60 a/o Au in an arc furnace. Au was 99.999% pure; Cr was Marz grade. Impurity levels in the alloys were <60 ppm. The samples were annealed at 1000°C for 6 days and furnace cooled. Figure 3 shows the experimental results. The curves are typical for itinerant electron antiferromagnets. The minimum in ϱ near 300°K is close to the transition temperature, T_N, from the antiferromagnetic to the paramagnetic state. Theoretical considerations suggest that the minimum in $d\varrho/dT$ is found at T_N. The change of residual resistivity with composition is 14 $\mu\Omega$cm/(a/o)Au.

Toth et al.[10] measured the resistivity of a Au_4Cr compound, appropriately heat treated to give an ordered structure (see Figure 4). T_N is the Neel temperature and T_o is the ordering temperature.

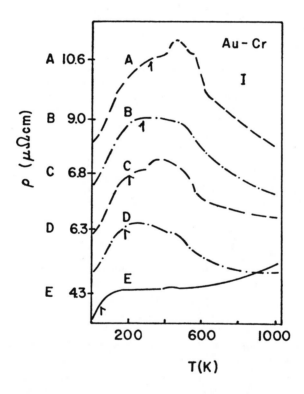

FIGURE 1.   Resistivity of Au-Cr alloys. Origin of curves are shifted. Bar, I, corresponds to 1 $\mu\Omega$cm. Arrow indicates Neel temperature(a/o Cr). (A) 25.6; (B) 23.7; (C) 16.9; (D) 15.8; (E) 4.6.

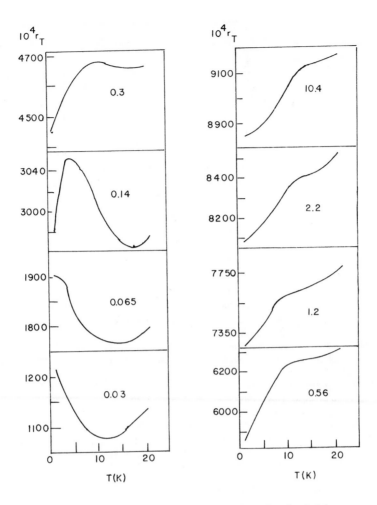

FIGURE 2.   Resistivity of $\underline{Cr}$Au alloys. r_T is a reduced resistivity.

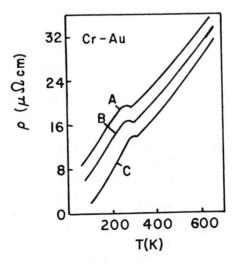

FIGURE 3.   Resistivity of $\underline{Cr}$Au alloys. (A) 0.6 a/o Au; (B) 0.27 a/o; (C) Cr.

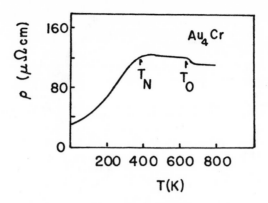

FIGURE 4.   Resistivity of Au_4Cr. T_N = Neel temperature; T_O = atomic ordering.

REFERENCES

1. Hansen, M. and Anderko, K., *Constitution of Binary Alloys,* McGraw-Hill, New York, 1958.
2. Eroglu, A., Arajs, S., Moyer, C. A., and Rao, K. V., *Phys. Status Solidi B,* 87, 287, 1978.
3. Ford, P. J., Babic, E., and Mydosh, J. A., *J. Phys. F,* 3, L75, 1973.
4. Shiozaki, Y., Nakai, Y., and Kunitomi, N., *Solid State Commun.,* 12, 429, 1973.
5. Kunitomi, N., Sakamoto, H., and Nakai, Y., *12th Int. Conf. on Low Temp. Physics,* Kana, E. Ed., Keigaku Publ. Co., Tokyo, 1971, 324, 347.
6. Shiozka, Y., *Solid State Commun.,* 12, 429, 1973.
7. Ford, P. J. and Mydosh, J. A., *Phys. Rev. B,* 14, 2057, 1970.
8. Inouo, K., Nakamura, Y., and Kunitomi, N., *J. Phys. Soc. Jpn.,* 27, 1159.
9. Davis, L. A. and Gordon, R. B., *Rev. Sci. Instrum.,* 38, 371, 1967.
10. Toth, R. S., Arrott, A., Shinuzaki, S., Werner, S. A., and Sato, H., *J. Appl. Phys.,* 40, 1373, 1969.
11. Birch, J. A., Kemp, W. R. G., Klemens, P. G., and Tainsh, R. J., *Intr. J. Phys.,* 12, 455, 1959.
12. Teutsch, W. B. and Love, W. F., *Phys. Rev.,* 105, 487, 1957.
13. Schulze, A. and Eicke, H., *Z. Angew. Phys.,* 4, 321, 1952.
14. Linde, J. O., *Nature (London),* 165, 645, 1950.
15. Schulze, A., *Phys. Z.,* 41, 121, 1940.
16. Thomas, J. L., *Bur. Stand. J. Res.,* 13, 681, 1934.

Au-Cu (Gold-Copper)

Cu and Au form a complete series of disordered solid solution alloys above 410°C. The order-disorder reaction occurs at the Cu_1Au_1 composition below 410°C and occurs at the Cu_3Au_1 composition below 390°C.

The resistivity of the Cu-Au alloy system has been extensively investigated, with particular attention focused on the resistivity changes accompanying the order-disorder reactions. The resistivity of the disordered concentrated alloys obeys Nordheim's rule (see Figure 1), and a systematic decreasing of the resistivity is evident when the Cu_1Au_1 and Cu_3Au_1 quenched alloys are heated to achieve an ordered structure.

The ordered state also influences the temperature dependence of the resistivity as shown for the Cu_1Au_1 alloys (see Figure 2). The Cu_1Au_3 composition has been suspected to experience the order-disorder reaction, but the electrical resistivity of this alloy behaves anomolously as this alloy is heat treated to realize the ordered state (see Figure 3). The resistivity appears to increase initially with heat treatment before decreasing. The resistivity of the final annealed state does not appear to be significantly different than the resistivity of the initial state suggesting that Cu_1Au_3 is not a order-disorder alloy.

The resistivity of the dilute CuAu alloys is linear with Au content (see Figure 4), giving K(CuAu) = 0.55 $\mu\Omega$cm/(a/o)Au. While the resistivity of dilute AuCu alloys gives K(AuCu) = 0.45 $\mu\Omega$cm/(a/o)Cu.

The DMR have been investigated for the dilute CuAu alloys. The deviation is dependent on the composition and temperature for temperature lower than 300°K (see Figure 5) and shows a maximum value at about 60°K. At room temperature and above, the deviation is relatively small and temperature insensitive. Deviations from Matthiessen's rule have also been determined for a dilute AuCu alloy and were found to be small and almost temperature independent.[12]

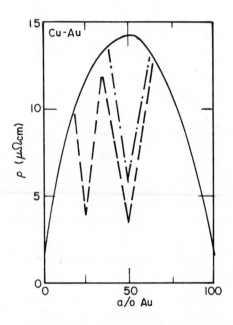

FIGURE 1.   Resistivity of Au-Cu at 200°C. Full line quenched from 600°C. Dash dot line cooled slowly. Dashed line tempered at 200°C for 120 hr.

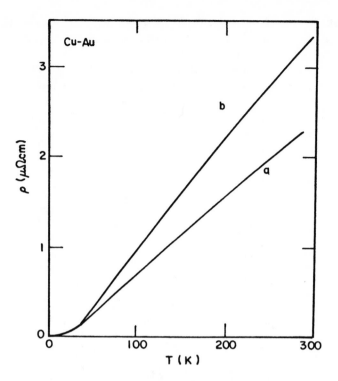

FIGURE 2.   Temperature-dependent part of resistance of CuAu. Curves a and b correspond to the state quenched from 500°C and the state cooled slowly from 350 to 200°C in 13 days, respectively.[2]

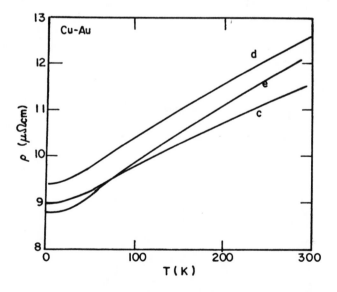

FIGURE 3.   Electrical resistivity of Cu_1Au_3 as a function of temperature. (c) Quenched from 450°C; (d) annealed for 11 days at 180°C; (e) annealed an additional 11 days at 150°C, then 10 days at 120°C and 2 days at 100°C.[2]

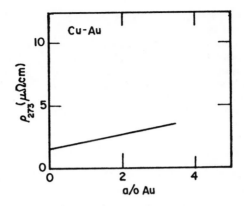

FIGURE 4.　Resistivity of dilute C̲u̲Au alloys at
273°K.[18]

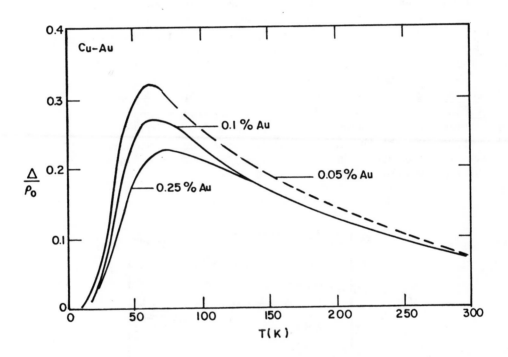

FIGURE 5.　Variation with temperature of $\Delta(c,T)/\varrho_o(c)$ for Cu-Au alloys.[8]

REFERENCES

1. Johansson, C. H. and Linde, J. O., *Ann. Phys.,* 25, 1, 1936.
2. Hirabayashi, M. and Muto, Y., *Acta Metall.,* 9, 497, 1961.
3. Passoglia, E. and Love, W. F., *Phys. Rev.,* 98, 1006, 1955.
4. Seemann, H. J., *Z. Phys.,* 62, 824, 1930.
5. Bowen, D. B., *Acta Metall.,* 2, 573, 1954.
6. Anguetil, M. C., *J. Phys. Paris,* 23, 113, 1962.
7. Rossiter, P. L. and Bykovec, B., *Philos. Mag.,* 38, 55, 1978.
8. Dugdale, J. S. and Basinski, Z. S., *Phys. Rev.,* 157, 52, 1962.
9. Lengeler, B., Schilling, W., and Wenzl, H., *J. Low Temp. Phys.,* 2, 59, 1970.
10. Stewart, R. G. and Heubener, R. P., *Phys. Rev.,* 1, 3323, 1970.
11. Damon, D. H., Mathur, M. P., and Klemens, P. G., *Phys. Rev.,* 176, 876, 1968.

12. Damon, D. H. and Klemens, P. G., *Phys. Rev. A*, 138, 1390, 1965.
13. Linde, J. O., *Ann. Phys.*, 10, 52, 1931.
14. MacDonald, D. K. G. and Pearson, W. B., *Acta Metall.*, 3, 392, 1955.
15. Rahim, C. A. and Barnard, R. D., *J. Phys. F*, 8, 1957, 1978.
16. Matthiessen, A., *Philos. Trans. R. Soc. London*, 150, 161, 1860.
17. Matthiessen, A. and Vogt, C., *Ann. Phys.*, 2, 19, 1864; *Philos. Trans R. Soc. London*, 154, 167, 1865.
18. Gerritsen, A. N., *Handbook of Physics*, Springer-Verlag, Berlin, 1957.

Au-Fe (Gold-Iron)

At 25°C, Fe appears to be modestly soluble in solid Au and the solubility limit increases rapidly with temperature. Au does not detectably dissolve in the Fe lattice until about 700°C. At 25°C, AuFe compositions with more than 10% Fe are two-phase alloys of α-Au and Fe.

The resistivity of several primary phase stable and metastable Au-Fe alloys quenched from 900°C was determined as a function of temperature (see Figure 1). The resistivity was found to decrease rapidly just below the Curie temperature in these alloys due to the advent of ferromagnetic ordering. Aging at 75°C for 1 week did not appreciably lower the resistivity of 12 and 24% Fe metastable alloys.

The dilute AuFe alloys are Kondo systems, and their resistivities decrease with increasing temperature at low temperatures (see Figure 2). The residual resistivity of the dilute AuFe alloys is linearly related to the iron concentration (see Figure 3) where $d\varrho$ (AuFe)/dc = 7.4 $\mu\Omega$cm/(a/o)Fe. The deviations from Matthiessen's rule have been determined for both dilute and concentrated Au-Fe alloys. The results (see Figures 4 to 6) show that the deviation increases with increasing temperature up to 300°K for the concentrated alloys. The maximum value for the DMR for the more dilute AuFe alloy generally occurs around 100 to 200°K, depending upon the composition.

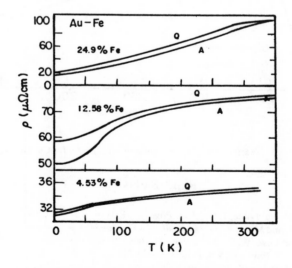

FIGURE 1.   Temperature dependence of ϱ for 4.53, 12.58, and 24.9% Fe alloys. The arrows refer to previously measured Curie temperatures. Q-quenched from 900°C. A-annealed.[1]

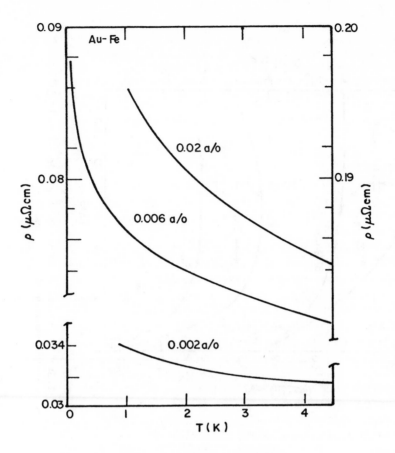

FIGURE 2.   Resistivities of A̲u̲Fe alloys.[2]

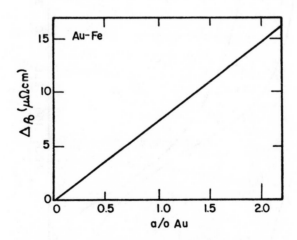

FIGURE 3.   Residual resistivity ϱ_o ($\mu\Omega$cm) vs. concentration c (a/o) for some of the more dilute A̲u̲Fe alloys studied.

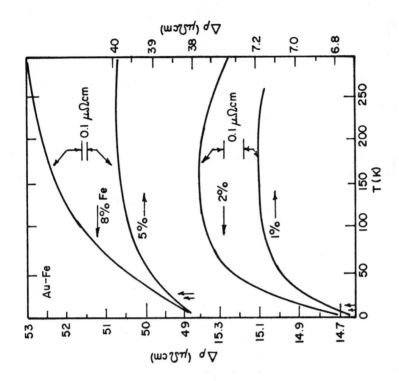

FIGURE 5.   Overall temperature variation of $\Delta\varrho$, ($\mu\Omega$cm) for $\underline{\text{Au}}$Fe alloys with concentrations of 1, 2, 5, and 8 a/o Fe.[4]

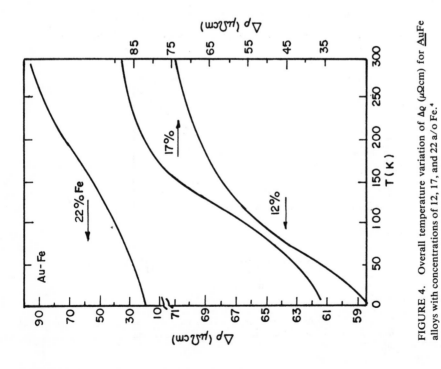

FIGURE 4.   Overall temperature variation of $\Delta\varrho$ ($\mu\Omega$cm) for $\underline{\text{Au}}$Fe alloys with concentrations of 12, 17, and 22 a/o Fe.[4]

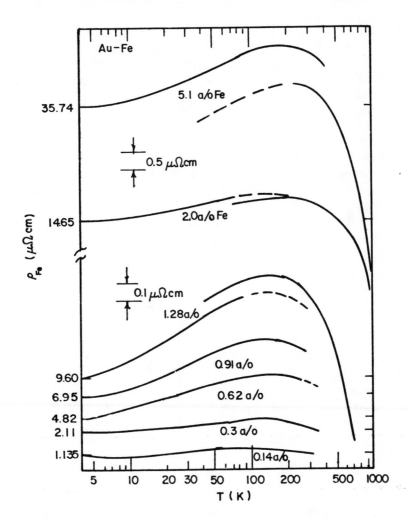

FIGURE 6.   Variation with temperature of the impurity contribution of the resistivity of <u>Au</u>Fe alloys.[7]

REFERENCES

1. Sundahl, R. C., Chen, T., Sivertsen, J. M., and Sato, Y., *J. Appl. Phys.*, 37, 1024, 1966.
2. MacDonald, D. K. C., Pearson, W. B., and Templeton, I. M., *Proc. R. Soc. London, Ser. A*, 266, 161, 1962.
3. Loram, J. W., Whall, T. E., and Ford, P. J., *Phys. Rev. B*, 2, 857, 1970.
4. Mydosh, J. A., Ford, P. J., Kawatra, H. P., and Whall, T. E., *Phys. Rev. B*, 10, 2845, 1974.
5. Gerritsen, A. N., *Physica*, 23, 1087, 1957.
6. Linde, J. O., *Ann. Phys.*, 10, 32, 1931.
7. Domenically, C. A. and Tristenson, F. A., *Phys. Rev.*, 95, 1134, 1954.

Au-Ga (Gold-Gallium)

Hansen and Anderko[1] show in the Au-Ga phase diagram four intermediate phases. The boundary of the Au-rich side is only approximately outlined. The solubility of Ga in Au is 11.15 a/o at 500°C, 11.8 a/o at 400°C, and 9.4 a/o at 300°C. Au is practically insoluble in Ga.

Michikami and Yamaguchi[2] prepared Au-Ga alloys from 99.99% Au and 99.99% Ga. These elements were melted together in evacuated sealed quartz tubes and then cast into quartz tubes. Figure 1 gives $\varrho(T)$ of three alloys. The pronounced humps are associated with phase transformation. Figure 2 gives $\varrho(20°C)$ as a function of composition. Data obtained by Rapp et al.[3] for $\varrho(295°K)$ and $d\varrho/dT$ at 260°K are given in Table 1. Au was 99.9999% w/o pure, Ga had 0.1 wt ppm Ag ad 0.3 wt ppm Cu as impurities. Alloys were arc melted in an Ar atmosphere, wrapped in Ta and Ar foil, sealed in a quartz tube with 1/2 atm He, annealed for 12 hr, and quenched to room temperature. Resistivity values obtained by Köster and Hauk[4] are slightly smaller than those by Rapp et al.[3]

Jan and Pearson[5] prepared $AuGa_2$ which has a cubic structure of the fluorite type by melting components in a sealed evacuated tube and passing several molten zones through it. This gave essentially a single crystal, except for a few globular inclusions of a second phase. Figure 3 gives $\varrho(T)$.

Dukin and Aleksandrov[6] gave a critical account of pseudopotential calculations and the resistivity increase due to 1% impurity. They gave $[\Delta\varrho/\Delta c]_{exp} = 16.0 \ \mu\Omega cm/(a/o)Au$ in $\underline{Ga}Au$.

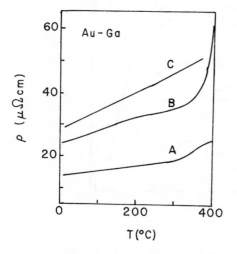

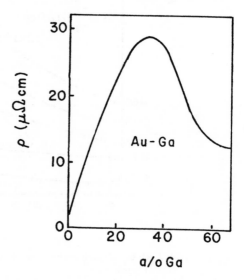

FIGURE 1.   Resistivity of Au-Ga alloys (a/o Ga). (A) 9.2; (B) 23.5; (C) 31.0.

FIGURE 2.   Room temperature (20°C) resistivity of Au-Ga alloys.

Table 1

(a/o)Ga	ϱ at 295°K ($\mu\Omega$cm)	$\partial\varrho/\partial T$ at 260°K (nΩcm/°K)
0	2.24	8.36
1.0	4.46	8.60
2.0	6.43	9.04
3.0	8.37	9.69
4.0	10.49	10.37
5.0	12.40	10.37
6.0	14.29	10.55
8.0	18.30	11.53
10.0	21.79	12.43

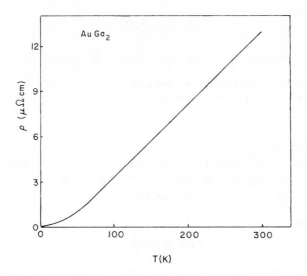

FIGURE 3.   Resistivity of AuGa₂.

REFERENCES

1. Hansen, M. and Anderko, K., *Constitution of Binary Alloys,* 2nd ed., McGraw-Hill, New York, 1958.
2. Michikami, O. and Yamaguchi, Y., *Jpn. J. Appl. Phys.,* 10, 660, 1971.
3. Rapp, O., Mota, A. C., and Hoyt, R. F., *Solid State Commun.,* 25, 855, 1978.
4. Koster, W. and Hauk, J., *Z. Metallkd.,* 56, 846, 1965.
5. Jan, J. P. and Pearson, W. B., *Philos. Mag.,* 8, 279, 1963.
6. Dukin, V. V. and Aleksandrov, B. N., *Pseudopotential Calculations of the Residual Resistivities of Dilute Solid Alloys Based on Normal Metals, Physics of Low Temperatures,* Akademia Nauk Uk SSR Academy of Science, USSR, Kharkov, 1978, 1.

Au-Ge (Gold-Germanium)

The phase diagram of Au-Ge shows one eutectic reaction, with the eutectic temperature of 356°C.[1] Ag is practically insoluble in Ge, but 3.2 a/o Ge will dissolve in Au at 356°C and 0.2 a/o Ge will dissolve at 200°C.

Köster and Hauk[2] measured the resistivity of disordered alloys with up to 2.5 a/o Ge. $\varrho(c)$ is a linear function of composition for all alloys. This implies that the solubility range is larger than reported.[1] Experimental results[2] are given in Table 1. Linde[3] gives for $\Delta\varrho/\Delta c$ 5.1 $\mu\Omega$cm/(a/o)Ge for <u>Au</u>Ge.

Table 1

(a/o)Ge	$\varrho(\mu\Omega$cm)
1	7.63
1.5	9.63
2	11.50
2.5	13.25

REFERENCES

1. Hansen, M. and Anderko, K., *Constitution of Binary Alloys,* 2nd ed., McGraw-Hill, New York, 1958.
2. Köster, W. and Hauk, J., *Z. Metallkd.,* 56, 846, 1965.
3. Linde, J. O., *Helv. Phys. Acta,* 41, 1013, 1968.

Au-Hg (Gold-Mercury)

Dukin and Aleksandrov[1] gave a critical account of pseudopotential calculations and the resistivity increase due to 1% impurity. They gave $[\Delta\varrho/\Delta c]_{exp} = 0.18$ $\mu\Omega$cm/(a/o)Au in HgAu. Linde[2] gives for $\Delta\varrho/\Delta c$ for dilute alloys a value of 0.4 $\mu\Omega$cm/(a/o)Au for AuHg.

REFERENCES

1. Dukin, V. V. and Aleksandrov, B. N., *Pseudopotential Calculations of the Residual Resistivities of Dilute Solid Alloys Based on Normal Metals, Physics of Low Temperatures,* Akademia Nauk Uk SSR Academy of Science, USSR, Kharkov, 1978, 1.
2. Linde, J. O., *Helv. Phys. Acta,* 41, 1013, 1968.

Au-In (Gold-Indium)

Au will dissolve up to 12.6 a/o In at 647°C and 10.4 a/o In at 400°C.[1] Au is practically insoluble in solid In. Four intermediate phases are shown to exist and are assumed to have average compositions corresponding to Au_4In, Au_3In, AuIn, and $AuIn_3$.

Damon et al.[2] measured the resistivity of AuIn alloys between 1.5 and 40°K. The samples in wire form were annealed at 750°K for 8 to 24 hr. The residual resistivities ϱ_o of the alloys are given in Table 1. Linde[4] gives for $\Delta\varrho/\Delta c$ 1.35 $\mu\Omega$cm/(a/o)In for AuIn. The temperature dependence of $\varrho - \varrho_o$ was determined as a function of temperature for alloys and pure Au. $(\varrho - \varrho_o)_{alloy} - (\varrho - \varrho_o)_{Au}$ was usually less than 10^{-8} $\mu\Omega$cm. The data were analyzed on the basis of two components, one due to phonon-assisted impurity scattering and the other on the basis of the "two band" effect.

Jan and Pearson[3] prepared a $AuIn_2$ compound which has a fluoride-type structure by melting components in a sealed evacuated tube and passing several molten zones through it. Figure 1 shows $\varrho(T)$ for this compound.

Table 1

Number	Samples	Residual resistivity ($\mu\Omega$cm)
1	Au	0.00779
2	Au	0.00910
3	Au	0.00907
4	Au + 0.29 a/o In	0.3309
5	Au + 0.58 a/o In	0.6530
6	Au + 0.86 a/o In	0.9693
7	Au + 1.15 a/o In	1.254

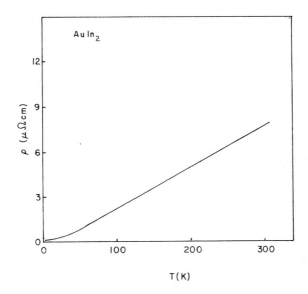

FIGURE 1.   Resistivity of AuIn₂.

REFERENCES

1. Hansen, M. and Anderko, K., *Constitution of Binary Alloys,* 2nd ed., McGraw-Hill, New York, 1958.
2. Damon, D. H., Mathur, M. P., and Klemens, P. G., *Phys. Rev.,* 176, 876, 1968.
3. Jan, J. and Pearson, W. B., *Philos. Mfg.,* 8, 279, 1963.
4. Linde, J. O., *Helv. Phys. Acta,* 41, 1013, 1968.

Au-Li (Gold-Lithium)

Dukin and Aleksandrov[1] gave a critical account of pseudopotential calculations and the resistivity increase due to 1% impurity. They gave $[\Delta\varrho/\Delta c]_{exp} = 4.8\ \mu\Omega$cm/(a/o)Au in LiAu.

REFERENCES

1. Dukin, V. V. and Aleksandrov, B. N., *Pseudopotential Calculations of the Residual Resistivities of Dilute Solid Alloys Based on Normal Metals, Physics of Low Temperatures,* Akademia Nauk Uk SSR Academy of Science, USSR, Kharkov, 1978, 1.

Au-Mg (Gold-Magnesium)

Up to 25 a/o Mg will dissolve in Au. The solubility limit is nearly temperature independent.[1] The intermediate phase, β', with average composition MgAu has a wide stability range. Three compounds have been detected: Mg₂Au, Mg₅Au₂, and Mg₃Au.

Köster and Hauk[2] measured the resistivity of AuMg alloys with up to 16 a/o Mg. Table 1 gives ϱ of alloys in the disordered, cold worked, or ordered states, probably at room temperature. Linde[3] gave for $\Delta\varrho/\Delta c$ for dilute alloys a value of 1.30 $\mu\Omega$cm/(a/o)Mg for AuMg.

Table 1

(a/o) Mg	Disordered alloy ϱ ($\mu\Omega$cm)	Cold worked alloy $\varrho(\mu\Omega$cm)	Ordered state $\varrho(\mu\Omega$cm)
1	3.40	3.46	3.4
2	4.56	4.68	4.59
4	7.77	7.48	7.36
8	11.86	12.16	12.01
12	15.90	16.28	16.10
16	27.20	27.54	

REFERENCES

1. Hansen, M. and Anderko, *Constitution of Binary Alloys,* 2nd ed., McGraw-Hill, New York, 1958; Mondolfo, L. F., *Aluminum Alloys Structure and Properties,* Butterworths, London, 1976.
2. Köster, W. and Hauk, J., *Z. Metallkd.,* 56, 846, 1965.
3. Linde, J. O., *Helv. Phys. Acta,* 41, 1013, 1968.

Au-Mn (Gold-Manganese)

The phase diagram of Au-Mn shows on the Au-rich side several phases; α'', η, Au_3Mn, χ, and Au_5Mn_2.[1-5] There may also exist a modulated stacking order at the composition range from 23 to 28 a/o Mn.[19] Resistivity measurements deal mostly with alloys on the Au-rich side of the phase diagram.[6-23]

Schilling et al.[22] studied the low-temperature resistivity of a Au-Mn alloy with .1 a/o Mn as a function of pressure (see Figure 1). They associated the resistivity minimum above 10°K and the maximum at about 2.5°K with the Kondo effect and spin glass behavior. Otter[7] prepared Au-Mn alloys from 99.99 + % pure Au and 99.9 + % pure Mn. Samples were melted in vacuum, homogenized for 10 hr at 800°C, swaged or drawn to wire, and annealed for about 1 hr at 500 to 550°C. Figure 2 shows $\varrho(T)$ of Au-rich alloys. $d\varrho/dc = 2.18$ $\mu\Omega$cm/(a/o)Mn. This agrees with results by Linde.[8-11]

The resistivity of Au_2Mn is given in Figure 3. This structure has a rather narrow range of homogeneity and exists in the ordered state below 730°C. It becomes antiferromagnetic below 90°C. Toth et al.[12] measured $\varrho(T)$ of Au_4Mn. Their results are given in Figure 4.

Yamamoto[6] determined the resistivity of Au-Mn alloys with 20 to 29 a/o Mn between room temperature and 800°C. $\varrho(T)$ is given in Figure 5. A linear correlation between $\varrho(700°C)$ and c shows that the alloys in this temperature and composition range are disordered. Deviations from such a relationship at 250°C indicate different states of order. Peaks in the $\varrho = f(c, 250°C)$ curve occur for the η-phase which is reported to be a hexagonal atomically disordered phase and the noncubic χ-phase.

Smith and Wells[13] investigated the intermetallic compound Au_5Mn_2. They prepared the sample from 99.99% Au and 99.995% Mn. The components were melted several times in an arc furnace. The ingot was sealed under Ar in a silica capsule, homogenized at 900°C for 3 days, and quenched in water. Specimens were spark cut and annealed under vacuum for 3 days at 400°C. The composition (71.4 a/o Au) was confirmed by chemical analysis. Figure 6 shows $\varrho(T)$ from 270 to 800°K. The rapid change in $d\varrho/dT$ near 350°K is associated with the change from the antiferromagnetic to paramagnetic state. $T_N < 353°K$. At $T_f = 450 \pm 2°C$, the sample transforms from face centered cubic above 450°C to monoatomic below 450°C. The resistivity curve near T_N is typical for a second order transformation.[14]

Smith[19] measured the resistivity of Au-Mn alloys with 44.8 to 52.9 a/o Au between room temperature and 600°C. Figure 7 gives $\varrho(T)$. Anomalies in the resistivity curves are associated with the double martensitic bcc → bct structure transformation.

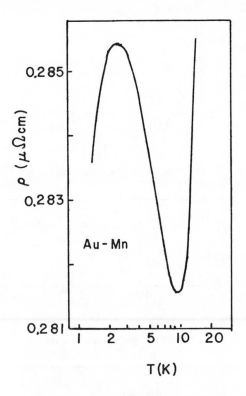

FIGURE 1.   Resistivity of Au + 0.1 a/o Mn.

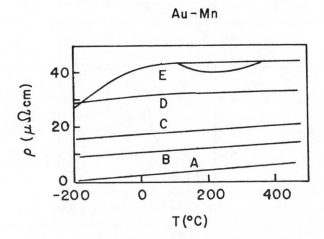

FIGURE 2.   Resistivity of AuMn alloys. (A) Au; (B) Au + 3.5 a/o Mn; (C) Au + 6.83 a/o Mn; (D) Au + 13.08 a/o Mn; (E) Au + 18.64 a/o Mn.

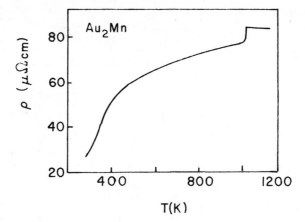

FIGURE 3.   Resistivity of Au₂Mn.

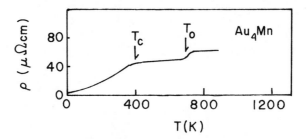

FIGURE 4.   Resistivity of Au₄Mn.

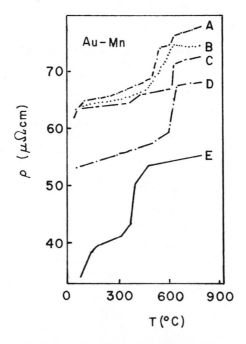

FIGURE 5.   Resistivity of Au-Mn alloys.
(A) 28.9 a/o Mn, Au₅Mn₂; (B) 27.8 a/o Mn,
X + Au₅Mn₂; (C) 26.7 a/o Mn, X + Au₃Mn;
(D) 25.2 a/o Mn, Au₃Mn; (E) 20.2 a/o Mn,
Au₄Mn.

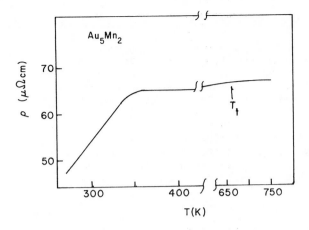

FIGURE 6. Resistivity of Au₅Mn₂.

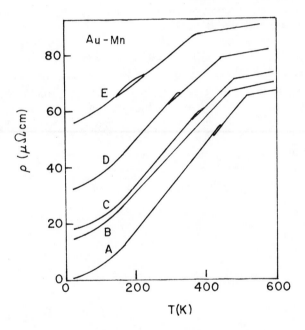

FIGURE 7. Resistivity of Au-Mn alloys (a/o Au). (A) 49.5;
(B) 52.9; (C) 48.0; (D) 46.6; (E) 44.8.

REFERENCES

1. Giansldati, A., Linde, J. O., and Borelius, G., *J. Phys. Chem. Solids,* 11, 46, 1959.
2. Watanbe, D., *J. Phys. Soc. Jpn.,* 15, 1030, 1960.
3. Humble, S. G., *Acta Crystallogr.,* 17, 1485, 1964.
4. Morris, D. P. and Huges, J. L., *Acta Crystallogr.,* 17, 10, 1964.
5. Morris, D. P. and Huges, J. L., *Acta Crystallogr.,* 15, 1062, 1962.
6. Yamamoto, K., *J. Phys. Soc. Jpn.,* 28, 1374, 1976.
7. Otter, F. A., Jr., *J. Appl. Phys.,* 27, 197, 1956.
8. Linde, J. O., *Ann. Phys.,* 10, 52, 1931.
9. Linde, J. O., *Ann. Phys.,* 14, 353, 1932.
10. Linde, J. O., *Ann. Phys.,* 15, 219, 1932.
11. Linde, J. O., *Ann. Phys.,* 30, 151, 1937.

12. Toth, R. S., Arrott, A., Shinozaki, S. S., Werner, S. A., and Sato, H., *J. Appl. Phys.*, 40, 1373, 1969.
13. Smith, J. H. and Wells, P., *J. Phys. C. Ser. 2*, 2, 356, 1969.
14. Fischer, M. E. and Langer, J. S., *Phys. Rev. Lett.*, 20, 665, 1968.
15. Bindloss, W., *Pys. Rev.*, 165, 725, 1968.
16. Yessik, M., Noakes, J., and Sato, H., *J. Appl. Phys.*, 41, 1234, 1970.
17. Adiatullen, R. and Fakidov, I. G., *Fiz. Trer. Tela.*, 12, 3152, 1970.
18. Smith, J. H. and Street, R., *Proc. Phys. Soc. London*, 71(4), 633, 1958.
19. Smith, J. H., *J. Appl. Phys.*, 39, 675, 1968.
20. Giansoldati, A. and Linde, J. O., *J. Phys. Radium*, 16, 341, 1955.
21. Crone, J. and Schilling, J., *Solid State Commun.*, 17, 791, 1975.
22. Schilling, J. S., Ford, P. J., Larsen, V., and Mydosh, J. A., *Phys. Rev. B*, 14, 4368, 1976.
23. Sato, H., Toth, R. S., Shirane, G., and Cox, D. E., *J. Phys. Chem. Solids*, 27, 413, 1966.

Au-Ni (Gold-Nickel)

Above 800°C, Au and Ni form a continuous series of solid solution alloys. Below 800°C, a miscibility gap with a critical point forms. At 25°C, the miscibility gap extends over almost the entire composition range and the terminal phases are essentially pure Ni and α-Au containing about 5% Ni.

The resistivity of concentrated metastable Au-Ni alloys obtained by rapidly quenching the high-temperature solid solution alloys were measured from 0 to 300°K. The result (see Figure 1) for the various metastable alloys displays a variety of features. The temperature coefficient of resistivity is very low for some alloys ($Ni_{37}Au_{63}$, $Ni_{42}Au_{58}$) and negative for others ($Ni_{45}Au_{55}$, $Ni_{50}Au_{50}$). A knee appears in the resistivity-temperature plots at the Curie temperature for the magnetic Ni-rich alloys.

The resistivity of the dilute NiAu alloys and the dilute AuNi alloys[2-3] has been measured. The concentration dependence of the residual resistivity of the dilute NiAu alloys yields $d\rho(NiAu)/dc = 1.2$ $\mu\Omega cm/(a/o)$ while the concentration coefficient of the residual resistivity for the dilute AuNi alloys is $d\rho/dc = 0.8$ $\mu\Omega cm/(a/o)Ni$.

135

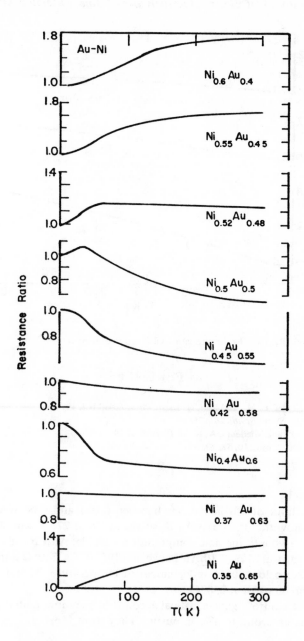

FIGURE 1A.   Resistance ratio as a function of temperature in Au-Ni.[1]

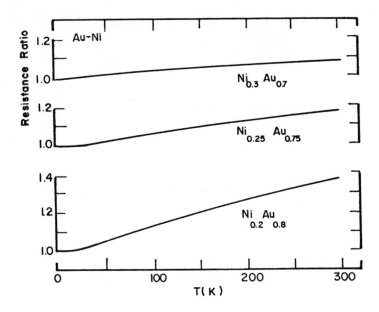

FIGURE 1B.   Resistance ratio as a function of temperature in Au-Ni.[1]

REFERENCES

1. Clinton, J. R., Tyler, E. H., and Luo, H. L., *J. Phys. F*, 4, 1162, 1974.
2. Arajas, S., *Phys. Status Solidi*, 28, 171, 1968.
3. Dorleijn, J. W. F. and Miedema, A. R., *J. Phys. F*, 5, 487, 1975.
4. Linde, J. O., *Ann. Phys.*, 10, 52, 1931.

Au-Pb (Gold-Lead)

The solubility limit of Pb in Au lies between 0.005 and 0.06 w/o Pb at 650°C. Diffusion tests show that 0.03 a/o Au dissolves in Pb at 770°C and 0.08 a/o Au dissolves at 200°C.[1] Two intermediate compounds are stable: Au_2Pb and $AuPb_2$.

Cohen et al.[2] measured ϱ of PbAu between −196 and +240°C and found $\Delta\varrho/c$ = 3.5 $\mu\Omega$cm/(a/o)Au below 80°C, independent of temperature. Then it increased gradually to 4.5 $\mu\Omega$cm/(a/o)Au at 250°C.

Dukin and Aleksandrov[3] gave a critical account of pseudopotential calculations and the resistivity increase due to 1% impurity. They gave $[\Delta\varrho/\Delta c]_{exp}$ = 3.22 $\mu\Omega$cm/(a/o)Au in PbAu.

REFERENCES

1. Hansen, M., and Anderko, K., *Constitution of Binary Alloys*, 2nd ed., McGraw-Hill, New York, 1958.
2. Cohen, B. M., Turnbull, D., and Warburton, W. K., *Phys. Rev. B*, 16, 2491, 1977.
3. Dukin, V. V. and Aleksandrov, B. N., *Pseudopotential Calculations of the Residual Resistivities of Dilute Solid Alloys Based on Normal Metals, Physics of Low Temperatures*, Akademia Nauk Uk SSR Academy of Science, USSR, Kharkov, 1978, 1.

Au-Pd (Gold-Palladium)

Au-Pd forms a continuous series of solid solution.[1] No evidence for compounds has been found.

The resistivity of Au-Pd alloys has been studied extensively by several authors.[2-24] The effect of short-range order was investigated, e.g., by Haas and Lücke,[5] Katsnelson,[9] Iveronova,[10] and Kim and Flanagan,[12,25] who studied the effect of plastic deformation. Hamaguchi[13] studied quenched in vacancies. A resistance minimum, found by March[2] and Edwards et al.,[7] was attributed by March[2] to magnetic impurities. The recent work by March[2] agrees well with the detailed study by Rowland et al.[4] These authors prepared Pd-Au alloys from spectroscopically pure Pd and 5N Au. Alloys had typically not more than 50 ppm of each impurity element in a given alloy. Samples were annealed at 800°C for a few hours and slowly cooled. Table 1 gives specimen data and $\varrho(4.2°K)$. Figure 1 gives $\varrho(T)$ for temperatures between 4 and 1300°K. The data are explained on the basis of a nonspherical Fermi surface, that the variation of the relaxation time over the Fermi surface is significant, and that N(E) is composition dependent. Figure 2 gives the residual resistivity of Pd-Au with data from Reference 4, from extrapolated data,[12] and the results from a calculation by Dugdale and Guenault,[18] who used a modified Mott model. Linde[26] gave for $\Delta\varrho/\Delta c$ for dilute alloys a value of 0.41 $\mu\Omega$cm/(a/o)Pd for AuPd. Kim and Flanagan[12,25] studied the effect of cold work on ϱ. They melted appropriate amounts of Pd and Au in a high-purity alumina crucible by induction heating under a vacuum of 10^{-5} to 10^{-6} Torr. The alloys were homogenized for 1 week at 900°C in a vacuum of 10^{-5} to 10^{-6} Torr. The alloys were then quenched to room temperature. The resistivity of these alloys after 40 or 90% deformation by a reduction in cross section is given in Figure 3. The mechanical deformation leads to a decrease of the resistivity for Au-Pd alloys with more than 20 a/o Au. Alloys with less than 20 a/o Au showed an increase in ϱ. Plastic deformation can lead to a decrease in ϱ for certain changes in short range order. The increase in plastic deformation should in most cases lead to an increase in ϱ, since it is associated with an increase in lattice defects.

Kim and Flanagan[12,25] measured also the Hall effect of these alloys. They extrapolated their data to that of complete random solid solutions. They could interpret their data on the basis of a two-band model and classical random scattering.

Table 1

Nominal concentration	Actual concentration (†±0.3 a/o, otherwise ±0.05 a/o)	$\varrho(4.2°K)$ ($\mu\Omega$cm)	$\dfrac{10^3\varrho(4.2°K)}{\varrho(273°K)}$
Pd (spec pure)	—	0.022	2.21
0.11% Au	0.12	0.115	11.79
0.5% Au	0.57	0.420	39.92
1.5% Au	1.52	1.059	96.27
5.0% Au	4.66	3.249	244.1
10% Au	9.3	6.68	410.6
20% Au	19.5	13.15	612.8
30% Au	30.5	19.58	735.1
40% Au	39.0	23.38	820.9
50% Au	48.5	20.72	878.0
55% Au	54.8	15.07	847.6
60% Au	59.3	12.76	831.8
70% Au	69.0	10.20	827.3
80% Au	79.5	7.01	792.1
90% Au	89.5	3.96	661.1
95% Au	95.67	.680	464.1
97% Au	97.30	1.023	317.7
99% Au	99.09	0.402	155.5
99.5% Au	99.49	0.169	71.98
Au (spec pure)		0.006	3.00

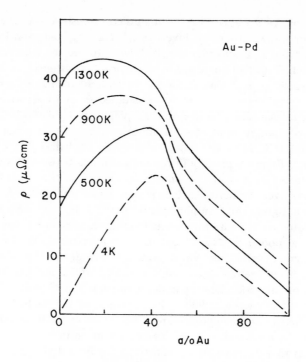

FIGURE 1.   Isotherms of the resistivity of Au-Pd alloys.

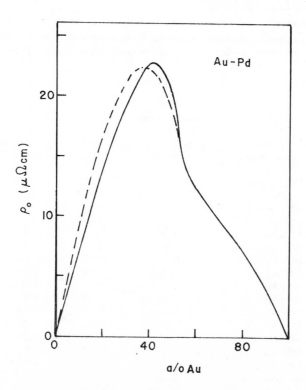

FIGURE 2.   Residual resistivity of Au-Pd. Full line: Rowland et al.,[4] extrapolated results; dashed line: calculated results.

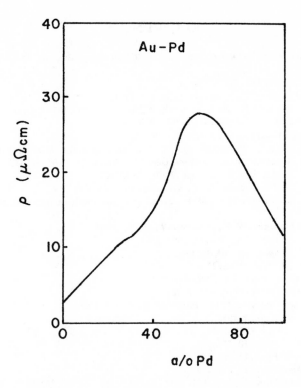

FIGURE 3A. Room temperature resistivity of 40% plastically deformed samples.

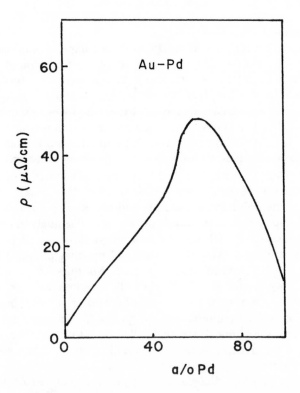

FIGURE 3B. Room temperature resistivity of 90% plastically deformed samples.

REFERENCES

1. Hansen, M. and Anderko, K., *Constitution of Binary Alloys,* 2nd ed., McGraw-Hill, New York, 1958.
2. March, J.-F., *Z. Metallkd.,* 69, 377, 1978.
3. Koster, W. and Halpern, T., *Z. Metallkd.,* 52, 821, 1961.
4. Rowland, T., Cusack, N. E., and Ross, R. G., *J. Phys. F,* 4, 2189, 1974.
5. Haas, H. and Lücke, K., *Scr. Metall.,* 6, 715, 1972.
6. Gimpl, M. L. and Fuochillo, N., *J. Met.,* 23, 39, 1971.
7. Edwards, I. R., Chen, C. W., and Legvold, S., *Solid State Commun.,* 8, 1403, 1970.
8. Maelano, A. J., Research Symp. Electronic Density of States, Gaithersburg, Md., 1969, 205.
9. Katsnelson, A. A., *Fiz. Met. Metalloved.,* 28(6), 1090, 1969.
10. Iveronova, V. I. and Katsnelson, A. A., *Kristallografiya,* 11, 576, 1966.
11. Lansitz, M. J. and Van der Meer, M. P., Proc. 7th Conf. on Thermal Conducitivity, Gaithersburg, Md., 1967, 325.
12. Kim, M. J. and Flanagan, W. F., *Acta Metall.,* 15, 735, 1967.
13. Hamaguchi, Y., *J. Phys. Soc. Jpn.,* 16, 1692, 1961.
14. Geibel, W., *Z. Anorg. Chem.,* 69, 43, 1911.
15. Otter, F. A., Jr., *J. Appl. Phys.,* 27, 197, 1956.
16. Conybeare, J. G. G., *Phys. Soc. Proc.,* 49, 29, 1937.
17. Sedstrom, E., Dissertation, Stockholm, 1924.
18. Dugdale, J. S. and Guenault, A. M., *Philos. Mag.,* 13, 503, 1966.
19. Vogt, E., *Ann. Phys.,* 14, 1, 1932.
20. Knook, B., *Physica,* 24, SIT4, 1958.
21. Linde, J. O., *Ann. Phys.,* 10, 52, 1931.
22. Linde, J. O., *Ann. Phys.,* 14, 353, 1932.
23. Linde, J. O., *Ann. Phys.,* 15, 219, 1932.
24. Linde, J. O., *Ann. Phys.,* 30, 151, 1937.
25. Kim, M. J. and Flanagan, W. F., *Acta Metall.,* 15, 747, 1967.
26. Linde, J. O., *Helv. Phys. Acta,* 41, 1013, 1968.

Au-Pt (Gold-Platinum)

Hansen and Anderko[1] report that Au-Pt forms a complete range of solid solutions at high temperatures. Below 1250°C, a two-phase region is found extending from about 20 to 95 a/o Pt at 700°C. There is also some evidence of the formation of $PtAu_3$.

March[2] prepared Au-Pt alloys from 99.999% pure elements, usually in an induction furnace (sometimes an arc furnace was used). Resistivities were essentially independent of sample preparation. The alloys were rolled and drawn to wires. Wires were finally annealed at 1000 to 1270°C and quenched. This should lead to stable or metastable single-phase alloys at low temperatures. Figure 1 gives the resistivity of the alloys at LHe temperature, together with data by Johansson and Linde.[16] The composition dependence of ϱ shows a similar behavior as found for other binary alloys where one element belongs to the noble metal elements and the other is a transition element adjacent in the periodic system to the noble metals. Possible analyses of $\varrho(c,T)$ are outlined in the introduction. Pt-rich and Au-rich alloys show the T^2 power dependence of the thermal contribution. This result differs from that obtained by Stewart and Huebner.[3] These authors studied deviations from Matthiessen's rule. They gave for the residual resistivity ϱ_o and $\delta\varrho_o/\delta c$ of AuPt alloys values given in Table 1 and for PtAu alloys in Table 2. Minor impurities were of the order of parts per million.

Damon et al.[17] measured the resistivity of AuPt alloys between 1.5 and 40°K. The samples in wire form were annealed at the 750°K for 8 to 24 hr. The residual resistivities ϱ_o of the alloys are given in Table 3. $\Delta\varrho/c$ is close to the values given.[3] $\varrho - \varrho_o$ was determined as a function of temperature for alloys and pure Au. $(\varrho - \varrho_o)_{alloy} - (\varrho - \varrho_o)_{Au}$ was usually less than 10^{-8} $\mu\Omega$cm. The data were analyzed on the basis of two components, one due to phonon assisted impurity scattering and the other on the basis of the "two-band" effect.

Otter[5] measured the resistivity of <u>Au</u>Pt alloys with up to 4.04 a/o Pt from −200 to +500°C. The elements had purities of 99.99 + %. The alloys were first superheated about 100°C above the melting point and then homogenized in vacuum for 24 hr at 900°C. Samples were then drawn or swaged to wires and reannealed at 500 to 550°C. Figure 2 gives results. $d\varrho/dc = 0.83$ $\mu\Omega$cm/(a/o)Pt. This agrees well with results by Linde,[6-9] but is lower than the more recent data by Stewart and Huebner[3] and Damon et al.[17] Resistivity measurements have been used to study decomposition processes and the shape of the phase diagram.

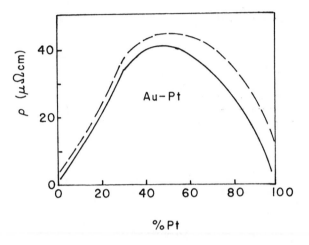

FIGURE 1. Resistivity of Au-Pt alloys. Full line, 4.2°K; dashed line, $\sim$295°K.

Table 1

Impurities concentration, c	Residual resistivity, ϱ_0	$\delta\varrho/\delta c$	Low-temperature power law, T^n
(a/o Pt)	($\mu\Omega$cm)	($\mu\Omega$cm/(cm))	(n)
0.10	0.111	1.11	2.8
0.50	0.480	0.96	3.2
1.03	0.980	0.95	3.2
4.98	4.82	0.97	3.2

Table 2

Impurities concentration, c	Residual resistivity, ϱ_0	$\delta\varrho/\delta c$	Low temperature power law, T^n
(a/o Au)	($\mu\Omega$cm)	($\mu\Omega$cm/(a/o))	(n)
0.09	0.432	4.8	3.4
0.45	0.960	2.13	3.4
0.88	1.57	1.79	3.4
4.84	6.64	1.37	3.4

Table 3

Sample number	Impurities concentration (a/o Pt)	Residual resistivity ($\mu\Omega$cm)	$\Delta\varrho/c$ ($\mu\Omega$cm/(a/o)Pt)
10	0.22	0.2230	1.0136
11	0.49	0.4943	1.0136
12	1	0.9239	0.9239
13	1.6	1.576	0.985

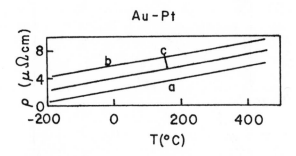

FIGURE 2.   Resistivity of Au and AuPt alloys. a, Au; b, 4.04 a/o Pt; c, 2.02 a/o Pt.

REFERENCES

1. Hansen, M. and Anderko, K., *Constitution of Binary Alloys*, 2nd ed., McGraw-Hill, New York, 1958; Mondolfo, L. F., *Aluminum Alloys Structure and Properties*, Butterworths, London, 1976.
2. March, J-F., *Z. Metallkd.*, 69, 377, 1978.
3. Stewart, R. G. and Huebener, R. P., *Phys. Rev. B*, 1, 3323, 1970.
4. Blood, P. and Greig, D., *J. Phys. F*, 2, 79, 1972.
5. Otter, F. A., Jr., *J. Appl. Phys.*, 27, 197, 1956.
6. Linde, J. O., *Ann. Phys.*, 10, 52, 1931.
7. Linde, J. O., *Ann. Phys.*, 14, 353, 1932.
8. Linde, J. O., *Ann. Phys.*, 15, 219, 1932.
9. Linde, J. O., *Ann. Phys.*, 30, 151, 1937.
10. Kralik, G., *Z. Metallkd.*, 61, 751, 1976.
11. Weise, J. and Gerold, N., *Z. Metallkd.*, 59, 904, 1968.
12. Birch, J. A., Kemp, W. R. G., Klemens, P. G., and Tainsh, R. J., *Aust. J. Phys.*, 12, 455, 1959.
13. Grube, G., Schneider, A., and Esch, O., *Festschrift aus Anlass des 100 Jahrigen Jubilaums der Firma W. C. Heraeus*, 1951, 20.
14. Wictorin, C. G., Dissertation, University of Stockholm. 1924
15. Hardy, H. K., *J. Inst. Met.*, 81, 599, 1952—1953.
16. Johansson, C. J. and Linde, J. O., *Ann. Phys.*, 5, 762, 1930.
17. Damon, D. H., Mathur, M. P., and Klemens, P. G., *Phys. Rev.*, 176, 876, 1968.
18. Greig, D. and Rowland, J. A., *J. Phys. F*, 4, 536, 1974.

Au-Rh (Gold-Rhodium)

Linde[1] gives for $\Delta\varrho/\Delta c$ for dilute alloys a value of 4.2 $\mu\Omega$cm/(a/o)Rh for AuRh.

REFERENCES

1. Linde, J. O., *Helv. Phys. Acta*, 41, 1013, 1968.

Au-Sb (Gold-Antimony)

Linde[1] gives for $\Delta\varrho/\Delta c$ for dilute alloys a value of 6.8 $\mu\Omega$cm/(a/o)Sb for AuSb.

REFERENCES

1. Linde, J. O., *Helv. Phys. Acta,* 41, 1013, 1968.

Au-Sn (Gold-Tin)

A total of 6.8 a/o Sn will dissolve in Au at the eutectic temperature of 498°C and 5.9 a/o Sn will dissolve at 400°C.[1] The maximum solubility of Au in solid Sn is about 0.2 a/o. A ξ-phase and the components AuSn, $AuSn_2$ and $AuSn_4$ are given.[1]

Bass[2] studied the quenched resistance of AuSn alloys with up to 0.3 a/o Sn. Most of the alloys were prepared from 5N Au and 4N Sn.

Köster and Hauk[3] measured the resistivity of AuSn alloys with up to 6 a/o Sn. Their results are given in Table 1. It is surprising that ordering leads only to very small increases in ϱ, and plastic deformation leads to large increases in ϱ. One may suspect that the disordered alloy may show partial order.

Linde[4] gives for $\Delta\varrho/\Delta c$ for dilute alloys 3.3 $\mu\Omega$cm/(a/o)Sn for AuSn.

Dukin and Aleksandrov[5] gave a critical account of pseudopotential calculations and the resistivity increase due to 1% impurity. They gave $[\Delta\varrho/\Delta c]_{exp}$ = 1.88 $\mu\Omega$cm/(a/o)Au in SnAu.

The *International Critical Tables*[6] give a summary of older data (see Tables 2 and 3).

Table 1

(a/o) Sn	Disordered $\varrho(\mu\Omega$cm)	Ordered $\varrho(\mu\Omega$cm)	Plastically deformed $\varrho(\mu\Omega$cm)
1	5.51	5.51	5.67
2	8.24	8.25	8.49
4	13.61	13.67	14.20
6	18.44	18.48	19.51

Table 2

Sn (w/o)	ϱ ($\mu\Omega$cm)	T (°C)
0.41	7.86	18.8
0.81	11.73	21.4
12.82	29.75	15.0
22.75	17.35	15.9
37.05	10.79	18.1
41.9	17.4	21.0
54.05	30.1	22.3
59.5	36.0	21.3
63.9	30.9	21.7
70.2	25.1	19.2
78.0	19.85	19.8
85.2	16.8	24.2
88.6	15.45	23.8
96.75	13.86	23.6

Note: For Ag, ϱ_o = 1.54.

Table 3

Sn (w/o)	ϱ(room temperature) (μΩcm)	
37.6	11.1	AuSn
54.6	33.3	AuSn$_2$
64.4	25	AuSn$_4$

REFERENCES

1. Hansen, M. and Anderko, K., *Constitution of Binary Alloys*, 2nd ed., McGraw-Hill, New York, 1958.
2. Bass, J., *Phys. Rev.*, 137, A765, 1965.
3. Köster, W. and Hauk, J., *Z. Metallkd.*, 56, 846, 1965.
4. Linde, J. O., *Helv. Phys. Acta*, 41, 1013, 1968.
5. Dukin, V. V. and Aleksandrov, B. N., *Pseudopotential Calculations of the Residual Resistivities of Dilute Solid Alloys Based on Normal Metals*, Physics of Low Temperatures, Akademia Nauk Uk SSR Academy of Science, USSR, Kharkov, 1978, 1.
6. *International Critical Tables*, Vol. 6, McGraw-Hill, New York, 1929.

Au-Ti (Gold-Titanium)

Hansen and Anderko[1] list four intermediate compounds in the Au-Ti system: TiAu$_6$, TiAu$_2$, TiAu, and Ti$_3$Au. More than 10 a/o Ti will dissolve in Au at 1115°C, and 1.8 a/o Ti will dissolve at 500°C. About 3 a/o Au will dissolve in Ti at 830°C.

The resistivity difference $\Delta\varrho = \varrho(_{alloy}) - \varrho(_{host})$ of two alloys with 0.22 and 1.06 a/o Ti[2] is given in Figure 1. It shows a weak resistance minimum. The low-temperature behavior can be described with an equation of the form $(1 - T/T_K)^2$. T_K is dependent on the impurity concentration.

White[2] finds that the low-temperature resistivity data between 4 and 80°K of AuTi alloys follow a curve of the form,

$$\rho(T) = \rho_o \left\{ 1 - (T/T_K)^2 \right\} + AT^3$$

with the following constants:

a/o Ti	T_K(K)	$\varrho_o(\mu\Omega cm)$	A (ΩcmK^{-3})
0.22	650 ± 100	2.6661	5×10^{-12}
1.06	400 ± 100	12.9475	8×10^{-12}

Toth et al.[3] measured the resistivity of Au$_4$V appropriately heat treated to obtain an ordered structure. $\varrho(T)$ is given in Figure 2.

Linde[4] gave for $\Delta\varrho/\Delta c$ for dilute alloys a value of 12.9 $\mu\Omega cm$/(a/o)Ti for AuTi.

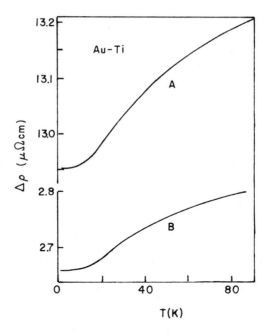

FIGURE 1. $\varrho_{(alloy)} - \varrho_{(host)}$ of $\underline{Au}$Ti alloys (a/o Ti). (A) 1.06; (B) 0.22.

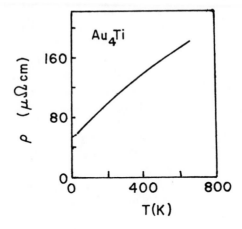

FIGURE 2. Resistivity of Au_4Ti.

REFERENCES

1. Hansen, M. and Anderko, K., *Constitution of Binary Alloys,* 2nd ed., McGraw-Hill, New York, 1958.
2. White, R. F., *J. Phys. F,* 2, 503, 1972.
3. Toth, R. S., Arrott, A., Shinozaki, S. S., Werner, S. A., and Sato, H., *J. Appl. Phys.,* 40, 1373, 1969.
4. Linde, J. O., *Helv. Phys. Acta,* 41, 1013, 1968.

Au-Tl (Gold-Thallium)

Linde[1] gives for $\Delta\varrho/\Delta c$ for dilute alloys a value of 1.7 $\mu\Omega cm/(a/o)$Tl for AuTl.

REFERENCES

1. Linde, J. O., *Helv. Phys. Acta,* 41, 1013, 1968.

Au-V (Gold-Vanadium)

The solubility of V in Au was determined by Summers-Smith.[1] He found that 17.5 a/o V will dissolve in Au at about 970°C, and that 14 a/o V will dissolve at 500°C. Resistivity measurements[2-12] are mostly concerned with scattering on low-concentration impurities and the properties of Au_4V.

Singh and Meaden[2] studied the resistivity of three AuV alloys with 0.4, 0.7, and 1 a/o V, prepared from elements of 99.999 + % purity. Figure 1 gives $\varrho(T)$. The low-temperature resistivity follows a T^2 law; the implied Kondo temperature (280°K) is higher than that obtained by other means.

Kume[3] studied the resistivity and susceptibility of AuV alloys with 0.3 to 2 a/o V between 1.4 and 1000°K. The resistivity of the solute varies logarithmically at high temperatures and follows a temperature dependence given by $[1-(T/T_o)^2]$, with $T_o \approx$ 280 to 310°K at the low-temperature limit. Results are analyzed on the basis of s-d interaction.

Sakamoto et al.[4] determined the temperature dependence of the excess electrical resistivity $\Delta\varrho = \varrho - \varrho_{host}$. They used the equation $\Delta\varrho = \varrho_o + aT - bT^2$.

Flukiger et al.[5] studied the phase diagram of Au-V and found a α-V primary solid solution (A2-type structure), ordered β-phase (A15), the extremely wide γ-Au terminal solid solution (A1), the γ'-ordered phase (VAu_2), and the γ''-ordered phase VAu_4. The nominal purity of the elements in the alloys were V, 99.9%, and Au, 99.999%. Figure 2 gives $\varrho(T)/\varrho(0°C)$ for $Au_{0.235}V_{0.765}$, $Au_{0.22}V_{0.78}$, and $Au_{21}V_{79}$.

Star[11] gives the electrical resistivity of AuV alloys with up to 1 a/o V in the liquid He temperature range. These alloys show the Kondo effect. All alloys were prepared by induction melting under an Ar atmosphere. Liquid droplets were quenched in water and drawn. These wires were not annealed. The residual resistivity calculated from these data is $d\varrho/dc = 15$ $\mu\Omega cm/(a/o)$V. Linde[13] gave for $\Delta\varrho/\Delta c$ for dilute alloys a value of 12.6 $\mu\Omega cm/(a/o)$V for AuV.

Toth et al.[6] measured the resistivity of Au_4V appropriately heat treated to obtain ordered structures. $\varrho(T)$ is given in Figure 3. T_c is the Curie temperature, and T_o is the atomic ordering temperature.

Fürnrohr and Gerold[7] studied the kinetics of the transformation from disordered fcc to ordered bct in Au_4V. They prepared foil specimens by cold rolling 99.99% pure Au_4V. Samples were heat treated in a high-vacuum furnace and oil quenched. The resistivity change with temperature is given in Figure 4. Cooling rates from 10 to 1°K/min were used. The sample was also quenched. ϱ(quenched) lies on the dashed curve (a) which is linearly extrapolated from the high-temperature data. The negative slope of $\varrho(T)$ is explained as being typical for the disordered state. The resistivity of an Au_4V sample annealed after quenching from 1000°C for 28 days at 300°C prior to $\varrho(T)$ measurements between 4 and 350°K was measured.[8] This sample was prepared by melting constituents, homogenized for 2 days at 1000°C, and then quenched. The change in $d\varrho/dT$ near 45 to 46°K is attributed to a ferromagnetic transition.

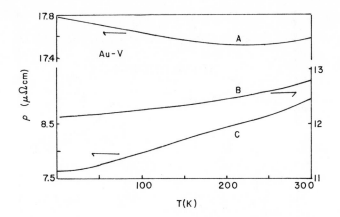

FIGURE 1.   Resistivity of $\underline{Au}V$ alloys (a/o V). (A) 1; (B) 0.7; (C). 0.4.

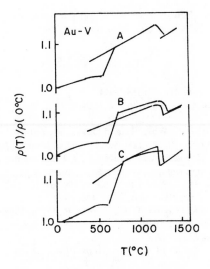

FIGURE 2.   Resistivity of Au-V alloys (a/o Au). (A) 23.5; (B) 22; (C) 21.

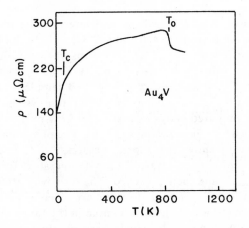

FIGURE 3.   Resistivity of Au_4V.

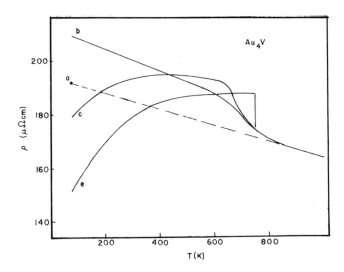

FIGURE 4.   Resistivity of Au₄V. (a) Quenched from 948°K; (b) cooling rate 10°K/min; (c) cooling rate 2.5°K/min.

REFERENCES

1. Summers-Smith, D., *J. Inst. Met.*, 83, 189, 1954.
2. Singh, R. L. and Meaden, G. T., *Phys. Rev. B*, 6, 2660, 1972.
3. Kume, K., *J. Phys. Soc. Jpn.*, 23, 1226, 1967.
4. Sakamoto, N., Yamaguchi, Y., Waki, S., and Ogawa, S., 12th Int. Conf. on Low Temps. Physics, Kyoto, Japan, September 4 to 10, 1970; (Kyoto, Japan Sci. Council, Japan, 1970), 332.
5. Flukiger, R., Sucz, Ch., Heiniger, F, and Muller, J., *J. Less Common Met.*, 40, 103, 1975.
6. Toth, R. S., Arrott, A., Shinozaki, S., Werner, S. A., and Sato, H., *J. Appl. Phys.*, 40, 1373, 1969.
7. Fürnrohr, P. and Gerold, V., *J. Phys. Chem. Solids*, 39, 167, 1978.
8. Maple, M. B. and Luo, H. L., *Phys. Lett. A*, 25, 121, 1967.
9. Ford, P. J., Whall, T. E., and Loram, J. W., *J. Phys. F.*, 4, 225, 1974.
10. Kozavzewski, B., *J. Phys. F*, 4, L35, 1974.
11. Star, W. M., *Physica*, 58, 623, 1972.
12. Van der Berg, G. J., 12th Int. Conf. on Low Temps. Physics, Kyoto, Japan, September 4 to 10 1970; (Kyoto, Japan Sci. Council, Japan, 1970), 671.
13. Linde, J. O., *Helv. Phys. Acta*, 41, 1013, 1968.

Au-Zn (Gold-Zinc)

The phase diagram of Au-Zn reveals several intermediate phases.[1] More than 30 a/o Zn will dissolve in Au between ∿420 and 642°C, and about 14 a/o Zn will dissolve at room temperature. A few percent of Au will be absorbed in solid Zn.

Köster and Hauk[2] measured the resistivity of Au-Zn alloys with up to 17.5 a/o Zn for disordered, ordered, and mechanically deformed samples. Their results are given in Table 1. The alloy with 17.5 a/o Zn may be a two-phase system.[1]

Köster and Storing[3] prepared alloys by casting 99.99% purity Au and Zn with intermitten anneals, rolling down these castings to foils. The phase diagram of Au-Zn given in this paper schematically shows two superstructures, Au₄Zn and Au₃Zn. The Au₃Zn compounds exist in a high-temperature and a low-temperature form. The samples were annealed for 1 hr at T_{anneal} and then quenched. The resistivity after the quench is measured at room temperature. Experimental results are given in Figure 1 for a sample in the solid solution range (14 a/o Zn), for a sample in the range in which Au₄Zn forms (15 a/o Zn and for the Au₃Zn sample. The last sample shows an unusual maximum at

300°C. Linde[7] gave for $\Delta\varrho/\Delta c$ for dilute alloys a value of 0.94 $\mu\Omega$cm/(a/o)Zn for AuZn.

Superlattice phases in Au_3Zn were studied by Predel and Schwermann[4] and Krompholz and Weiss.[5] Dukin and Aleksandrov[6] gave a critical account of pseudopotential calculations and the resistivity increase due to 1% impurity. They gave $[\Delta\varrho/\Delta c]_{exp}$ = 1.48 $\mu\Omega$cm/(a/o)Au in ZnAu.

Table 1

(a/o)Zn	ϱ Ordered ($\mu\Omega$cm)	ϱ Disordered ($\mu\Omega$cm)	ϱ Deformed ($\mu\Omega$cm)
1	2.98	2.97	3.04
2	3.98	3.95	4.08
4	5.82	5.72	5.99
8	9.57	9.09	9.54
12	—	17.8	—
14	—	14.6	—
17.5	—	18.0	—

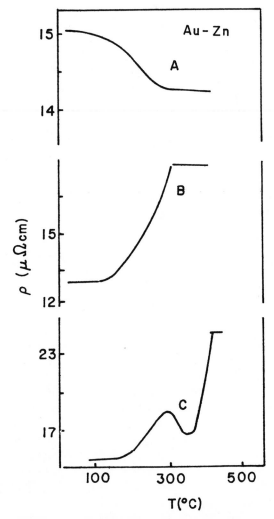

FIGURE 1. Resistivity of annealed AuZn alloys at room temperature (a/o Zn). T = temperature of anneal. (A) 14; (B) 15; (C) 25.

REFERENCES

1. **Hansen, M. and Anderko, K.,** *Constitution of Binary Alloys,* 2nd ed., McGraw-Hill, New York, 1958.
2. **Köster, W. and Hauk, J.,** *Z. Metallkd.,* 56, 846, 1965.
3. **Köster, W. and Storing, R.,** *Z. Metallkd.,* 57, 34, 1966.
4. **Predel, B. and Schwermann, W.,** *Z. Metallkd.,* 62, 517, 1971.
5. **Krompholz, K. and Weiss, A.,** Ber. Bunsenges Phys. Chem. Meeting of German Bunsenges, Phys. Chem. 82, Konigstein, Germany, 29, (October 5 to 7, 1977) 1978, 334.
6. **Dukin, V. V. and Aleksandrov, B. N.,** *Pseudopotential Calculations of the Residual Resistivities of Dilute Solid Alloys Based on Normal Metals, Physics of Low Temperatures,* Akademia Nauk Uk SSR Academy of Science, USSR, Kharkov, 1978, 1.
7. **Linde, J. O.,** *Helv. Phys. Acta,* 41, 1013, 1968.

BARIUM (Ba)

Ba-Pb (Barium-Lead)

Grube and Dietrich[1] studied the phase diagram of Ba-Pb, with resistivity measurements using 98.7% pure Ba. Single-phase samples according to this study were $BaPb_3$, $BaPb$, and Ba_2Pb. The phase diagram given by Hansen and Anderko[2] indicates that the stability range of the intermediate phase near Ba_2Pb extends over several percent. $BaPb$ and $BaPb_3$ are compounds which show a slightly expanded stability range of a few percent.

REFERENCES

1. Grube, G. and Dietrich, A., *Z. Electrochemie*, 44, 755, 1938.
2. Hansen, M. and Anderko, K., *Constitution of Binary Alloys*, 2nd ed., McGraw-Hill, New York, 1958.

BERYLLIUM (Be)

Be-Ti (Beryllium-Titanium)

Matthias et al.[1] studied $TiBe_2$ which crystallizes in a cubic Laves phase. $\varrho(T)$ of this compound is given in Figure 1. This material is antiferromagnetic, with $T_N \simeq 10°K$.

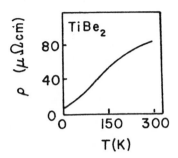

FIGURE 1.   Resistivity of $TiBe_2$.

REFERENCES

1. Matthias, B. T., Giorji, A. L., Struebing, V. O., and Smith, J. L., *Phys. Lett. A,* 69, 221, 1978.

BISMUTH (Bi)

Bi-Cd (Bismuth-Cadmium)

The solubilities of Bi and Cd in each other are very restricted.[1] The phase diagram gives one eutectic reaction.

Gabe and Evans[2] prepared Bi-Cd alloys from 99.9% pure Cd and 99.9+% Bi. Alloys were cast in the form of plates, annealed for days at 125°C, and furnace cooled. Results of resistivity measurements at 0°C are given in Table 1. All alloys are mechanical mixtures of essentially pure Bi and Cd phases. ϱ was a smooth function of composition in these experiments.

Table 1

(w/o)Bi	Before anneal ($\mu\Omega$cm)	After anneal ($\mu\Omega$cm)
0	7.07	6.83
14.9	12.56	10.03
30.0	16.67	14.67
45.0	20.13	20.16
55.0	25.57	25.11
64.0	33.90	32.49
75.0	48.38	44.90
90.0	71.04	73.01
100.0	107.40	106.80

REFERENCES

1. Hansen, M. and Anderko, K., *Constitution of Binary Alloys,* 2nd ed., McGraw-Hill, New York, 1958.
2. Gabe, S. and Evans, E. J., *Philos. Mag.,* 19, 773, 1935.

Bi-In (Bismuth-Indium)

Hansen and Anderko[1] report a very limited solubility of In in solid Bi. A few percent Bi will dissolve at room temperature in In. The compound InBi and the intermediate phase, In_3Bi, are stable to temperatures of 110 and 89°C, respectively.

Lomonos et al.[2] studied deviations from Matthiessen's rule on InBi alloys. Samples were prepared from high purity In (RRR = 0.3×10^5). They obtained the following data (see Table 1) for ϱ(LHe temp)/ϱ(290°K).

Dukin and Aleksandrov[5] gave a critical account of pseudopotential calculations and the resistivity increase due to 1% impurity. They gave $[\Delta\varrho/\Delta c]_{exp} = 1.7 \mu\Omega$cm/(a/o)Bi in InBi.

Hashimoto[3] measured the resistivity of InBi. More recently Mori and Saito[4] prepared a single crystal of InBi from 99.999% pure Bi and 99.9999% pure In by melting constituents in a glass tube and then passing a molten zone. $\perp$ or $\parallel$ refer to the orientation of the crystal. $\varrho(0°K) \parallel = 3.40 \mu\Omega$cm, and $\varrho(0°K) \perp = 1.75 \mu\Omega$cm. $\varrho\perp \propto T^{2.5}$, $\varrho\parallel \propto T^2$ for 20°K < T < 70°K.

Table 1

(a/o)Bi	ϱ(LHe temp)/ ϱ(290°K)	ϱ(290°K) ($\mu\Omega$ cm)
0.027	0.55×10^{-2}	9.00
0.110	2.20	9.14
0.164	3.18	9.28
0.275	5.270	9.53

REFERENCES

1. Hansen, M. and Anderko, K., *Constitution of Binary Alloys,* 2nd ed., McGraw-Hill, New York, 1958.
2. Lomonos, O. I., Aleksandrov, B. N., and Chernov, A. P., *Sov. Phys. JETP,* 40(3), 552, 1975.
3. Hashimoto, K., *J. Phys. Soc. Jpn.,* 12, 1423, 1957.
4. Mori, K. and Saito, Y., *J. Phys. Soc. Jpn.,* 44, 944, 1978.
5. Dukin, V. V. and Aleksandrov, B. N., *Pseudopotential Calculations of the Residual Resistivities of Dilute Solid Alloys Based on Normal Metals, Physics of Low Temperatures,* Akademia Nauk Uk SSR Academy of Science, USSR, Kharkov, 1978, 1.

Bi-Li (Bismuth-Lithium)

Grube et al.[1] studied the phase diagram of Li-Bi with resistivity measurements. They found the compounds LiBi and Li_3Bi. The samples were two-phase alloys except for pure Bi, pure Li, and these two compounds.

REFERENCES

1. Grube, G., Vosskühler, H., and Schecht, H., *Z. Elektrochem.,* 40, 270, 1934.

Bi-Mg (Bismuth-Magnesium)

A total of 1.12 a/o Bi will dissolve in Mg at 551°C and 0.04 a/o Bi will dissolve at 300°C.[1] Mg is essentially insoluble in solid Bi. The intermediate phase, Mg_3Bi_2, has a stability range of a few percent. It undergoes a transformation near 700°C.

Grube et al.[2] melted 99.9% Bi and 99.93% Mg under a protective mixture of KCl and LiCl. Samples were annealed for several days at a temperature ≈30°C below the eutectic temperature. The conductivity of the alloys is shown in Table 1 as a function of composition. All alloys should contain two-phase systems except for alloys in a narrow composition range near Mg_3Bi_2 (about 35 to 40 a/o Bi) and for some alloys very close to Bi (up to 1.5 a/o Mg). Surprisingly, the alloys with 0.1 a/o Bi has a higher conductivity than that of the Mg sample.

Table 1
ELECTRICAL CONDUCTIVITY (10^4 OHM^{-1}cm^{-1}) OF Bi-Mg ALLOYS

Bi (a/o)	Temperature (°C)										
	50	100	150	200	250	300	350	400	450	500	550
0.0	20.04	17.11	15.03	13.35	12.07	10.99	9.97	9.25	8.56	7.95	7.39
1.0	19.36	16.52	14.50	12.83	11.59	10.46	9.33	8.22	7.13	6.15	5.41
2.5	17.81	15.22	13.35	11.80	10.73	9.70	8.65	7.56	6.59	5.71	4.81
5.0	15.28	13.13	11.52	10.33	9.32	8.37	7.37	6.41	5.58	4.88	4.18
10.0	11.46	9.98	8.64	7.73	6.94	6.26	5.53	4.88	4.24	3.69	3.15
20.0	4.78	4.08	3.58	3.21	2.90	2.58	2.29	2.05	1.80	1.57	1.32
30.0	1.42	1.23	1.10	0.998	0.914	0.840	0.761	0.692	0.614	0.530	0.361
32.0	0.537	0.472	0.434	0.408	0.386	0.360	0.334	0.309	0.288	0.256	0.184
32.5	0.248	0.217	0.200	0.187	0.176	0.164	0.153	0.143	0.134	0.124	—
33.0	0.257	0.225	0.205	0.189	0.175	0.162	0.149	0.138	0.131	0.123	0.117
35.0	0.338	0.284	0.247	0.217	0.192	0.171	0.152	0.138	0.125	0.115	0.109
38.5	0.446	0.377	0.329	0.288	0.259	0.232	0.208	0.187	0.169	0.152	0.138
40.0	0.499	0.421	0.364	0.320	0.283	0.252	0.222	0.197	0.179	0.161	0.146
40.5	0.411	0.352	0.311	0.278	0.262	—	—	—	—	—	—
41.5	0.368	0.313	0.273	0.241	0.214	—	—	—	—	—	—
42.0	0.352	0.300	0.262	0.232	0.207	—	—	—	—	—	—
50.0	0.371	0.320	0.281	0.251	0.231	—	—	—	—	—	—
55.0	0.389	0.338	0.293	0.255	0.225	—	—	—	—	—	—
60.0	0.425	0.367	0.318	0.276	0.238	—	—	—	—	—	—
70.0	0.495	0.432	0.375	0.324	0.285	—	—	—	—	—	—
75.0	0.513	0.451	0.391	0.334	0.286	—	—	—	—	—	—
80.0	0.502	0.447	0.391	0.338	0.292	—	—	—	—	—	—
90.0	0.536	0.471	0.407	0.347	0.295	—	—	—	—	—	—
97.0	0.637	0.554	0.471	0.393	0.329	—	—	—	—	—	—
100.0	0.708	0.608	0.520	0.437	0.365	—	—	—	—	—	—

REFERENCES

1. Hansen, M. and Anderko, K., *Constitution of Binary Alloys,* 2nd ed., McGraw-Hill, New York, 1958.
2. Grube, G., Mohr, L., and Bornhak, R., *Z. Electrochem.,* 40, 143, 1934.

Bi-Pb (Bismuth-Lead)

Hansen and Anderko[1] report that more than 16 a/o Bi will dissolve in Pb at room temperature, but less than 1 a/o Pb will dissolve in Bi. Thomas and Evans[2] prepared Bi-Pb alloys from 99.9% pure Bi and 99.99 + % Pb. by casting from the melt. Samples were annealed for several days at 80, 60, and finally 40°C. The resistivity at 0°C is a smooth function of composition, with a maximum for the alloy with 97.3 w/o Bi. Results are given in Table 1.

Table 1

Composition by weight (%)		Resistivity at 0°C before annealing ($\mu\Omega cm^3$)	Resistivity at 0°C after annealing ($\mu\Omega cm^3$)
100	Pb	19.97	19.90
5	Bi	24.82	24.16
10	Bi	29.32	29.43
16	Bi	35.26	34.17
20	Bi	39.85	36.89
25	Bi	42.11	42.31
32	Bi	50.95	49.22
42	Bi	66.37	64.51
50	Bi	87.16	81.41
58	Bi	102.0	102.13
65	Bi	137.0	132.02
70	Bi	154.9	157.01
80	Bi	109.9	211.90
90	Bi	281.8	280.05
96	Bi	315.2	318.4
97	Bi	301.0	306.2
98	Bi	264.2	275.4
99	Bi	230.1	236.5
99.7	Bi	183.1	179.1
100	Bi	113.5	108.4

REFERENCES

1. Hansen, M. and Anderko, K., *Constitution of Binary Alloys,* 2nd ed., McGraw-Hill, New York, 1958.
2. Thomas, W. and Evans, E. J., *Philos. Mag.,* 16, 329, 1933.

Bi-Sb (Bismuth-Antimony)

Bi and Sb form a complete series of solid solutions.[1] Jain[2] studied in detail the resistivity and the Hall coefficient of single crystals with up to 41 a/o Sb. The data can be explained on the basis of a semimetal-semiconducting transition at about 5 a/o Sb in the low temperature range. Figure 1 shows $\varrho(T)/\varrho(300°K)$ as a function of 1/T for alloys with 12.5 to 41 a/o Sb. At low temperatures, ϱ decreases with increasing temperature, typical for a semiconductor. At high temperatures, $d\varrho/dT$ is positive. However, $\delta\varrho/\delta T$ of pure Bi and Bi + 4.8 a/o Sb is always positive, as seen in Figure 2. The resistivity at 4.2 and 300°K is given as a function of composition in Figure 3. The room temperature data follow a behavior predicted for metals. The energy gap calculated is given in Figure 4 is a function of composition for the semiconductors (5 a/o < c < 40 a/o). The overlap of conduction and valency band is given for the semimetals with less than 5 a/o Sb. Data are analyzed in the basis of shifting energy bands.

Negative slopes of $\varrho(T)$ have been found by Ivanov et al.[3] for alloys with 2 to 12 a/o Sb in the temperature range from −200 to 0°C. Tanuma[4] measured the resistivity and the Hall coefficient of Bi-Sb alloys over the complete composition range at 80°K.

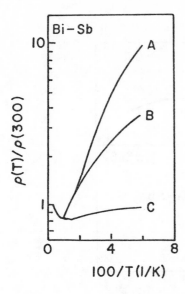

FIGURE 1. Resistivity ratio of Bi-Sb alloys (%Sb). (A) 12.5; (B) 17.5; (C) 41.

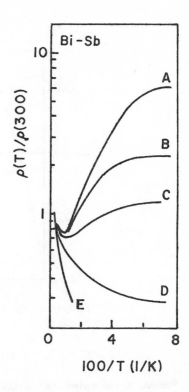

FIGURE 2. Resistivity ratio of Bi-Sb alloys (%Sb). (A) 10; (B) 7.2; (C) 6.2; (D) 4.8; (E) pure Bi.

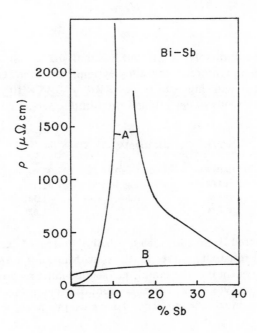

FIGURE 3. Resistivity of Bi-Sb alloys (°K). (A) 4.2; (B) 300.

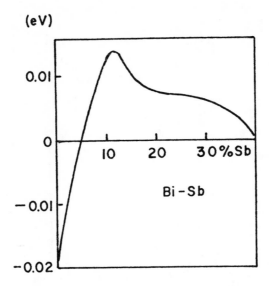

FIGURE 4.   Energy gap (+) or band overlap (−) in Bi-
Sb alloys.

REFERENCES

1. Hansen, M. and Anderko, K., *Constitution of Binary Alloys,* 2nd ed., McGraw-Hill, New York, 1958.
2. Jain, A. L., *Phys. Rev.,* 114, 1518, 1959.
3. Ivanov, G. A., Popov, A. M., and Chistyakov, B. I., *Phys. Met. Metallogr. (USSR),* 16(2), 23, 1963.
4. Tanuma, S., *J. Phys. Soc. Jpn.,* 16, 2349, 1961.

Bi-Sn (Bismuth-Tin)

Up to 13.1 a/o Bi will dissolve in Sn, but less than 0.2 a/o Sn will dissolve in Bi at 139°C, the eutectic temperature of this alloy system.[1] Lomonos et al.[2] prepared poly-crystalline Sn-Bi alloys with high-purity Sn (RRR = 0.45 × 10^5). They obtained the following results in the study of deviations from Matthiessen's rule.

(a/o) Bi	ϱ(LHe temp)/ϱ(290°K)	ϱ(290°K) ($\mu\Omega$cm)
0.057	0.41 × 10^{-2}	11.44
0.114	0.88	11.51
0.170	1.24	11.54
0.294	2.10	11.62

Ivanov et al.[3] prepared Bi-Sn alloys from V-000 Bi and Sn of chemical purity. Figure 1 gives the measured resistivity curves of alloys with up to 0.4 a/o Sn. Negative curvatures imply in this system that the sample has semiconducting properties.

Dukin and Aleksandrov[4] gave a critical account of pseudopotential calculations and the resistivity increase due to 1% impurity. They gave $[\Delta\varrho/\Delta c]_{exp} = 0.87 \ \mu\Omega$cm/(a/o)Bi in SnBi.

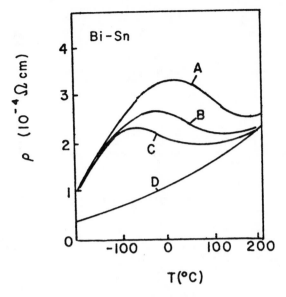

FIGURE 1.   Resistivity of Bi-Sn alloys. (A) 0.4 a/o Sn;
(B) 0.2 a/o Sn; (C) 0.1 a/o Sn; (D) Bi.

REFERENCES

1. Hansen, M. and Anderko, K., *Constitution of Binary Alloys,* 2nd ed., McGraw-Hill, New York, 1958.
2. Lomonos, O. I., Aleksandrow, B. N., and Zhernov, A. P., *Sov. Phys. JETP,* 40, 552, 1975.
3. Ivanov, G. A., Popov, A. M., and Chistyakov, B. I., *Phys. Met. Metallogr. (USSR),* 16(2), 23, 1963.
4. Dukin, V. V. and Aleksandrov, B. N., *Pseudopotential Calculations of the Residual Resistivities of Dilute Solid Alloys Based on Normal Metals, Physics of Low Temperatures,* Akademia Nauk Uk SSR Academy of Science, USSR, Kharkov, 1978, 1.

Bi-Te (Bismuth-Tellurium)

Hansen and Anderko[1] report that up to 2.4 a/o Te will dissolve in Bi at the eutectic temperature of 266°C. Bi is practically insoluble in Te. An intermediate phase is stable in the composition range from 53 to 64 a/o Te. Its nominal composition is Bi_2Te_3.

Ivanov et al.[2] prepared Bi-Te alloys from V-000 grade Bi and chemically pure Te. Figure 1 gives experimental results on pure Bi and an alloy with 0.2 a/o Te.

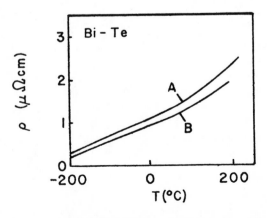

FIGURE 1.   Resistivity of Bi-Te alloys. (A) Bi; (B) Bi + 0.2 a/o Te.

REFERENCES

1. Hansen, M. and Anderko, K., *Constitution of Binary Alloys,* 2nd ed., McGraw-Hill, New York, 1958.
2. Ivanov, G. A., Popov, A. M., Chistyakov, B. I., *Phys. Met. Metallogr. (USSR),* 16(2), 23, 1963.

Bi-Tl (Bismuth-Thallium)

Dukin and Aleksandrov[1] gave a critical account of pseudopotential calculations and the resistivity increase due to 1% impurity. They gave $[\Delta \varrho / \Delta c]_{exp} = 5.32$ $\mu\Omega$cm/(a/o)Bi in T̲lBi.

REFERENCES

1. Dukin, V. V. and Aleksandrov, B. N., *Pseudopotential Calculations of the Residual Resistivities of Dilute Solid Alloys Based on Normal Metals, Physics of Low Temperatures,* Akademia Nauk Uk SSR Academy of Science, USSR, Kharkov, 1978, 1.

CALCIUM (Ca)

Ca-Cd (Calcium-Cadmium)

Dukin and Aleksandrov[1] gave a critical account of pseudopotential calculations and the resistivity increase due to 1% impurity. They gave $[\Delta\varrho/\Delta c]_{exp} = 0.47$ ($\mu\Omega$cm/(a/o)Ca in $\underline{Cd}$Ca.

REFERENCES

1. Dukin, V. V. and Aleksandrov, B. N., *Pseudopotential Calculations of the Residual Resistivities of Dilute Solid Alloys Based on Normal Metals, Physics of Low Temperatures,* Akademia Nauk Uk SSR Academy of Science, USSR, Kharkov, 1978, 1.

CADMIUM (Cd)

Cd-Cu (Cadmium-Copper)
Dukin and Aleksandrov[1] gave a critical account of pseudopotential calculations and the resistivity increase due to 1% impurity. They gave $[\Delta\varrho/\Delta c]_{exp} = 0.28$ $\mu\Omega$cm/(a/o)Cu in CdCu. Linde[2] gives for $\Delta\varrho/\Delta c$ for dilute alloys a value of 0.2 $\mu\Omega$cm/(a/o)Cd for CuCd.

REFERENCES

1. Dukin, V. V. and Aleksandrov, B. N., *Pseudopotential Calculations of the Residual Resistivities of Dilute Solid Alloys Based on Normal Metals, Physics of Low Temperatures*, Academia Nauk Uk SSR Academy of Science, USSR, Kharkov, 1978, 1.
2. Linde, J. O., *Helv. Phys. Acta*, 41, 1013, 1968.

Cd-Ga (Cadmium-Gallium)
Dukin and Aleksandrov[1] gave a critical account of pseudopotential calculations and the resistivity increase due to 1% impurity. They gave $[\Delta\varrho/\Delta c]_{exp} = 0.61$ $\mu\Omega$cm/(a/o)Cd in GaCd.

REFERENCES

1. Dukin, V. V. and Aleksandrov, B. N., *Pseudpotential Calculations of the Residual Resistivities of Dilute Solid Alloys Based on Normal Metals, physics of Low Temperatures*, Akademia Nauk Uk SSR Academy of Science, USSR, Kharkov, 1978, 1.

Cd-Ge (Cadmium-Germanium)
Dukin and Aleksandrov[1] gave a critical account of pseudopotential calculations and the resistivity increase due to 1% impurity. They gave $[\Delta\varrho/\Delta c]_{exp} = 1.47$ $\mu\Omega$cm/(a/o)Ge in CdGe.

REFERENCES

1. Dukin, V. V. and Aleksandrov, B. N., *Pseudopotential Calculations of the Residual Resistivities of Dilute Solid Alloys Based on Normal Metals, Physics of Low Temperatures*, Adademia Nauk Uk SSR Academy of Science, USSR, Kharkov, 1978, 1.

Cd-Hg (Cadmium-Mercury)
Dukin and Aleksandrov[1] gave a critical account of pseudopotential calculations and the resistivity increase due to 1% impurity. They gave $[\Delta\varrho/\Delta c]_{exp} = 0.24$ $\mu\Omega$cm/(a/o)Hg in CdHg.

REFERENCES

1. Dukin, V. V. and Aleksandrov, B. N., *Pseudopotential Calculations of the Residual Resistivities of Dilute Solid Alloys Based on Normal Metals, Physics of Low Temperatures*, Akademia Nauk Uk SSR Academy of Science, USSR, Kharkov, 1978, 1.

Cd-In (Cadmium-Indium)

The Cd-In alloy system[1] shows one eutectic reaction at 123°C, in which the liquid phase with 74 a/o In decomposed into nearly pure Cd and a InCd alloy with about 83 a/o In. The solubility of Cd in In decreases rapidly with decreasing temperature for T <123°C.

Predel and Sandig[2] measured the resistivity of In-Cd over the complete composition range. Both liquid and solid alloys were studied. Samples were prepared from 99.9998% Cd and 99.9995% In. Resistivities were measured at 0.1 Torr. Figure 1 gives $\varrho(c)$ at constant temperature. It is surprising that $\varrho(c)$ at 400°C shows a complex composition dependence, since all alloys are liquid at this temperature. At 50 and 100°C, cubic and tetragonal fcc phases exist for InCd alloys with 10 to 15 a/o Cd. Then a two-phase regime follows, extending to Cd with a few percent In.

Dukin and Aleksandrov[3] gave a critical account of pseudopotential calculations and the resistivity increase due to 1% impurity. They gave $[\Delta\varrho/\Delta c]_{exp} = 0.34$ $\mu\Omega$cm/(a/o)Cd in InCd and $[\Delta\varrho/\Delta c]_{exp} = 0.54$ $\mu\Omega$cm/(a/o)In in CdIn.

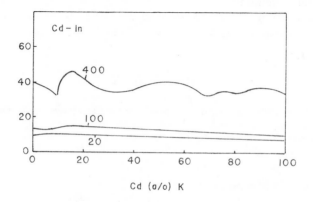

FIGURE 1.   Resistivity of Cd-In; numbers, °C.

REFERENCES

1. Hansen, M. and Anderko, K., *Constitution of Binary Alloys,* 2nd ed., McGraw-Hill, New York, 1958.
2. Predel, B. and Sandig, H., *J. Less Common Met.,* 21, 71, 1970.
3. Dukin, V. V. and Aleksandrov, B. N., *Pseudopotential Calculations of the Residual Resistivities of Dilute Solid Alloys Based on Normal Metals, Physics of Low Temperatures,* Akademia Nauk Uk SSR Academy of Science, USSR, Kharkov, 1978, 1.

Cd-Li (Cadmium-Lithium)

Grube et al.[1] studied the phase diagram of Cd-Li. They used high-purity Cd (Kahlbaum) and 99% pure Li. Table 1 gives the conductivity (in $10^4\Omega^{-1}$cm^{-1}). The CdLi phase extends to about 3 to 5 a/o Li. The LiCd intermediate phase is stable from about 45 a/o Li, in good agreement with the more recent phase diagram by Hansen and Anderko.[2] That phase diagram shows that a χ-phase is stable between 9 a/o In and 30 a/o In, and a LiCd$_3$ intermediate phase is stable between 14.5 and 29 a/o In. The phase diagram is more complicated than the σ(c)-curves suggest.

Dukin and Aleksandrov[3] gave a critical account of pseudopotential calculations and the resistivity increase due to 1% impurity. They gave $[\Delta\varrho/\Delta c]_{exp} = 2.52$ $\mu\Omega$cm/(a/o)Cd in LiCd.

Table 1
ELECTRICAL CONDUCTIVITY ($10^{-4}\Omega^{-1}cm^{-1}$)

(a/o) Li	Temperature (°C)							
	50	100	150	200	250	300	350	400
0.0	12.02	10.25	8.83	7.71	6.83	6.04	—	—
2.2	11.20	9.68	8.29	7.25	6.43	5.76	—	—
4.6	10.05	8.48	7.48	6.68	6.04	5.45	—	—
4.8	9.56	7.92	7.06	6.37	5.76	5.21	—	—
5.6	9.34	7.69	6.73	6.05	5.52	5.03	—	—
6.0	9.00	7.68	6.74	6.06	5.53	5.03	—	—
11.3	7.02	5.93	5.08	4.59	4.31	4.09	—	—
18.0	5.89	5.20	4.49	4.13	3.85	3.61	3.38	3.16
23.4	9.06	7.99	7.10	6.34	5.70	4.98	3.37	3.11
26.2	9.06	7.99	7.02	6.24	5.58	4.89	4.13	2.87
28.0	8.24	7.32	6.54	5.85	5.27	4.66	4.03	3.12
32.5	5.91	5.44	5.01	4.64	4.28	3.97	3.65	3.34
42.0	5.01	4.61	4.25	4.01	3.82	3.50	3.06	2.81
51.4	9.74	8.27	7.13	6.08	5.24	4.46	3.80	3.16

REFERENCES

1. Grube, G., Vosskuhler, H., and Vogt, H., *Z. Elektrochem.*, 38, 869, 1932.
2. Hansen, M. and Anderko, K., *Constitution of Binary Alloys*, 2nd ed., McGraw-Hill, New York, 1958.
3. Dukin, V. V. and Aleksandrov, B. N., *Pseudopotential Calculations of the Residual Resistivities of Dilute Solid Alloys Based on Normal Metals, Physics of Low Temperatures*, Akademia Nauk UK SSR Academy of Science, USSR, Kharkov, 1978, 1.

Cd-Pd (Cadmium-Lead)

Dukin and Alekasndrov[1] gave a critical account of pseudopotential calculations and the resistivity increase due to 1% impurity. They gave $[\Delta\varrho/\Delta c]_{exp} = 1.85\ \mu\Omega cm/(a/o)Cd$ in PbCd and $[\Delta\varrho/\Delta c]_{exp} = 4.17\ \mu\Omega cm/(a/o)Pb$ in CdPb.

REFERENCES

1. Dukin, V. V. and Aleksandrov, B. N., *Pseudopotential Calculations of the Residual Resistivities of Dilute Solid Alloys Based on Normal Metals, Physics of Low Temperatures*, Akademia Nauk Uk SSR Academy of Science, USSR, Kharkov, 1978, 1.

Cd-Tl (Cadmium-Thallium)

Dukin and Aleksandrov[1] gave a critical account of pseudopotential calculations and the resistivity increase due to 1% impurity. They gave $[\Delta\varrho/\Delta c]_{exp} = 1.65\ \mu\Omega cm/(a/o)Cd$ in TlCd and $[\Delta\varrho/\Delta c]_{exp} = 1.30\ \mu\Omega cm/(a/o)Tl$ in CdTl.

REFERENCES

1. Dukin, V. V. and Aleksandrov, B. N., *Pseudopotential Calculations of the Residual Resistivities of Dilute Solid Alloys Based on Normal Metals, Physics of Low Temperatures*, Akademia Nauk Uk SSR Academy of Science, USSR, Kharkov, 1978, 1.

Cd-Zn (Cadmium-Zinc)

The phase diagram of Cd-Zn shows one eutectic reaction, in which the liquid phase with 26.5 a/o Zn decomposes at 266°C into C̲dZn with 5 a/o Zn and Z̲nCd with 2.3 a/o Cd.

LeBlanc and Schöpel[2] measured the conductivity of Cd-Zn alloys between 20 and 240°C. Samples were heated to 200°C, then cooled every 2 days by 15°C to obtain equilibrium states. Table 1 gives $\varrho(c)$.

Dukin and Aleksandrov[3] gave a critical account of pseudopotential calculations and the resistivity increase due to 1% impurity. They gave $[\Delta\varrho/\Delta c]_{exp} = 0.08 \ \mu\Omega cm/(a/o)Zn$ in C̲dZn, and $[\Delta\varrho/\Delta c]_{exp} = 0.25 \ \mu\Omega cm/(a/o)Cd$ in Z̲nCd.

Table 1
ELECTRICAL CONDUCTIVITY ($10^4 \Omega^{-1} cm^{-1}$)

Zn (a/o)	$K \cdot 10^{-4}$°C					
	20	50	100	150	200	240
100	16.83	15.07	12.80	11.17	9.82	8.92
99	15.63	14.00	11.90	10.32	9.00	8.06
97	16.31	14.76	12.60	10.92	9.53	8.56
95	16.22	14.58	12.37	10.77	9.40	8.42
90	15.86	14.30	12.18	10.52	9.20	8.24
80	15.40	13.82	11.70	10.11	8.79	7.89
60	14.77	13.30	11.28	9.66	8.38	7.48
40	14.28	12.81	10.79	9.22	7.92	7.06
20	13.94	12.46	10.51	8.94	7.66	6.77
5	13.74	12.24	10.32	8.82	7.54	6.71
3	13.64	12.20	10.28	8.73	7.51	6.73
1	13.54	12.08	10.15	8.72	7.62	6.84
0.5	13.40	12.00	10.14	8.77	7.60	6.83
0	14.18	12.74	10.81	9.35	8.14	7.30

REFERENCES

1. Hansen, M. and Anderko, K., *Constitution of Binary Alloys,* 2nd ed., McGraw-Hill, New York, 1958.
2. LeBlanc, M. and Schöpel, H., *Z. Elektrochem.,* 39, 695, 1933.
3. Dukin, V. V. and Aleksandrov, B. N., *Psuedopotential Calculations of the Residual Resistivities of Dilute Solid Alloys Based on Normal Metals, Physics of Low Temperatures,* Akademia Nauk Uk SSR Academy of Science, USSR, Kharkov, 1978, 1.

COBALT (Co)

Co-Cu (Cobalt-Copper)

At room temperature, Cu is insoluble in the solid Co lattice, while α-Co will dissolve up to 10 a/o Cu. A miscibility gap involving just α-Co and essentially pure Cu extends over the entire composition range between Co-10 a/o Cu and pure Cu with no evidence of any intermediate phases or intermetallic compounds.

The resistivity of the Co-Cu alloy system[1] at 25°C is shown in Figure 1, where it is evident that the resistivity of the concentrated alloys in the two-phase region is a complex function of composition and is not a simple volume average of the resistivity of the two terminal phases.

The resistivity of the dilute CuCo alloys at 0°C increases linearly with Co concentration[2] (see Figure 2). $d\varrho/dc = 6.9$ $\mu\Omega$cm/(a/o)Co. The resistivity of the very dilute CuCo alloys as a function of temperature has been investigated.[4-7] A Kondo minimum was observed. The DMR have also been determined for these dilute CuCo alloys. The deviation is both composition and temperature dependent (see Figure 3). The DMR shows a peak at low temperatures followed by a minimum at slightly higher temperatures. After the minimum, the DMR increases rapidly with increasing temperature, suggesting that the deviation will be sizeable at room temperature. The DMR for the dilute CoCu alloys gradually increases with increasing temperature up to 70°K (see Figure 4).

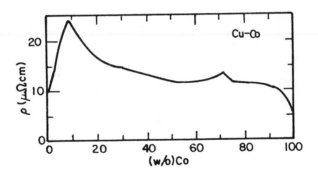

FIGURE 1.   Resistivity of Cu-Co at 25°C.[9]

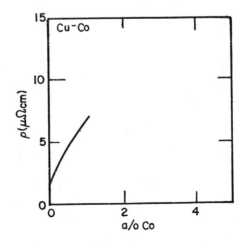

FIGURE 2.   Residual resistivity of dilute CoCu alloys.[3]

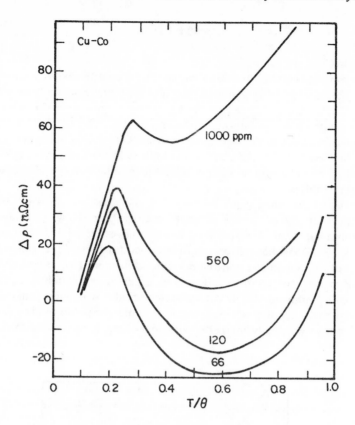

FIGURE 3.   The deviation from Matthiessen's rule as a function of temperature for several dilute C̲uCo alloys.

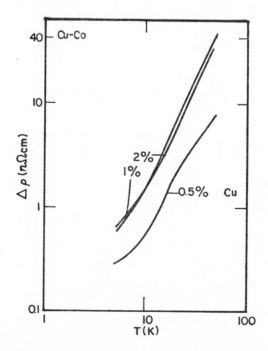

FIGURE 4.   Deviation from Matthiessen's rule for C̲oCu alloys.[8]

REFERENCES

1. Reichardt, G., *Ann. Phys.* 6, 832, 1901.
2. Gerritsen, A. N. and Linde, J. O., *Physica,* 18, 877, 1952; *Physica,* 17, 573, 1951.
3. Gerritsen, A. N., *Handbook of Physics,* Springer-Verlag, Berlin, 1957.
4. Dreyfull-Bourguard, A., Gautier, F., and Loegel, B., *J. Phys.,* 32, CI510, 1971.
5. Domenicali, C. A. and Christenson, E. L., *J. Appl. Phys.,* 32, 2450, 1961.
6. Loegel, B., *J. Phys. F,* 3, L106, 1973.
7. Jacobs, I. S. and Schmitt, R. W., *Physica,* 24, S174, 1958.
8. Hugel, J., *J. Phys. F,* 3, 1723, 1973.
9. *International Critical Tables,* Vol. 6, McGraw-Hill, New York, 1929.

Co-Fe (Cobalt-Iron)

Fe-Co exists as single phase BCC Solid solution alloys from 0 to 75 a/o Co at temperatures from 25 to 800°C. The Fe_1Co_1 composition in this phase field orders below 730°C. A two-phase region exists at room temperature for alloy compositions between 75 and 90% Co.

The composition dependence of the resistivity of Fe-Co alloys at temperatures between −253 and 800°C shows peaks in the resistivity at approximately the 15 and the 80 a/o Co compositions at all temperatures (see Figure 1). A resistivity minimum is also apparent at approximately the 55 a/o Co composition for all temperatures, including 800°C where the minimum can not be associated with the order-disorder reaction in the Fe_1Co_1 Alloy

The residual resistivity of the FeCo alloys as a function of the Co concentration is shown in Figure 2, and the resulting coefficient is about 0.9 $\mu\Omega$cm/(a/o) Co. The DMR for the dilute FeCo alloys at 296°K is relatively independent of cobalt concentration (see Figure 3), while the deviation of a 1a/o alloy is relatively independent of temperature up to 400°K (see Figure 4). The DMR in dilute CoFe is comparatively small at room temperature.

Yokoyama et al.[10,11] studied the order-disorder transition in FeCo. They measured ϱ as a function of annealing temperature and annealing times.

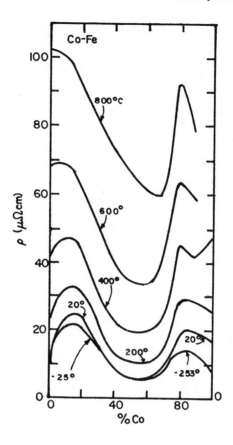

FIGURE 1.   Electrical resistivity and its temperature coefficient β for Fe-Co alloys.[2]

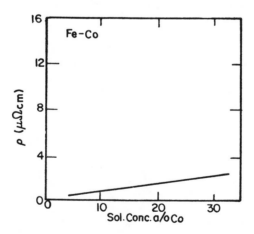

FIGURE 2.   Electrical resistivities of Co-Fe alloys
at 4.2°K.[3]

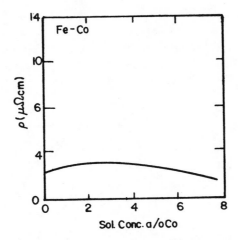

FIGURE 3. Apparent deviations from Matthiesson's rule at room temperature for Co-Fe alloys.[3]

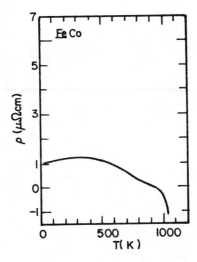

FIGURE 4. Temperature dependence of the solute resistivity for Fe + 1 a/o Co. No corrections were applied for changes in Curie temperature, Θ, with solute additions.[4]

REFERENCES

1. Kussmann, A., Scharnow, B., and Schulze, A., *Z. Tech. Phys.* 13, 449, 1932.
2. Bozorth, R. M., *Ferromagnetism,* Van Nostrand, Princeton, N.J., 1951.
3. Arajs, S., Schwerer, F. C., and Fisher, R. M., *Phys. Status Solidi,* 33, 731, 1969.
4. Schwerer, F. C. and Cuddy, L. J., *J. Appl. Phys.,* 41, 1419, 1970.
5. Fert, A. and Campbell, I. A., *J. Phys. F,* 6, 849, 1976.
6. Dorleijn, J. W. F., *Philips Res. Rep.,* 31, 287, 1976.
7. Dorleijn, J. W. F. and Miedema, A. R., *J. Phys. F,* 7, L23, 1977.
8. Majumdar, A. K. and Berger, L., *Phys. Rev. B,* 7, 4203, 1973.
9. Durand, J. and Gautier, F., *J. Phys. Chem. Solids,* 31, 2773, 1970.
10. Yokoyama, T. and Takezawa, T., *J. Phys. Soc. Jpn.,* 27, 509, 1969.
11. Yokoyama, T., Takezawa, T., and Higashida, Y., *Trans. Jpn. Inst. Met.,* 12, 80, 1971.

Co-Mo (Cobalt-Molybdenum)

Sarachik and Knapp[1] prepared Mo-Co alloys in an arc furnace. The MoCo alloy contained 0.90 a/o Co. Figure 1 shows the excess resistivity per percent Co referred to its value at 20°K as a function of log T. The $\varrho(T)$ curves showed a knee. This behavior was associated with the Kondo temperature.

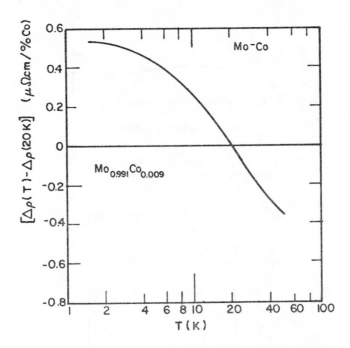

FIGURE 1.   Excess resistivity per percent Co referred to the value of 20°K, as a function of temperature, for Co dissolved in Mo. The quantity $\Delta\varrho = (\varrho - \varrho_o)/(a/o)$ Co, where ϱ and ϱ_o are the resistivities of the alloys with and without Co, respectively.

REFERENCES

1. Sarachik, M. P. and Knapp, G. S., *J. Appl. Phys.*. 40, 1105, 1969.

Co-Ni (Cobalt-Nickel)

Above 420°C, solid Co and Ni combine to form a complete series of face-centered cubic randomly structured solid solution alloys. The Co_1Ni_3 composition is suspected to be an order-disorder alloy. At 420°C, Co experiences an α(FCC) to ε(HCP) allotropic transformation, and as a result, a two-phase region appears in the phase diagram below 420°C to separate the hexagonally structural CoNi ε-primary phase from the fcc NiCo α-primary phase. At 25°C, alloys with compositions between 20 and 30 a/o Ni are a two-phase mixture of these α and ε-phases.

Several studies were made of the resistivity of the Co-Ni alloy systems[1-3] near room temperature (see Figure 1), and the agreement between the various studies has been poor, even when allowances are made for the slightly different temperatures of each study.

A considerable amount of recent work has focussed on the resistance of the dilute NiCo alloys. The residual resistivity of the dilute NiCo alloys is linear with concentration (see Figure 2) and the resulting coefficient at 4.2°K is $d\varrho/dc = 1.8\ \mu\Omega cm/$ (a/o)Co.

The DMR has also been evaluated for the dilute N̲i̲Co alloys (see Figure 3). The deviations become progressively larger with increasing temperature and are appreciable at room temperature.

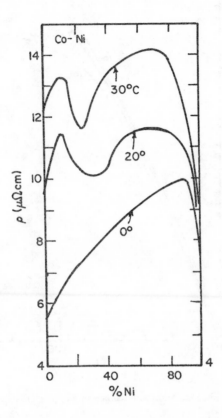

FIGURE 1. Electrical resistivity of Co-Ni alloys at 0 to 30 °C according to several reports.[3]

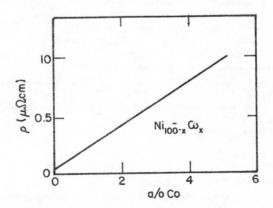

FIGURE 2. Residual resistivity of dilute N̲i̲Co alloys of various concentrations.[6]

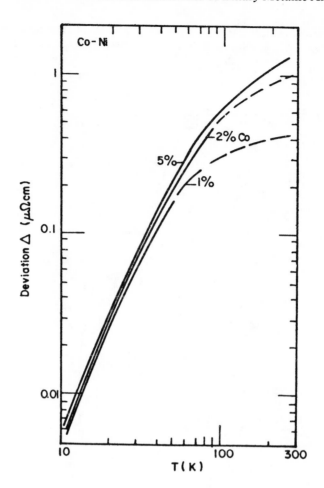

FIGURE 3.   Variation with temperature of Δ (c, T) for various NiCo alloys.[9]

REFERENCES

1. Masumoto, H., *Sci. Rep. Tohoku Imp. Univ.,* 16, 34, 1927.
2. Broniewski, W. and Pietrek, W., *C. R.,* 201, 206, 1935.
3. Bozorth, R. M., *Ferromagnetism,* Van Nostrand, Princeton, N.J., 1951.
4. Schulze, A., *Z. Tech. Phys.,* 8, 423, 1927; Phys. Z., 28, 669, 1927.
5. Dorleijn, J. W. F. and Miedema, A. R., *J. Phys. F,* 5, 487, 1975.
6. Dorleijn, J. W. F., *Philips Res. Rep.,* 31, 287, 1976.
7. Fert, A. and Campbell, I. A., *J. Phys. F,* 6, 849, 1976.
8. Leonard, P., Cadeville, M. C., and Durand, J., *J. Phys. Chem. Solids,* 30, 2169, 1969.
9. Farell, T. and Grieg, D., *J. Phys. C,* 1, 1359, 1968.
10. Caderville, M. C., Gautier, F., Robert, C., and Rossel, J., *Solid State Commun.,* 1, 1701, 1968.
11. Hugel, J., *J. Phys. F,* 3, 1723, 1973.
12. Farrell, T. and Grieg, D., *J. Phys. C,* 1, 1359, 1968.

Co-Pd (Cobalt-Palladium)

The phase diagram of Pd-Co above 408°C shows a complete range of solid solutions. Colp and Williams[2] measured $\varrho(T)$ of alloys with 2 to 7.5 a/o Co between 1.4 and 77°K. Their results for the difference between the resistivity of the alloy and that of pure Pd, $\Delta\varrho = \varrho$alloy $(T) - \varrho_{Pd} (T)$ is shown in Figures 1 and 2 where the difference follows a temperature dependence of $\Delta\varrho = A + BT^2$ in this temperature range. A is a linear function of composition, with $A = c\,(1.27 \pm 0.03)\ \mu\Omega\text{cm}$ and $B\ \alpha\ c^n$, with $n = -0.75 \pm 0.05$.

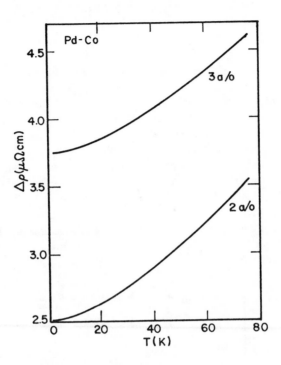

FIGURE 1. $\Delta\varrho = \varrho_{alloy}\,(T) - \varrho_{Pd}\,(T)$ of PdCo alloys. Number gives Co percentage.[2]

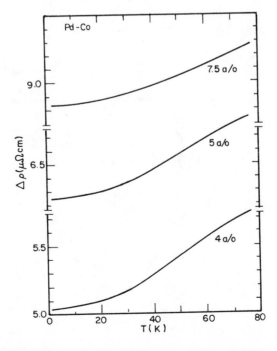

FIGURE 2. $\Delta\varrho = \varrho_{alloy}\,(T) - \varrho_{Pd}\,(T)$ of PdCo alloys. Number gives Co percentage.[2]

REFERENCES

1. Hansen, M. and Anderko, K., *Constitution of Binary Alloys,* McGraw-Hill, New York, 1958.
2. Colp, M. E. and Williams, G., *Phys. Rev. B,* 5, 2599, 1972.

Co-Pt (Cobalt-Platinum)

Shen et al.[1] studied $\varrho(T)$ of PtCo alloys. These authors used a 0.07 a/o Co in Pt alloy. Their experimental results are given in Figure 1. The impurity resistivity, $\Delta\varrho$, increases linearly with ln T. $\Delta\varrho = A + B \ln T$ with $0.3 < \ln T < 3$. The data can be explained by both Kondo's theory, or Knapp's phenomenological model.

Williams et al.[2] studied the effect of magnetic fields on a Pt-0.061 a/o Co wire prepared from 99.999% pure Pt and Co sponge by induction melting in a water-cooled Cu boat levitation furnace. After cold rolling, the sample was annealed for 24 hr at 1000°C in a vacuum of 10^{-5} Torr. $\Delta\varrho(T)$ due to impurities was found to be given by $\Delta\varrho = A + B \ln[(T^2 + \Theta^2)^{1/2}]$ in zero magnetic field. This equation also describes the experimental results in a magnetic field, if T^2 is replaced by $T^2 + (g M_B/k_B)^2 H^2$. The authors suggests that $\Delta\varrho$ is due to localized spin fluctuations. $\Delta\varrho(T)$ values of alloys with up to 5 a/o Co for temperatures to 60°K were obtained[3] on samples prepared by arc-melting (see Figure 2).

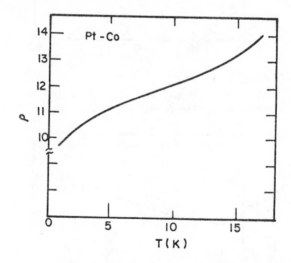

FIGURE 1.   Resistivity of Pt + 0.07 a/o Co.

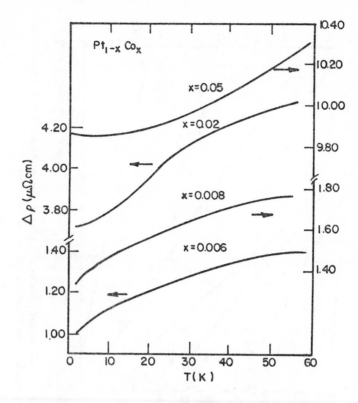

FIGURE 2. Temperature dependence of the impurity resistivity of $Pt_{1-x}Co_x$.

REFERENCES

1. Shen, L., Schreiber, D. S., and Arko, A. J., *Phys. Rev.*, 179, 512, 1969.
2. Williams, G., Swallow, G. A., and Loram, J. W., *Phys. Rev. B*, 11, 344, 1975.
3. Rao, L. V., Rapp, O., Johanneson, Ch., Buduick, J. I., Burch T. J., and Cannella, V., AIP Conf. Proc., 346, 1976.

Co-Rh (Cobalt-Rhodium)

α-Co and Rh form a continuous series of FCC solid solution alloys at elevated temperatures. For compositions greater than 50 a/o Rh, these FCC alloys are stable to temperatures below 25°C. The compositions containing less than 48% exist at 25°C as single-phase disordered HCP alloys.

Much of the resistivity work on the Co-Rh alloy system has been confined to low temperatures where the alloys can exist in the paramagnetic, spinglass, or ferromagnetic states, depending upon the composition and temperature. The temperature dependence of the resistivity shows a bend (see Figure 1) at the temperature boundary separating these magnetic regions and shows a slope at low temperatures that is a function of the magnetic state of the alloy (see Figure 2). The DMR for dilute CoRh is relatively small at room temperature.[4]

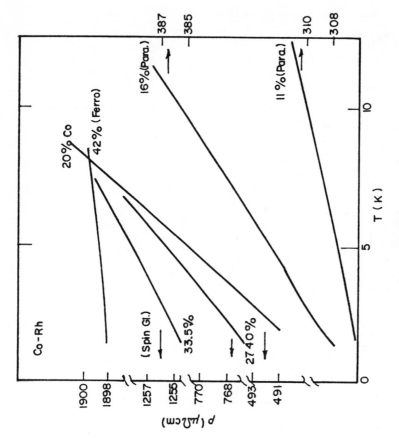

FIGURE 2.   The resistivity-temperature plots of some of the alloys in the concentration range investigated. Number gives Co concentration (in %).[1]

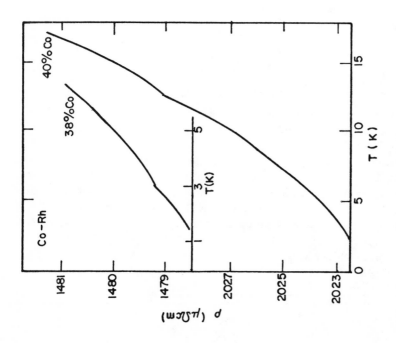

FIGURE 1.   The resistivity-temperature plots of two samples showing a resistance anomaly occuring at the boundary separating the spin glass and ferromagnetic regions.[1]

REFERENCES

1. Tari, A., *J. Phys. F,* 6, 1313, 1976.
2. Rusby, R. L., Ph.D. thesis, University of London, 1973.
3. Coles, B. R., Mozumber, S., and Rusby, R. L., *Proc. 12th Int. Conf. Low Temperature Physics,* Kanda, E., Ed., Academic Press, New York, 1971, 737.
4. Durand, J. and Gautier, F., *J. Phys. Chem. Solids,* 31, 2773, 1970.

Co-Sb (Cobalt-Antimony)

At 25°C, the solid Cu lattice can dissolve up to 3 a/o Sb. Three intermediate phases with limited ranges of homogeneity exist near the Cu_1Sb_1, Co_1Sb_2, and Co_1Sb_3 compositions. All other compositions are two-phase alloys. The DMR for the primary phase CoSb alloys has been determined at temperatures to 300°K (see Figure 1) and was found to continually increase with temperature. The resistance at room temperature of the intermediate CoSb phase is dependent on the Sb concentration (see Figure 2).

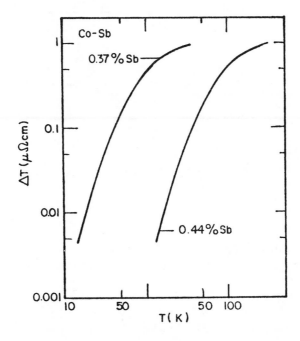

FIGURE 1.   Deviations from Matthiessen's rule for CoSb alloys.[3]

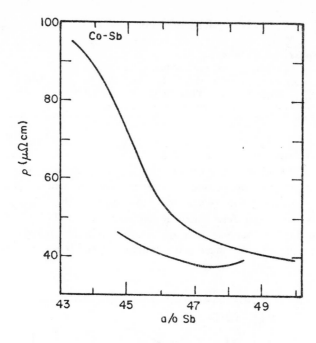

FIGURE 2.   Resistivity vs. Sb content of the intermediate phase (CoSb) alloys: top line (Reference 1); bottom line: resistivity data of Makarov given by Reference 1.

REFERENCES

1. Penn, J. and Miller, E., *J. Appl. Phys.*, 44, 177, 1973.
2. Makorov, E. S., *Dokl.Akad.Nauk SSSR*, 40, 191, 1943.
3. Ross, R. N., Price, D. C., and Williams, G., *J. Phys. F*, 8, 2367, 1978.

Co-Ti (Cobalt-Titanium)

Sherbakov et al.[1] determined ϱ of T̲iCo. Results are given in Figure 1. Alloys with higher impurity levels show negative $d\varrho/dT$ values. The alloy with 6 a/o shows a resistivity minimum.

The resistivity of CoTi has been measured by Butler et al.,[2] Arajs et al.,[3] and Hilscher and Gratz.[4] Arajs et al.[3] used a sample prepared by the electron beam melting method. Figure 2 shows $\varrho(T)$ from 300 to 1400°K. The measurements indicate no phase transitions.

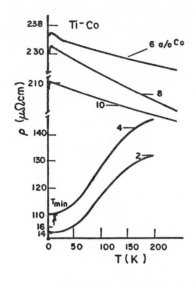

FIGURE 1. Resistivity of T̲iCo alloys. Number gives (a/o)Co.

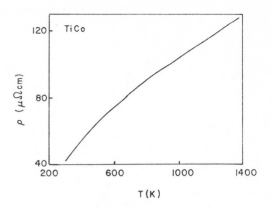

FIGURE 2. Resistivity of TiCo.

REFERENCES

1. Sherbakov, Prekul, A. F., Volkenshtein, N. V., and Nikolaev, A. L., *Sov. Phys. Solid State,* 21, 398, 1979.
2. Butler, S. R., Hanlon, J. E., and Wasilewski, R. J., *J. Phys. Chem. Solids,* 30, 281, 1969.
3. Arajs, S., Stelmach, A. A., and Martin, M. C., *J. Less Common Met.,* 32, 178, 1973.
4. Hilscher, G. and Gratz, E., *Phys. Status Solidi A,* 48, 473, 1978.

Co-Zr (Cobalt-Zirconium)

Hossain et al.[1] prepared $Zr_{50}Co_{50}$ alloys by nonconsumable arc-melting in a pregettered Ar atmosphere. ZrCo has the ordered bcc structure (B2-type). $\varrho(T)$ of these samples show a small hysteresis effect. Figure 1 gives the resistivity.

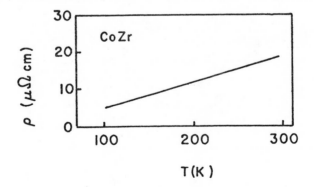

FIGURE 1. Resistivity of CoZr.

REFERENCES

1. Hossain, D., Harris, I. R., and Barradough, K. G., *J. Less Common Met.,* 37, 35, 1974.

CHROMIUM (Cr)

Cr-Cu (Chromium-Copper)

Cu and Cr show no evidence of mutual solid solubility at temperatures below 1000°C and alloys over the entire range of compositions appear to be a two-phase mixture of essentially pure Cu and pure Cr. Resistivity measurements have been primarily confined to the very dilute CuCr alloys. The variation of the resistivity with chromium concentration in these alloys at 0°C is $d\varrho/dc = 4.0 \, \mu\Omega cm/(a/o)Cr$ (see Figure 1).

The temperature dependence of the resistivity of the very dilute CuCr alloys exhibits a Kondo response (see Figure 2) characteristic of magnetic atom solutes in a noble metal lattice. The DMR at various temperatures between 0 and 300°K for CuCr alloys shows a maximum at about 60°K (see Figure 3) and is dependent on the Cr concentration. The deviation is small at room temperature. The Cr impurity contribution to the resistivity of these dilute CuCr alloys as a function of temperature is presented in Figure 4.

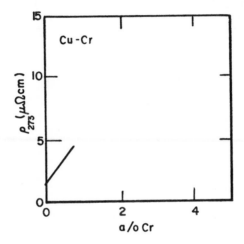

FIGURE 1. Resistivity of CuCr alloys at 273°K.[1]

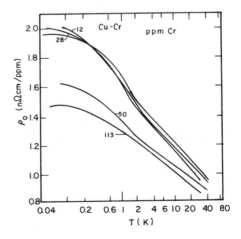

FIGURE 2. The low-temperature resistivity of CuCr for various Cr concentrations.[3]

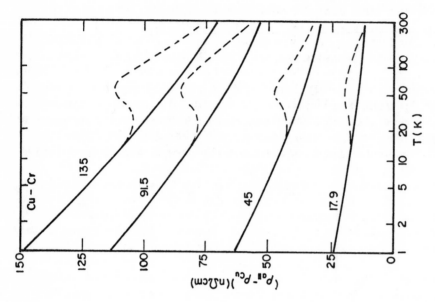

FIGURE 4. The Cr impurity contribution to the resistivity of CuCr alloy at various temperatures. The numbers designate Cr (ppm) in CuCr.[5] The solid curve above ~20K is an extrapolation of the Kondo contribution; the dashed line above ~20K and the full line below gives experimental results.

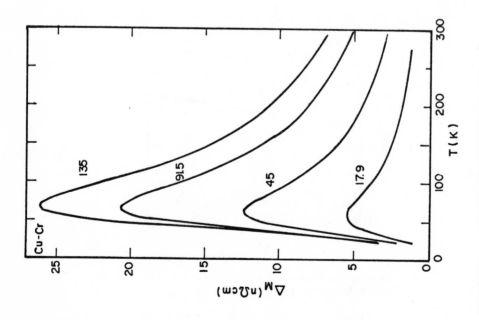

FIGURE 3. The deviations from Matthiessen's rule in CuCr at various temperatures. Numbers give impurity concentration (in ppm).[5]

REFERENCES

1. Gerritsen, A. N., *Handbook of Physics,* Springer-Verlag, Berlin, 1956.
2. Legvold, S., Vyrostek, T. A., Schafer, J. A., Burgardt, P., and Peterson, D., *Solid State Commun.,* 16, 477, 1975.
3. Daybell, M. D. and Steyert, W. A., *Phys. Rev. Lett.,* 20, 195, 1968.
4. Vreys, H., DeSmedt, E., Pitsi, G., and Dupre, A., *Physica,* 86, 88B, 455, 1977.
5. Haen, P., Souletie, J., and Texeira, J., *J. Low Temp. Phys.,* 23, 191, 1976.
6. Read, M. and Guenault, A. M., *J. Phys. F,* 4, 94, 1974.
7. Linde, J. O., *Low Temp. Phys. Chem.,* 402, 1948.

Cr-Fe (Chromium-Iron)

Except for a small γ-loop at elevated temperatures for the Fe-rich alloys and a σ-phase forming below 815°C at the equiatomic composition, Fe and Cr form a series of disordered body-centered cubic solid solution alloys that at room temperature exist at all compositions, except between approximately 40 to 60 a/o Cr.

The electrical resistivity of the Fe-Cr alloys has been thoroughly investigated as a function of temperature and composition.[1-5] In general, the resistivity of FeCr alloys at room temperature will increase as a function of Cr concentration as shown in Figure 1. The temperature dependence of the resistivity has been of particular interest for the CrFe alloys because of the anomolous resistance behavior associated with the antiferromagnetic order occurring below the Neel temperature in these alloys (see Figure 2). The more concentrated CrFe alloys do not order antiferromagnetically so the resistivity increases smoothly with temperature and shows no unusual features (see Figure 3).

The residual resistivities of dilute FeCr and CrFe alloy have been determined[7-14] as a function of solute concentration. The concentration coefficient of the residual resistivity is $d\varrho/dc = 2\ \mu\Omega cm/(a/o)Cr$ for the dilute FeCr alloys and $d\varrho/dc = 6\ \mu\Omega cm/(a/o)Fe$ for the dilute CrFe alloys. The residual resistivity of the CrFe alloys increases with Fe concentration to a maximum at the 10 a/o Fe composition (see Figure 4) as shown by Loegel.[14,15]

The DMR for the dilute FeCr alloys is both dependent on temperature and concentration.[10] The temperature dependence shows a maximum at 500°K.[11]

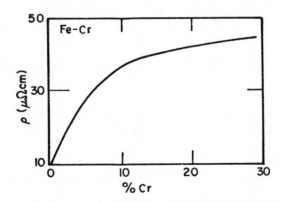

FIGURE 1. Room-temperature electrical resistivities of Fe-Cr alloys in w/o.[6]

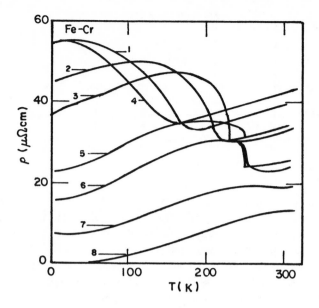

FIGURE 2. Electrical resistivity of Cr-Fe alloys below 320°K. (1) 9.5 a/o Fe; (2) 6.5 a/o Fe; (3) 4.9 a/o Fe; (4) 11.2 a/o Fe; (5) 3.3 a/o Fe; (6) 2.3 a/o Fe; (7) 0.9 a/o Fe; (8) Cr.[2]

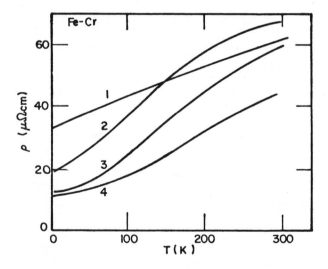

FIGURE 3. Electrical resistivities vs. temperature for Cr-Fe alloys with 29 to 78 a/o Fe (a/o Fe). (1) 20; (2) 29.1; (3) 37.5; (4) 78.[4]

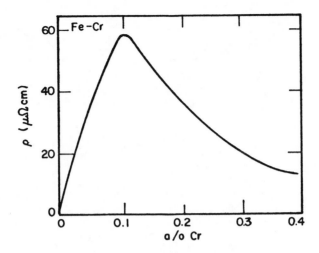

FIGURE 4.   Residual resistivity of C̲r̲Fe alloys as a function of composition.[14,15]

REFERENCES

1. Adock, F., *J. Iron Steel Inst. (London)*, 124, 99, 1931.
2. Arajs, S. and Dunmyre, G. R., *J. Appl. Phys.*, 37, 1017, 1966.
3. Cox, J. E. and Lucke, W. H., *J. Appl. Phys.*, 38, 3851, 1967.
4. Rajan, N. S., Waterstrat, R. M., and Beck, P. A., *J. Appl. Phys.*, 31, 731, 1960.
5. Schroder, K., Yessik, M. J., and Baum, N. P., *J. Appl. Phys.*, 37, 731, 1960.
6. Bozorth, R. M., *Ferromagnetism*, Van Nostrand, Princeton, N.J., 1951.
7. Dorleijn, J. W. F. and Miedema, A. R., *J. Phys. F*, 7, L23, 1977.
8. Dorleijn, J. W. F., *Philips Res. Rep.*, 31, 287, 1976.
9. Majumdar, A. K. and Berger, L., *Phys. Rev. B*, 7, 4203, 1973.
10. Fert, A. and Campbell, I. A., *J. Phys. F*, 6, 849, 1976.
11. Schwerer, F. C. and Cuddy, L. J., *J. Appl. Phys.*, 41, 1419, 1970.
12. Arajs, S., Schwerer, F. C., and Fisher, R. M., *Phys. Status Solidi*, 33, 731, 1969.
13. Biolluz, A., Thesis, Strasbourg, 1978.
14. Loegel, B., Ph.D. thesis, Strasbourg.
15. Loegel, B., *J. Phys. F*, 3, L106, 1973.

Cr-Ge (Chromium-Germanium)

Hansen and Anderko[1] report that Wallbaum[2] found these compounds in the Cr-Ge system: Cr_3Ge with the β-W structure, Cr_3Ge_2 which is isomorphous with Cr_3Si_2, and CrGe with the FeSi (B20)-type structure. Fakidov and Grazhdaukina[3] prepared Cr-Ge alloys from pure Ge (room temperature resistivity: 1.4 Ωcm) and Cr (impurities are Fe, 0.02%; O_2, 0.0013%; Si, 0.03%; H_2, 0.00019%; N, 0.0145%) by melting powders with more than 40 a/o Ge in evacuated quartz ampoules. They were kept at 1140°C for not less than 4 hr. Samples were furnace cooled. Table 1 shows the resistivities of the alloys at room temperature. It shows a sharp maximum at 75 a/o Ge, a composition which does not coincide with the composition of the compounds given by Wallbaum. Reference 3 determined $\varrho(t)/\varrho(77)$ of alloys with 50, 60, or 70 a/o Ge, respectively.

Table 1

Composition of alloy				
a/o		w/o		$\varrho_a \cdot 10^{-3}$
Cr	Ge	Cr	Ge	(Ωcm)
50.0	50.0	42.7	57.3	0.19
46.0	54.0	48.0	62.0	1.52
40.0	60.0	32.3	67.7	3.50
54.0	66.0	27.0	73.0	5.02
30.0	70.0	23.5	76.5	7.47
25.0	75.0	19.3	80.7	13.20
20.0	80.0	15.2	84.8	18.42
15.0	85.0	11.2	88.8	17.00
10.0	90.0	7.4	98.6	15.32
5.0	95.0	3.6	96.4	13.90

REFERENCES

1. Hansen, M. and Anderko, K., *Constitution of Binary Alloys,* McGraw-Hill, New York, 1958.
2. Wallbaum, H. J., *Naturwissenschaften,* 32, 76, 1944.
3. Fakidov, I. G. and Grazhdaukina, N. P., *Phys. Met. Metallogr. (USSR),* 6, 62, 1958.

Cr-Mn (Chromium-Manganese)

A total of 25 a/o Mn will dissolve in Cr at 500°C.[1] The maximum solubility of Mn in Cr is 71.4 a/o at 1310°C. The intermetallic σ-phase is found between 72 and 85 a/o Mn. Several percent of Cr will dissolve in the different allotropic modifications of Mn.

Alloys of the Cr-rich solid solutions were studied by Maki and Adachi.[2] They melted appropriate amounts of Cr (5N) and Mn (4N) several times in a plasma jet furnace in an Ar atmosphere. The samples were homogenized at 1000°C for 24 to 72 hr in an evacuated silica tube. Then the samples were quenched in water. X-ray and neutron diffraction studies showed the samples to be disordered.

Figure 1 gives $\varrho(T)$ of alloys with up to 60 a/o Mn. All alloys show a characteristic minimum, associated by the authors with the Neel temperature, and also a broad maximum in $\varrho(T)$ below T_N. The resistivity is analyzed with the model of a gap type antiferromagnet. $\varrho(T) = [\varrho_i + \varrho_1(T) + \varrho_m(T)]/[1 - g_m(T)]$, where ϱ_i is residual resistivity and $\varrho_1(T)$ is the phonon resistivity term ($\propto$ Grüneisen-Bloch function times T); $\varrho_m(T)_i \propto (1 - m^2(T))$, where m(T) is the normalized sublattice magnetization; g is the truncation factor. Sousa et al.[4] studied in detail $\varrho(T)$ of Cr-Mn alloys with 58 and 64 a/o Mn and attempted to fit $d\varrho/dT$ with a power function of the form $(1/\varrho)(d\varrho/dT) = A|(T - T_N)/T_N|^{-\lambda} + B$, where A, B, and λ are constants. Should this power function be the right approach to explain the antiferromagnetic paramagnetic transition, then one can obtain accurate values of T_N($T_N = 723$°K for $Cr_{36}Mn_{64}$, $T_N = 736$°K for $Cr_{42}Mn_{58}$).

Low-temperature $\varrho(T)$ measurements by Arajs[8] (see Figure 2) show no anomaly. Nagasawa and Senba[9] gave $d\varrho/dc = 80 \mu\Omega$cm/(a/o)Cr in Mn-rich alloys. They found a T^2 temperature dependence at low temperatures ($\varrho = \varrho_0 + A T^2$). The coefficient A is very large for Mn: $A = 0.125 \mu\Omega$cm/K^2. A changes rapidly with alloying. It is 0.5 $\mu\Omega$cm/K^2 for an alloy with 1 a/o Cr. The T^2 dependence is associated with spin fluctuations.

Williams and Standford[10] measured the resistivity of Mn rich allows between 4.2°K and room temperature. Samples were arc-melted and then sliced by spark-erosion. A vacuum annealed at 620°C for 6 hr followed. Figures 3 to 5 gives the experimental results. The minimum in $\varrho(T)$ is associated with antiferromagnetic ordering. T_N is depressed if Cr is added to Mn.

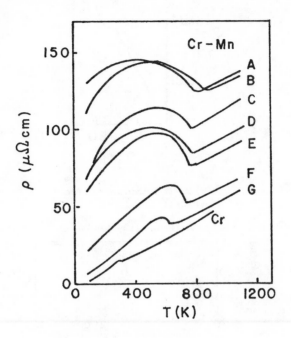

FIGURE 1. Resistivity of Cr-Mn alloys (a/o Mn). (A) 50; (B) 60; (C) 40; (D) 30; (E) 20; (F) 10; (G) 5.

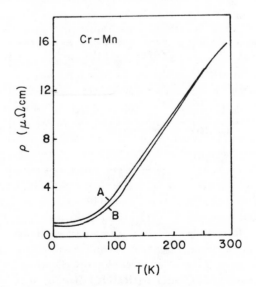

FIGURE 2. Resistivity of Cr-Mn alloys (a/o Mn). (A) 4; (B) 2.1.

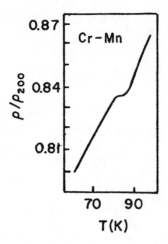

FIGURE 3.   Resistivity of Cr-Mn
alloys; Mn, 1 a/o Cr.

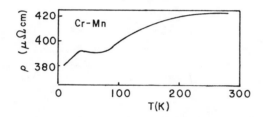

FIGURE 4.   Resistivity of Cr-Mn alloys; Mn—1.9
a/o Cr.

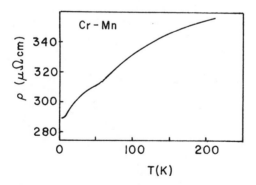

FIGURE 5.   Resistivity of Cr-Mn alloys; Mn—2.7
a/o Cr.

REFERENCES

1. **Hansen, M. and Anderko, K.,** *Constitution of Binary Alloys,* 2nd ed., McGraw-Hill, New York, 1958.
2. **Maki, S. and Adachi, K.,** *J. Phys. Soc. Jpn.,* 46, 1131, 1979.
3. **Suzuki, T.,** *J. Phys. Soc. Jpn.,* 21, 442, 1966.
4. **Sousa, J. B., Chaves, M. R., Braga, M. E., Reis, M. M., Pinheiro, M. F., and Crisan, M.,** *J. Phys. F,* 5, L155, 1975.

5. Hamaguchi, Y. and Kuniomo, N., *J. Phys. Soc. Jpn.*, 19, 1849, 1964.
6. Yasui, M. and Shimizu, M., *J. Phys. Soc. Jpn.*, 31, 378, 1971.
7. Elliott, R. J. and Wedgewood, F. A., *Proc. Phys. Soc.*, 81, 846, 1963.
8. Arajs, S., *Phys. Status Solidi*, 1, 499, 1970.
9. Nagasawa, H. and Senba, *J. Phys. Soc. Jpn.*, 39, 70, 1975.
10. Williams, W., Jr. and Stanford, J. L., *Phys. Rev. B*, 7, 3244, 1973; *AIP Conf. Proc.*, 15(1), 521, 1971.

Cr-Mo (Chromium-Molybdenum)

Cr and Mo form a complete series of solid solutions.[1] There is no evidence of compound formation.

Meaden et al.[2] prepared CrMo alloys with 0.6 and 5.1 a/o Mn. Their $\varrho(T)$ curves, given in Figures 1 and 2, confirm the measurements by Arajs.[3] They show a sharp minimum, attributed to the Neel temperature T_N. T_N decreases with increasing Mo concentration.

Arajs et al.[4] measured the resistivity of CrMo with up to 8 a/o Mo. Their $\varrho(4.2°K)$ curve is shown in Figure 3. It is in most other alloy systems possible to use the $\varrho(4.2°K, C)$ data to obtain the impurity resistivity. This is not possible in Cr alloys with their complex electronic structure and resistivity, as seen in Figure 4.

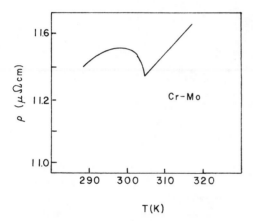

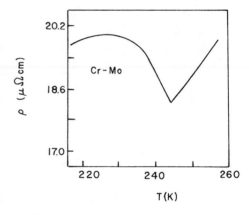

FIGURE 1.   Resistivity of Cr-Mo alloys, Cr + 0.6 a/o Mo.[2]

FIGURE 2.   Resistivity of Cr-Mo alloys, 5.1 a/o Mo.[2]

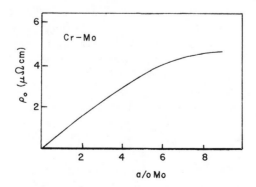

FIGURE 3.   Resistivity of Cr-Mo alloys, ϱ at 4.2°K.[4]

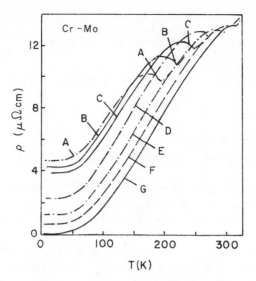

FIGURE 4.   Resistivity of Cr-Mo alloys. (A) 8.8 a/o Mo; (B) 6.9 a/o Mo; (C) 5.1 a/o Mo; (D) 3.2 a/o Mo; (E) 1.6 a/o Mo; (F) 0.6 a/o Mo; (G) 0 a/o Mo.[4]

REFERENCES

1. Hansen, M. and Anderko, K., *Constitution of Binary Alloys,* 2nd ed., McGraw-Hill, New York, 1958.
2. Meaden, G. T., Rao, K. V., and Tee, K. T., *Phys. Rev. Lett.,* 25, 359, 1970.
3. Arajs, S., *J. Appl. Phys.,* 39, 673, 1968.
4. Arajs, S., Anderson, E. E., and Rao, K. V., *J. Less Common Met.,* 26, 157, 1967.

Cr-Ni (Chromium-Nickel)

The Cr-Ni phase diagram is suggested to be a simple eutectic whose α-Cr primary solid solution phase field is capable of dissolving only a limited amount of Ni ($\sim$1 w/o) at 25°C, while the α-Ni primary solid solution can accommodate more than 30 w/o Cr at 25°C. The alloy range between Cr-1 w/o Ni and Cr-70 w/o Ni is a two-phase mixture of α-Cr and α-Ni.

The resistivity of the α-NiCr primary phase alloys has been frequently investigated.[1-6] The temperature dependence of the resistivity for various concentrated NiCr alloys is presented in Figure 1A. The resistivity of the alloys whose compositions are near 25% Cr are relatively temperature insensitive above 600°C. Anomolous resistivity behavior is apparent just below 600°C in these concentrated alloys and this is possibly associated with the onset of the Cr_1Ni_3 ordering reaction at approximately 540°C.[15] The influence of the ordering phenomena is also manifest in the annealing dependence of the resistivity of the concentrated alloys (see Figure 1B).

A more detailed investigation of the resistivity of α-NiCr alloys below 300°K (see Figure 2) indicates that alloys near 15% Cr have a very low-temperature coefficient of resistivity suitable for resistor applications. Also, the resistivity behaves anomalously with temperature for NiCr alloy compositions ($\sim$10% Cr) close to the critical value for ferromagnetic ordering.

Considerable attention has also been directed to the resistivity of the dilute NiCr alloy.[7-14] The residual resistivity at 4.2°K for various dilute NiCr alloy (see Figure 3) gives the concentration coefficient of resistivity as $d\rho(NiCr)/dc = 5$ $\mu\Omega cm/(a/o)Cr$, while the DMR is concentration dependent and increases with increasing temperature (see Figure 4).

195

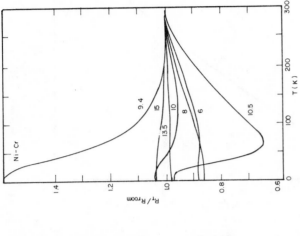

FIGURE 2. The resistance ratio, relative to room temperature, as a function of temperature for NiCr alloys. The numbers accompanying each set of data indicate the Cr concentration (a/o).[6]

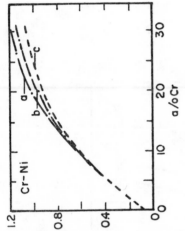

FIGURE 1B. Resistivity of Ni-Cr alloys at room temperature.[1,3] (a) Slowly cooled from 800°C; (b) quenched from 800°C; (c) hard drawn.

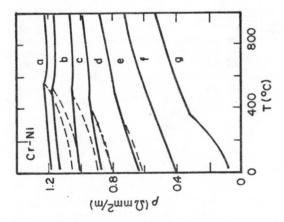

FIGURE 1A. Temperature dependence of ϱ (a/o Cr). (a) 31.4; (b) 26.3; (c) 20.0; (d) 16.0; (e) 10.3; (f) 5.4; (g) 0.

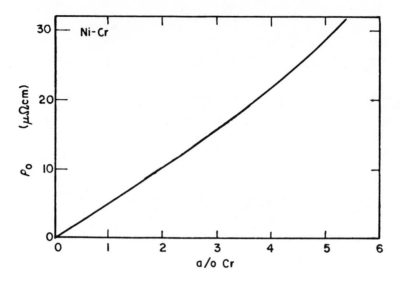

FIGURE 3.   Resistivity at 4.2°K, $\varrho_o(c)$, vs. solute concentration for N̲i̲Cr alloys.[14]

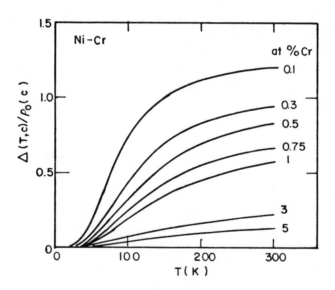

FIGURE 4.   Normalized deviations $[\Delta(T,c)/\varrho_o(c)]$ for N̲i̲Cr alloys. Numbers indicate approximate Cr concentrations.[14]

REFERENCES

1. Thomas, H., *Z. Phys.*, 129, 219, 1951.
2. Mooij, J. H., *Phys. Status Solidi A*, 17, 521, 1973.
3. Köster, W. and Rochell, P., *Z. Metallkd.*, 48, 485, 1957.
4. Arajs, S., *Phys. Stat. Solidi A*, 1, 499, 1970.
5. Taylor, A. and Hinton, K. G., *J. Inst. Met.*, 81, 169, 1952.
6. Smith, T. F., Tainsh, R. J., Shelton, R. N., and Gardner, W. E., *J. Phys. F*, 5, L96, 1975.
7. Farrell, T. and Grieg, D., *J. Phys. C*, 1, 1359, 1968.
8. Fert, A. and Campbell, I. A., *J. Phys. F*, 6, 849, 1976.
9. Cadeville, M. C., Gautier, F., Robert, C., and Rossel, J., *Solid State Commun.*, 7, 1701, 1968.
10. Hugel, J., *J. Phys. F*, 3, 1723, 1973.
11. Leonard, P., Cadeville, M. C., and Durand, J., *J. Phys. Chem. Solids*, 30, 2169, 1969.
12. Dorleijn, J. W. F., *Philips Res. Rep.*, 31, 287, 1976.
13. Dorleijn, J. W. F. and Miedema, A. R., *J. Phys. F*, 5, 487, 1975.
14. Schwerer, F. C. and Conroy, J. W., *J. Phys. F*, 1, 877, 1971.
15. Nordheim, R. and Grant, N. J., *J. Inst. Met.*, 82, 440, 1954.
16. Coles, B. R., *Adv. Phys.*, 7—8, 52, 1958.

Cr-Os (Chromium-Osmium)

Os absorbs up to 51 a/o Cr in solid solution.[1] The tetragonal Cr_2Os phase is stable at elevated temperatures; the Cr_3Os phase has a narrow stability range. About 9.5 a/o Os will dissolve in Cr.

Arajs et al.[2] prepared Cr-Os alloys with Cr of Iochrome purity. The alloys were melted repeatedly, then sealed in quartz capsules filled with Ar (150 Torr at 300°K) homogenized at 1400°K for 150 hr and furnace cooled. Figure 1 gives $\varrho(T)$ for these alloys. $\varrho(T)$ shows the characteristic minimum associated with the onset of antiferromagnetic ordering. T_N increases with increasing Os concentration, typical for impurities with higher outer electron/atom concentrations than Cr. Table 1 gives T_N and $\varrho(4.2°K)$ of the alloys.

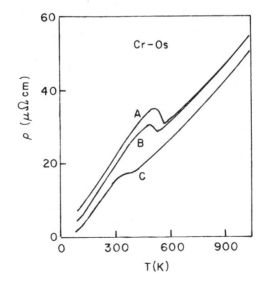

FIGURE 1. Resistivity of Cr-Os alloys (a/o Os).
(A) 2.0; (B) 1.1; (C) 0.3.[2]

Table 1

C(a/o Os)	T_N (K)	$\varrho(4.2°K)$ ($\mu\Omega$cm)
0.3	359	0.81
0.6	465	1.81
1.1	533	2.85
2.0	566	5.26

REFERENCES

1. Raub, E., *Z. Metallkd.*, 48, 53, 1957.
2. Arajs, S., deYoung, T. F., and Anderson, E. E., *J. Appl. Phys.*, 41, 1426, 1970.

Cr-Pd (Chromium-Palladium)

Hansen and Anderko show that $\sim$5 a/o Pd will dissolve in Cr at 1320°C.[1] Resistivity measurements[2] show that up to 60 a/o Cr (Cr_3Pd_2) is soluble in Pd. Cr_3Pd_2 and Cr form a two-phase regime.

Roshko and Williams[3] measured the resistivity of Pd-Cr alloys with 11 to 18 a/o Cr from 1.4 to 300°K. They found that the system resembled a canonical spin glass. $\Delta\varrho \propto$ $- \log c \cdot T^{3/2}$ at low temperatures. $\Delta\varrho$ shows a maximum. The alloys were prepared in an arc furnace from 5N Pd and 5N Cr. Figures 1 and 2 give the experimental results for alloys with more than 11 a/o Cr. Resistivities for alloys with 1 to 4 a/o Cr are given in Figure 3. At T > 20°K, $\varrho(T)/c = A + B \ln [(T^2 + \Theta^2)^{1/2}]$ for alloys with 4 to 6 at/o Cr. The spin fluctuation temperature Θ ranges from 25 to 33°K.

Low-concentration alloys with up to 0.3 at/o Cr were studied by the same authors.[5] The incremental resistivity $\Delta\varrho = \varrho(T)_{alloy} - \varrho(T)_{Pt}$ appears to contain two contributions. One is the conventional Matthiessen's rule deviation which is essentially temperature independent. The second contribution decreases monotomically with temperature and should follow a temperature dependence of the form $\ln[(T^2 + \Theta^2)^{1/2}]$.

Star et al.[6] studied $\varrho(T)$ of annealed PdCr alloys below room temperature. Figure 4 gives experimental results. $\delta\varrho/c$ is in good agreement with the value of 4.6 $\mu\Omega$cm/(a/o)Cr given by Nagasawa.[7] Star et al.[6] fitted the experimental data to a curve of the form $\varrho(T) = \varrho(T = 0) + AT^2 + BT^5$ and obtained good agreement with data for 1.2 < T < 10°K.

Grube and Knabe[2] prepared Pd-Cr alloys from pure Pd and electrolyte Cr. Figure 5 gives the electrical conductivity as a function of composition, and Table 1 gives the complete reuslts in tabulated form. Alloys with 5 to 40 a/o Pd are two-phase materials.

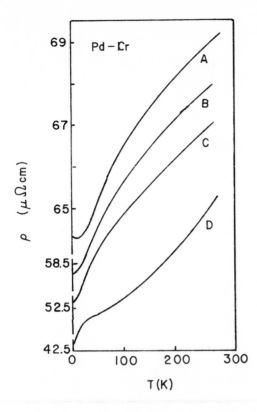

FIGURE 1.   Resistivity of Cr-Pd alloys (a/o Pd).
(A) 18; (B) 16; (C) 14; (D) 11.

FIGURE 2.   Resistivity of Cr-Pd alloys (a/o Pd).
(A) 82.7; (B) 49; (C) 23.

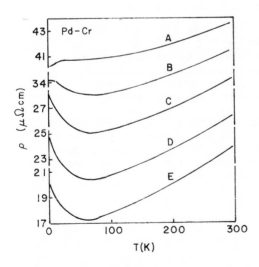

FIGURE 3.   Resistivity of Cr-Pd alloys (a/o Cr).
(A) 1.01; (B) 8.1; (C) 6.05; (D) 4.05.

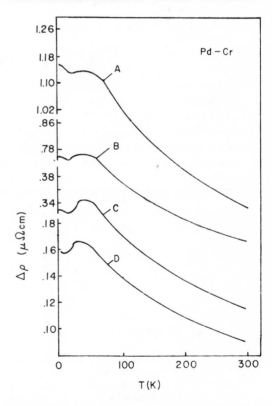

FIGURE 4.   Impurity resistivity of Cr-Pd alloys (a/
o Cr). (A) 0.3; (B) 0.2; (C) 0.1; (D) 0.05.

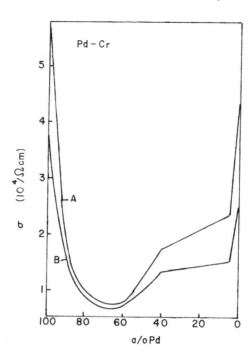

FIGURE 5.   Resistivity of Cr-Pd alloys (°C). (A)
200; (B) 500.

Table 1
ELECTRICAL CONDUCTIVITY OF Cr-Pd ALLOYS (10⁴Ohm⁻¹ cm⁻¹)

Pd (a/o)	Temperature (°C)													
	100	200	300	400	500	600	700	800	900	1000	1100	1200	1300	1400
100.00	7.143	5.747	4.762	4.115	3.650	3.289	3.012	2.809	2.639	2.488	2.348	2.227	2.119	2.016
92.50	2.342	2.222	2.114	2.016	1.927	1.845	1.779	1.712	1.653	1.600	1.548	1.495	1.447	1.401
82.70	1.112	1.089	1.065	1.044	1.024	1.003	1.977	0.951	0.928	0.904	0.879	0.757	0.732	0.711
71.50	0.818	0.800	0.781	0.763	0.745	0.729	0.710	0.695	0.677	0.662	0.645	0.632	0.618	—
62.50	0.750	0.740	0.730	0.717	0.704	0.689	0.674	0.664	0.656	0.649	0.641	0.630	0.620	—
52.50	1.230	1.114	1.048	0.982	0.939	0.901	0.861	0.818	0.768	0.725	0.688	0.661	0.629	—
49.00	1.452	1.284	1.178	1.102	1.052	1.012	0.932	0.869	0.805	0.762	0.722	0.685	0.649	—
40.50	2.000	1.765	1.584	1.450	1.345	1.252	1.178	1.085	0.979	0.905	0.818	0.850	—	—
32.75	2.174	1.880	1.669	1.504	1.379	1.235	1.149	1.062	0.980	0.903	0.833	0.768	0.700	—
23.00	2.415	2.041	1.776	1.587	1.433	1.289	1.183	1.079	0.988	0.910	0.840	0.778	0.725	—
12.00	2.615	2.242	1.921	1.700	1.472	1.305	1.170	1.055	0.972	0.899	0.832	0.773	0.719	—
5.00	2.857	2.358	1.984	1.721	1.520	1.339	1.196	1.081	0.988	0.909	0.840	0.777	0.725	—
2.00	4.657	3.654	3.000	2.532	2.185	1.895	1.656	1.482	1.321	1.198	1.088	0.995	0.912	0.840
Cr	5.714	4.545	3.745	3.125	2.667	2.310	2.041	1.818	1.618	1.439	1.290	1.153	1.042	0.949

REFERENCES

1. Hansen, M. and Anderko, K., *Constitution of Binary Alloys*, 2nd ed., McGraw-Hill, New York, 1958.
2. Grube, G. and Knabe, R., *Z. Elektrochem.*, 42, 793, 1936.
3. Roshko, R. M. and Williams, G., *Phys. Rev. B*, 16, 1503, 1977.
4. Kao, F. C. C. and Williams, G., *Phys. Rev. B*, 7, 267, 1973.
5. Roshko, R. M., and Williams, G., *J. Appl. Phys.*, 50, 1740, 1978.
6. Star, W. M., de Vroede, E., and van Baarle, C., *Physica*, 59, 128, 1972.
7. Nagasawa, H., *J. Phys. Soc. Jpn.*, 28, 1171, 1971.

Cr-Pt (Chromium-Platinum)

About 70 a/o Cr will dissolve in Pt.[1] A few percent of Pt will dissolve in Cr. Cr_3Pt exists in between these two solid solution ranges.

Star et al.[2] studied the resistivity of annealed P_tCr alloy wires with up to 3 a/o Cr at liquid He temperatures. $\Delta\varrho/c = 4.4 \ \mu\Omega cm/(a/o)Cr$ if c is the nominal composition, but $\Delta\varrho/c$ ranges from 3.8 to 4.4 $\mu\Omega cm/(a/o)Cr$ if one uses the Cr concentrations obtained analytically.

Roshko and Williams[3] measured $\varrho(T)$ of Pt-Cr alloys with 5 to 12 a/o Cr. Samples were prepared from 99.9995% pure Pt and 99.999% pure Cr by arc-melting. The authors plot the impurity resistivity $\Delta\varrho$ of the alloys from LHe temperatures to 300°K. The $\Delta\varrho$ data were fitted to the equation $\Delta\varrho/c = A - B \ ln[(T^2 + \Theta^2)^{1/2}]$ with the following constants:

Alloy (a/o Cr)	A ($\mu\Omega cm/(a/o)Cr$)	B	$\Delta\varrho(o)/c$
5	7.15	0.61	4.76
6	7.04	0.59	4.75
8	6.71	0.54	4.60
10	6.31	0.46	4.44
12	5.76	0.42	4

Typically $\Delta\varrho$ decreases by about 20 to 30% from LHe temperatures to 300°K. $\Delta\varrho(o)/c$ is also given by these authors. The fit to the data is very good above 50°K; below 50°K an error of up to 1% can be observed.

Roshko et al.[4,5] prepared P_tCr alloys with 13 to 18 a/o Cr from 5N Pt and 5N Cr in an arc furnace. Samples were melted several times to ensure homogeneity. Figure 1 gives $\varrho(T)$ of the alloys. The curves show minima in the temperature range from 38 (13 a/o Cr) to 10°K (18 a/o Cr). The incremental resistivity $\Delta\varrho(T)$ due to impurities can be described with the same equation as above, with $B \propto cos\{\pi\Delta Z/10\}$, where ΔZ is the effective impurity-host nuclear charge difference and $\Theta = T_s$ the single impurity characteristic temperature. T_s can be related to a local enhancement factor of the form $T_s \propto [1 - UN(E_F)]$. The authors suggest that magnetic freezing of the spin glass type occurs at higher impurity concentration.

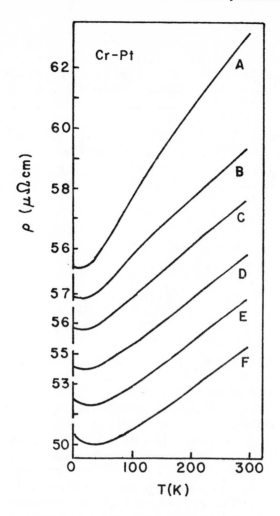

FIGURE 1.   Resistivity of Cr-Pt alloys (a/o Cr). (A) 18; (B) 17; (C) 16; (D) 15; (E) 14; (F) 13.

REFERENCES

1. Hansen, M. and Anderko, K., *Constitution of Binary Alloys,* 2nd ed., McGraw-Hill, New York, 1958.
2. Star, W. M., de Vroede, E., and van Baarle, C., *Physica,* 59, 128, 1962.
3. Roshko, R. M. and Williams, G., *Phys. Rev. B,* 9, 4945, 1974.
4. Roshko, R. M., Martense, I., and Williams, G., *J. Phys. F,* 7, 1811, 1977.
5. Roshko, R. M. and Williams, G., *Physica,* 86-88B, 829, 1977.

Cr-Rh (Chromium-Rhodium)

Coles et al.[1] prepared Rh-base alloy wires from metals of 99.99% purity. The experimental results are shown in Figure 1 for an alloy with nominally 0.5 a/o Cr.

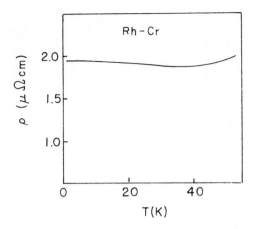

FIGURE 1.   Resistivity of Rh + 0.5 a/o Cr.

REFERENCES

1. Coles, B. R., Mozumder, S., and Rusby, R., *12th Int. Conf. On Low Temp. Physics,* Kanda, E., Ed., Keigaku Publ. Co., Tokyo, 1971, 737.

Cr-Ru (Chromium-Ruthenium)

Kao and Williams[1] prepared Ru̲Cr from 99.999% pure Ru and 99.999% pure Cr by arc-melting. $\varrho(T) - \varrho(1.45°K)$ of pure Ru is given in Figure 1. $\Delta\varrho(T)$ due to Cr impurities is given in Figure 2. $\Delta\varrho$ in Ru̲Cr alloys is at low temperatures temperature independent.

DeYoung et al.[2] measured the resistivity of Cr-Ru alloys with up to 14.0 a/o Ru from about 100 to 1000°K. Figure 3 gives experimental results. $\varrho(T)$ curves show the characteristic minimum which is close to the Neel temperature of the alloys. Table 1 gives T_N as a function of composition of the studies by DeYoung[2] and Arajs et al.[3,4]

Arajs et al.[4] prepared Cr̲Ru alloys with Cr of Iochrome purity. The alloys were melted repeatedly, then sealed in quartz capsules filled with Ar (150 Torr at 300°K) homogenized at 1400°K for 150 hr and furnace cooled. Figure 4 gives $\varrho(T)$ for the alloys. The resistivity shows the characteristic minimum at T_N. The sample changes from the antiferromagnetic state below this temperature T_N to the paramagnetic state above T_N. T_N usually increases with C_{Ru}, typical for impurities with higher outer electron/atom numbers than Cr. Table 2 gives the correlation between impurity concentration, T_N, and $\varrho(4.2°K)$.

Table 1			Table 2		
Ru (a/o)	T_N (°K)		Concentration (a/o)	T_N (°K)	$\varrho(4.2°K)$ ($\mu\Omega$cm)
0.9	496 ± 2		0.9	507	3.14
2.1	520 ± 2		2.1	530	8.39
3.0	557 ± 2		3.0	565	11.48
4.8	549 ± 2		4.8	558	19.89
6.6	522 ± 2				
8.3	479 ± 2				
10.1	440 ± 2				
11.4	392 ± 2				
14.0	276 ± 2				

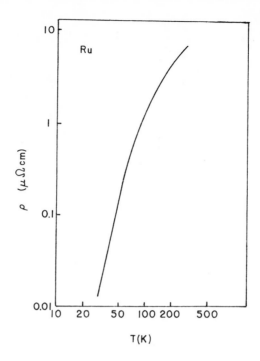

FIGURE 1.   Thermal part of the resistivity of pure Ru.

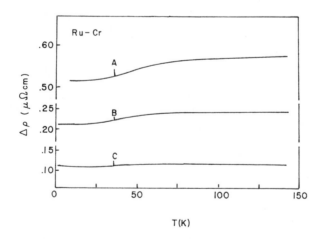

FIGURE 2.   Resistivity in RuCr due to Cr impurities (a/o Cr). (A) 0.8; (B) 0.4; (C) 0.2.

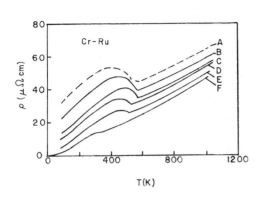

FIGURE 3.   Resistivity of Cr-Ru alloys. (A) 6.6 a/o Ru; (B) 4.8 a/o Ru; (C) 3.0 a/o Ru; (D) 2.1 a/o Ru; (E) 0.9 a/o Ru; (F) pure Ru.

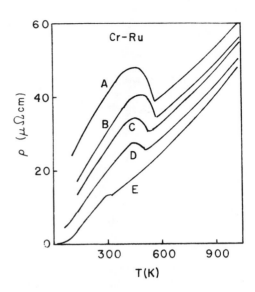

FIGURE 4.   Resistivity of Cr-Ru alloys. (A) 4.8 a/o Ru; (B) 3.0 a/o Ru; (C) 2.1 a/o Ru; (D) 0.9 a/o Ru; (E) Cr.

REFERENCES

1. Kao, F. C. C. and Williams, G., *J. Phys. F,* 4, 419, 1974.
2. DeYoung, T. F., Arajs, S., and Anderson, E. E., *J. Less Common Met.,* 32, 165, 1973.
3. Arajs, S. and Dunmyre, G. R., *J. Appl. Phys.,* 38, 1893, 1967.
4. Arajs, S., DeYoung, T. F., and Anderson, E. E., *J. Appl. Phys.,* 41, 1426, 1970.

Cr-Sb (Chromium-Antimony)

The phase diagram of Cr-Sb shows that two compounds are stable, CrSb and CrSb$_2$.[1] Several percent Sb can dissolve in Cr.

Fakinov and Afanasev[2] measured the resistivity of the CrSb compound which forms directly from the melt and has the NiAs-type structure (the second compound CrSb$_2$ is a semiconductor). $\varrho(T)$ shows a peak near 430°C (see Figure 1). CrSb is below 450°C antiferromagnetic,[3,4] an anomaly in the thermal expansion indicates that $T_N = 420 \pm 10$°C.

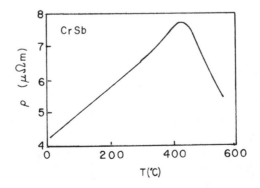

FIGURE 1.   Resistivity of CrSb.

REFERENCES

1. Hansen, M. and Anderko, K., *Constitution of Binary Alloys,* 2nd ed., McGraw-Hill, New York, 1958.
2. Fakinov, I. G. and Atanasev, A. Ya., *Phys. Met. Metallogr.,* 6, 160, 1958.
3. Snow, A. I., *Rev. Mod. Phys.,* 25, 127, 1953.
4. Snow, A. I., *Phys. Rev.,* 85, 365, 1952.

Cr-Se (Chromium-Selenium)

Masumoto et al.[1] determined the conductivity of CrSe. The samples were annealed at 250°C for 1 week. The conductivity increased by a factor of 10 at 305°C on heating, but it does not show a similar change on cooling. The quenched sample (Curve 3 in Figure 1) shows a high conductivity at room temperature which first decreases, then increases with increasing temperature. Preliminary X-ray investigations showed that the quenched samples and the annealed samples have the same crystal structure (NiAs type). It is supposed that the CrSe lattice slightly deviates from the pure NiAs structure below 305°C due to the Jahn-Teller effect. The distortion is too small to be detected by their X-ray measurements.

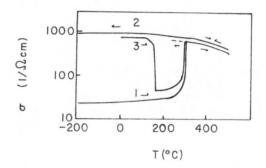

FIGURE 1.   Conductivity of CrSe.

REFERENCES

1. Masumoto, K., Hiara, T., and Kamigaichi, T., *J. Phys. Soc. Jpn.,* 17, 1209, 1962.

Cr-Ti (Chromium-Titanium)

The high-temperature β-phase of Ti forms a continuous solid solution with Cr,[1] but less than 0.5 a/o Cr will dissolve in α-Ti.

Mikheyev and Aleksashin[2] measured $\varrho(T)$ of Ti-rich Ti-Cr alloys. Ti type TG-O and electrolytic Cr (99.8% purity) were melted in a vacuum arc furnace with consumable electrodes. The ingots were forged at 950 to 1100°C. Then the samples were annealed in evacuated quartz tubes in the following way: 1100°C for 10 hr, 1000°C for 25 hr, 900°C for 50 hr, 800°C for 100 hr, 700°C for 200 hr, 600°C for 300 hr, and 550°C for 75 hr. This procedure should lead to a state close to equilibrium. $\varrho(T)$ of Ti-rich alloys is given in Figure 1. The kinks in $\varrho(T)$ at high temperature should be associated with α + γ → α + β transformation for the alloy with 5 w/o Cr. The 1.5 a/o Cr alloy is probably single phase below 800°C; the kinks in $\varrho(T)$ near 880°C should be associated with the α → α + β transformation. Alloys with up to 20 w/o Cr were studied.

Chiu et al.[3] studied $\varrho(T)$ of Cr-rich Cr-Ti alloys. They melted Cr(<30 ppm impurities) and 99.9% purity Ti by levitation in an Ar atmosphere at a pressure of 1.2 atm. The melt was poured into a Cu mold. The resistivity of the alloys is given in Figure 2. $\varrho(T)$ shows a minimum or at least a noticeable change in slope which should be associated with a change from the low-temperature antiferromagnetic state to the high-temperature paramagnetic state. The residual resistivity change with Ti concentration is $d\varrho/dc = 0.50 \,\mu\Omega cm/(a/o)Ti$.

The Neel temperature decreases with increasing Ti concentration. This is typical for Cr-alloys in which the outer electron/atom ratio of the impurity atom (el/at = 4 for Ti) is smaller than that for Cr (el/at = 6). Chiu et al.[3] analyze their data on the basis of a s-d scattering mechanism and separated $\varrho(T)$ into several components.

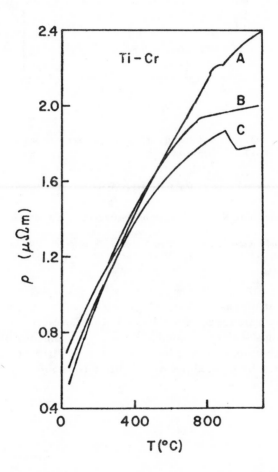

FIGURE 1.   Resistivity of Ti-Cr alloys. (A) 1.5% Cr; (B) 5% Cr; (C) Ti.

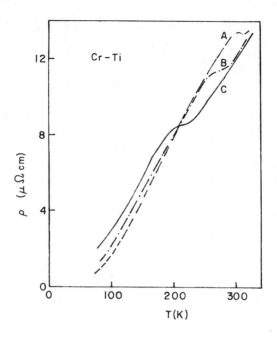

FIGURE 2.   Resistivity of C̲rTi alloys. (A) Cr; (B) 0.17 a/o Ti; (C) 0.53 a/o Ti.

REFERENCES

1. Hansen, M. and Anderko, K., *Constitution of Binary Alloys,* 2nd ed., McGraw-Hill, New York, 1958.
2. Mikheyev, V. S. and Aleksashin, V. S., *Phys. Met. and Metallogr. (USSR),* 14(2), 62, 1962.
3. Chiu, C. H., Jericho, M. H., and March, R. H., *Can. J. Phys.,* 49, 3010, 1971.

Cr-V (Chromium-Vanadium)

V-Cr forms a complete range of solid solutions.[1] The Neel temperature decreases rapidly with increasing V concentration. This leads to a shifting of the minimum in $\varrho(T)$ to lower temperature, as found by Giannuzzi et al.[2] and Moller et al.[3] Figure 1 gives experimental results.[3] ϱ of alloys with higher V concentration are given in Figure 2.[2]

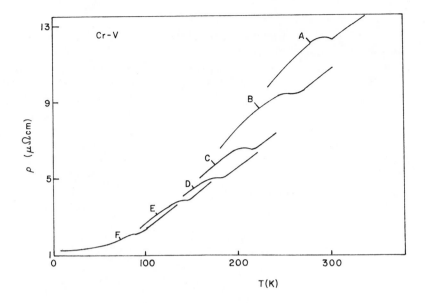

FIGURE 1.   Resistivity of C̲r̲V alloys (a/o V). (A) 0.1; (B) 0.5; (C) 1.0; (D) 1.5; (E) 2.0; (F) 3.0.

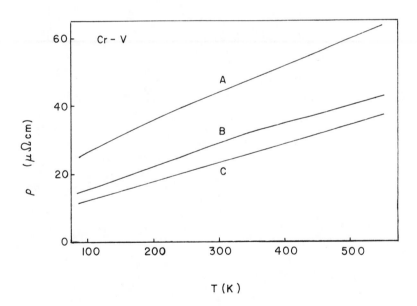

FIGURE 2.   Resistivity of Cr-V(a/o V). (A) 50; (B) 77; (C) 20.

REFERENCES

1. Hansen, M. and Anderko, K., *Constitution of Binary Alloys*, McGraw-Hill, New York, 1958.
2. Giannuzzi, A., Tomaschke, H., and Schroder, K., *Philos. Mag.*, 21, 479, 1970.
3. Moller, H. B., Trego, A. L., and Makintosh, A. R., *Solid State Commun.*, 3, 137, 1965.

Cr-W (Chromium-Tungsten)

Cr and W form at elevated temperatures a complete series of solid solutions.[1] An immiscibility gap shows up below 1495°C.

Arajs et al.[2,3] measured the resistivity of C̲rW alloys with up to 3.4 a/o W. $\varrho(4.2°K, c)$ is a linear function of c and $d\varrho(4.2°K)/dc = 0.8$ $\mu\Omega$cm/(a/o)W. It is in most alloy systems possible to estimate the impurity resistivity at other temperatures from 4.2°K data. However, the complex temperature dependence of the electronic structure in Cr alloys makes it impossible to estimate the impurity resistivity in this way. $\varrho(T)$ of alloys is given in Figure 1. All the curves show a minimum in $\varrho(T)$ except the curve for the alloy with 3.4 a/o W which shows only a change in slope. The temperature at ϱ_{min} is close to the temperature of the transition from the low temperature antiferromagnetic state to the paramagnetic high temperature state.

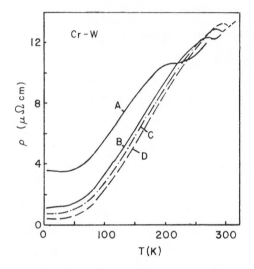

FIGURE 1.   Resistivity of Cr-W alloys (a/o W).
(A) 3.14; (B) 1.0; (C) 0.7; (D) 0.3.

REFERENCES

1. Hansen, M. and Anderko, K., *Constitution of Binary Alloys,* 2nd ed., McGraw-Hill, New York, 1958.
2. Arajs, S., Anderson, E. E., and Rao, R. V., *J. Less Common Met.,* 26, 157, 1971.
3. Arajs, S., *J. Appl. Phys.,* 39, 675, 1968.

Cr-Zn (Chromium-Zinc)

At temperatures around 250°C, 25 a/o Zn will dissolve in Cr, but not more than 0.04 a/o Cr will dissolve in solid Zn at 415°C. There exists one intermediate phase with ∿5.6 a/o Cr, with the possible formula $CrZn_{17}$ (288 atoms per unit cell).[1]

Resistivity measurements deal mostly with the study of impurity states.[2-5] Bell and Caplin[2] studied the effect of isolated Cr atoms in Zn. Samples were prepared from 6N Zn and the transition elements impurities by melting elements in vacuum. The residual resistivity ratio per atomic percent solute being measured as equal to 3.30 $\mu\Omega$cm/(a/o)Cr.

Ford et al.[3] prepared Zn alloys by melting pure Zn with a master alloy under pure Ar in a Pyrex® container. This was evacuated and the molten alloy was cast. Even with these preparations, annealing led frequently to signs of precipitates. In all alloys, the resistivity shows a $\log_{10}T$ dependence, but at very low temperatures a T^2 dependence can be observed. The residual resistivities can be explained on the basis of a magnetized Friedel virtual bound state.

These measurements confirmed earlier studies between 1.3 and 4.2°K which showed for ZnCr alloys with 0.01 to 0.1 a/o Cr a minimum in ϱ.[4] A sample with 20 ppm impurities with possibly some precipitation gave a $\log_{10}T$ behavior between 0.4 to 8°K[5] and ZnFe alloys with 28 and 1000 ppm impurities revealed a small temperature-dependent term.[6]

REFERENCES

1. Hansen, M. and Anderko, K., *Constitution of Binary Alloys,* 2nd ed., McGraw-Hill, New York, 1958.
2. Bell, A. E. and Caplin, A. P., *J. Phys. F,* 5, 143, 1974.
3. Ford, P. J., Rizzuto, C., and Salamoni, E., *Phys. Rev. B,* 6, 1851, 1972.
4. Muto, Y., *J. Phys. Soc. Jpn.,* 15, 2119, 1960.
5. Boato, G., Rizzuto, C., and Vrig, J., *Proc. 10th Int. Conf. on Low Temp. Phys.,* Vimiti Publishing House, Moscow, 1967, 4, 292, 1956.

CESIUM (Cs)

Cs-K (Cesium-Potassium)

Bauhofer and Simon[1] used a contactless measuring system to determine ϱ of $K_{1-x}Cs_x$ alloys. Figure 1 gives $\varrho(c)$ for T = 78, 190, and 373°K. At 373°K all alloys should be liquid; at the other temperatures, the samples should be solid. Anomalies in $\varrho(c)$ of 78°K indicate that intermetallic compounds form. They have the composition K_2Cs_5 and K_7Cs_6. The alloy samples used in this investigation should at 78°K contain two phases, the composition is not exactly given by K_2Cs_5 and K_7Cs_6. All samples may be single-phase at 180°K.

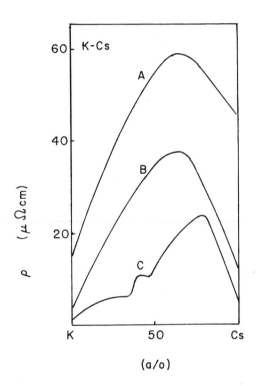

FIGURE 1.   Resistivity of K-Cs (°K). (A) 373; (B) 190; (C) 78.

REFERENCES

1. Bauhofer, W. and Simon, A., *Z. Naturforsch. Teil A*, 32(11), 1275, 1977.

COPPER (Cu)

Cu-Fe (Copper-Iron)

Below 830°C, the equilibrium Cu-Fe alloys exist as an immiscible mixture of primary solid solutions α-Cu and α-Fe over almost the entire composition range. At 25°C, the solid solubility of Cu in α-Fe and Fe in α-Cu are both less than 1%.

The resistivity of the two-phase CuFe alloys has been determined and appears to be a simple well-behaved function of composition (see Figure 1) and also appears to be relatively independent of the thermal-mechanical treatment of the alloy except when the alloy is quenched.

Resistivity measurements have been made on both the primary dilute solid solution alloys.[1-7] The room temperature resistivity as a function of composition in the dilute Cu and Fe alloys (see Figure 2) gives the $d\varrho$(CuFe, 273°C)/dc = 9.3 $\mu\Omega$cm/(a/o)Fe. The concentration coefficient of the residual resistivity in FeCu alloys is $d\varrho$/dc = 6.8 $\mu\Omega$cm/(a/o)Cu. The resistivity as a function of temperature (see Figure 3) suggests that dilute CuFe alloys are Kondo systems.

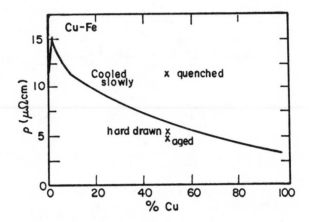

FIGURE 1. Electrical resistivities of Fe-Cu alloys and their change with heat treatment.[10]

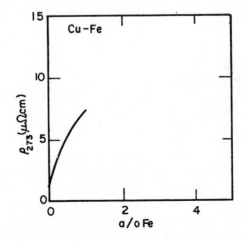

FIGURE 2. Resistivity of CuFe alloys at 0°C.[8]

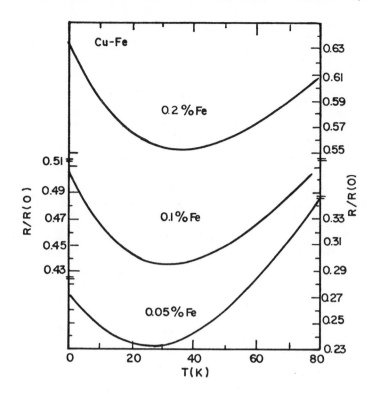

FIGURE 3.   The low-temperature resistivity of several CuFe alloys.[4]

REFERENCES

1. Linde, J. O., *Ann. Phys.*, 15, 219, 1932.
2. Domenicali, C. A. and Christenson, E. L., *J. Appl. Phys.*, 32, 2450, 1961.
3. Daybell, M. D. and Steyert, W., *Phys. Rev. Lett.*, 18, 398, 1967.
4. Manchester, F. D. and Martin, D. L., *Proc. R. Soc. London Ser. A*, 263, 494, 1961.
5. Kos, J. F. and Moussouros, P. K., *Solid State Commun.*, 18, 1501, 1976.
6. Kos, J. F. and Moussouros, P. K., *J. Phys. Chem. Solids*, 39, 239, 1978.
7. Rumbo, E. R., *J. Phys. F*, 3, L9, 1973.
8. Gerritsen, A. N., *Handbook of Physics*, Springer-Verlag, Berlin, 1929.
9. Star, W. M., Basters, F. B., Nap, G. M., deVroede, E., and van Baarle, C., *Physica*, 58, 585, 1972.
10. Bozorth, R. M., *Ferromagnetism*, Van Nostrand, Princeton, N.J., 1951, 231.

Cu-Ga (Copper-Gallium)

The Cu-Ga phase diagram is a complex system that at room temperature has an extensive primary α-Cu phase field that can accommodate up to 18% Ga. Four narrow intermediate phases apparently exist in the 18 to 44 a/o Ga composition range. Between 44 and 100 a/o Ga, the equilibrium alloys exist as predominantly a mixture of two phases involving a narrow intermediate phase at approximately Cu-66 a/o Ga. Cu appears to be insoluble in solid Ga. The resistivity of Cu-Ga as a function of composition at several temperatures for the α-Cu solid solution alloys is shown in Figure 1.

A linear response with $d\varrho(\underline{Cu}Ga)/dc = 0.4\ \mu\Omega cm/(a/o)Ga$. Ga is only evident when the composition of the alloys do not exceed approximately 4 a/o Ga.

The DMR has also been determined (see Figure 2) for the α-Cu solid solution alloys and is at maximum about 70°K and is relatively small at room temperature. The DMR for the more concentrated α-Cu Ga solid solution alloys does not show a well-defined hump and generally increases with increasing temperature so the DMR is appreciable at room temperature.

Dukin and Aleksandrov[7] gave a critical account of pseudopotential calculations and the resistivity increase due to 1% impurity. They gave $[\Delta\varrho/\Delta C]_{exp} = 8.0\ \mu\Omega cm/(a/o)Cu$ in $\underline{Ga}Cu$.

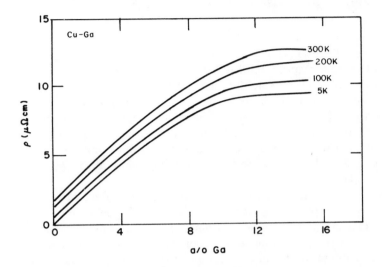

FIGURE 1. Resistivity of $\underline{Cu}$Ga alloys vs. concentration of Ga at selected temperatures.[1]

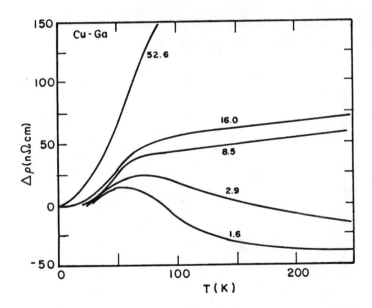

FIGURE 2. Deviations from Matthiessen's rule for $\underline{Cu}$Ga alloys from the data of Crisp et al.[1] Numbers designate Ga concentration (in a/o).[4]

REFERENCES

1. Crisp, R. S., Henry, W. G., Schroeder, P. A., and Wilson, R. W., *J. Chem. Eng. Data*, 11, 556, 1966.
2. Linde, J. O., *Ann. Phys.*, 15, 219, 1932.

3. MacDonald, D. K. C. and Pearson, W. B., *Acta Metall.,* 3, 403, 1955.

4. Loegel, B., Ph.D. thesis, Strasbourg.

5. Loegel, B., *J. Phys. F,* 3, L106, 1973.

6. Kierspe, W., *Z. Metallkd.,* 58, 895, 1967.

7. Dukin, V. V. and Aleksandrov, B. N., *Pseudopotential Calculations of the Residual Resistivities of Dilute Solid Alloys Based on Normal Metals, Physics of Low Temperatures,* Akademia Nauk Uk SSR Academy of Science, USSR, Kharkov, 1978, 1.

Cu-Ge (Copper-Germanium)

The Cu-Ge phase diagram indicates that a primary phase field exists at the Cu end, but not at the Ge end. The solubility limit of Ge in the α-Cu primary phase is 9.5 a/o at 25°C. Two intermediate phases, ζ and ε, exist at room temperature. The ζ-phase is the product of a peritetic reaction and exists as a single phase region at 25°C over a moderately large range (11.4 to 16.9 a/o Ga) of compositions. The narrow ε_1-phase occurs near the Cu-25 a/o Ga composition and results from a peritectoid reaction at 635°C. Between 25.5 and 100 a/o Ga, Cu-Ga alloys exist at 25°C as a two-phase mixture of ε_1 and Ga.

The concentration dependence of the resistivity[1-3] and the DMR[3-7] have been determined for the α-Cu primary-phase alloys. The resistivity as a function of composition for several temperatures in the α-Cu primary phase is shown in Figure 1.

The resistivity at 17°C is linear with composition to approximately 2 a/o Ge (see Figure 1) and the resulting concentration coefficient for the residual resistivity $d\varrho/dc$ is 3.7 $\mu\Omega$cm/(a/o)Ge.[1] The DMR as a function of temperature for the dilute CuGe alloys below 300°K displays a slight maximum at about 85°K, while for the more concentrated alloys, the deviation generally increases with increasing temperature and does not exhibit any pronounced features. The deviations are also strongly concentration dependent. Above 300°K, the DMR generally increases with increasing temperature and again is composition sensitive (Figure 2).

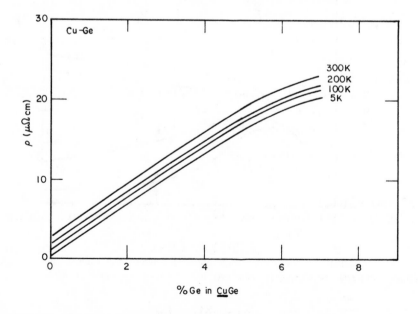

FIGURE 1.   Resistivity of CuGe alloys at selected temperatures.[1]

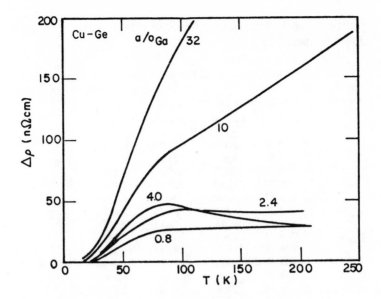

FIGURE 2. Deviation from Matthiessen's rule as a function of temperature for CuGe alloys from the data of Crisp et al.[1] Numbers designate (a/o) Ga concentration.[6]

REFERENCES

1. Crisp, R. S., Henry, W. G., Schroeder, P. A., and Wilson, R. W., *J. Chem. Eng. Data*, 11, 556, 1966.
2. Linde, J. O., *Ann. Phys.*, 15, 219, 1932.
3. Dugdale, J. S. and Basinski, Z. S., *Phys. Rev.*, 157, 552, 1967.
4. Loegel, B., Ph.D. thesis, Strasbourg.
5. Rahim, C. A. and Barnard, R. D., *J. Phys. F*, 8, 1957, 1978.
6. Loegel, B., *J. Phys. F*, 3, L106, 1973.
7. Black, J. E., *Can. J. Phys.*, 52, 345, 1974.

Cu-In (Copper-Indium)

Linde[1] gives for $\Delta\varrho/\Delta c$ for dilute alloys a value of 1.05 $\mu\Omega$cm/(a/o)In for CuIn.

REFERENCES

1. Linde, J. O., *Helv. Phys. Acta*, 41, 1013, 1968.

Cu-Ir (Copper-Iridium)

Linde[1] gives for $\Delta\varrho/\Delta c$ for dilute alloys a value of 5.7 $\mu\Omega$cm/(a/o)Ir for CuIr.

REFERENCES

1. Linde, J. O., *Helv. Phys. Acta*, 41, 1013, 1968.

Cu-Mg (Copper-Magnesium)

Up to 7 a/o Mg is soluble in Cu at 722°C and about 3 a/o is soluble at 400°C. Cu is practically insoluble in Mg. One intermediate phase ($MgCu_2$) with a stability range of a few percent and an intermediate compound (Mg_2Cu) exists.[1] Table 1 gives the resistivity at room temperature.[2]

Table 1

Mg (w/o)	ϱ (μΩcm)	Mg (w/o)	ϱ (μΩcm)
0.56	2.115	26.7	6.146
0.65	3.25	27.55	6.83
0.77	3.339	31.0	7.134
1.53	4.185	34.4	7.155
2.97	4.784	36.35	8.111
3.28	5.061	36.8	7.937
6.75	6.214	38.15	9.272
8.76	6.641	41.11	8.265
10.06	5.925	41.12	10.31
11.65	5.411	43.3	11.94
14.53	5.803	41.5	8.529
15.38	5.746	45.0	7.08
15.47	5.717	48.0	6.508
16.05	5.588	52.9	6.258
16.28	5.16	53.05	6.072
16.46	5.145	60.2	6.020
17.71	5.500	68.8	5.786
22.66	6.069	77.5	5.171
23.7	6.371	87.8	4.923

REFERENCES

1. Hansen, M. and Anderko, K., *Constitution of Binary Alloys,* McGraw-Hill, New York, 1958.
2. *International Critical Tables,* Vol. 6, McGraw-Hill, New York, 1929, 156.

Cu-Mn (Copper-Manganese)

A complete series of Cu-Mn solid solution alloys forms at elevated temperatures. Mn experiences several alloy tropic transformations on cooling, so at room temperature the Cu-Mn phase diagram consists of a α-Cu primary solid solution from 0 to 25 a/o Mn and a two-phase mixture of α-Cu and Mn from 25 to 100 a/o Mn.

The resistivity of several CuMn primary phase alloys have been investigated in the temperature interval from −191 to 438°C (see Figure 1). The resistivity of the more concentrated CuMn alloys is seen to have a low value of the temperature coefficient of resistance at 25°C. At lower temperatures, these alloys show a Kondo minimum when the Mn concentration is small (see Figure 2).

The concentration dependence of the room temperature resistivity of these dilute CuMn alloys is linear to at least 4 a/o Mn (see Figure 3) with an associated coefficient of $d\varrho/dc$ = 2.9 μΩcm/(a/o)Mn. The DMR has been investigated for both dilute and concentrated CuMn alloys. The DMR as a function of temperature for the dilute CuMn alloy increases with temperature to a maximum value at 75°K (see Figure 4) and then decreases rapidly with increasing temperature to a relatively small value at 25°C. A similar behavior is also apparent in the temperature dependnce of the DMR for more concentrated CuMn alloys (see Figure 5), except that the temperature of the maximum progressively increases with Mn concentration.

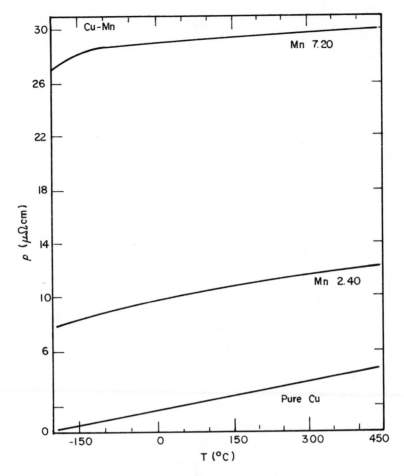

FIGURE 1.   Electrical resistivities of $\underline{Cu}$Mn alloys.

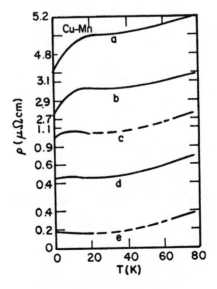

FIGURE 2A.   Resistivities of $\underline{Cu}$Mn alloys (a/o Mn). (a) 1.8; (b) 1.0; (c) 0.4; (d) 0.2; (e) 0.05.

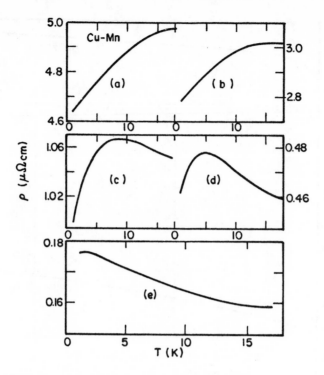

FIGURE 2B.   Resistivities of CuMn alloys (a/o Mn). (a) 1.8; (b) 1.0; (c) 0.4; (d) 0.2; (e) 0.05.

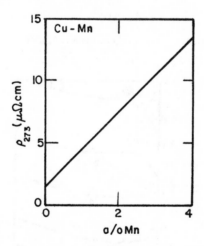

FIGURE 3.   Resistivity of CuMn alloys at 273°K.[16]

223

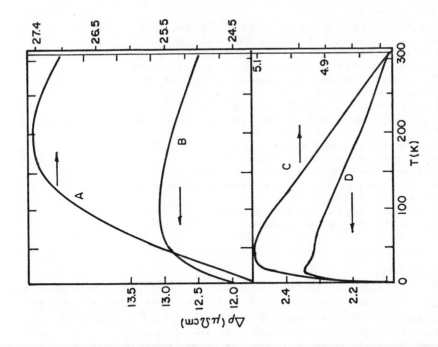

FIGURE 5. Overall temperature variation in ϱ for C̲u̲Mn alloys
(a/o Mn). (A) 9.7; (B) 4.5; (C) 1.5; (D) 0.7.

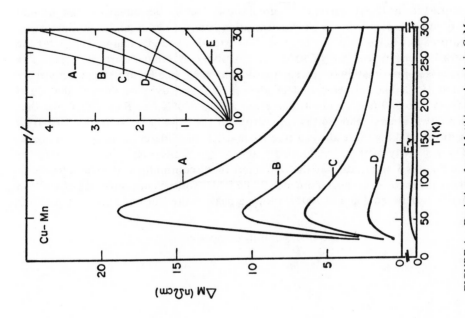

FIGURE 4. Deviations from Matthiessen's rule in C̲u̲Mn
(ppm Mn). (A) 197; (B) 95.5; (C) 47.8; (D) 10.5.

REFERENCES

1. Norbury, A. L. and Kuwada, F., *Philos. Mag.*, 4, 1338, 1927.
2. Linde, J A., *Ann. Phys.*, 15, 219, 1932.
3. Domenicali, C. A. and Christenson, E. L., *J. Appl. Phys.*, 32, 2450, 1961.
4. Nakamura, A. and Kinoshita, N., *J. Phys. Soc. Jpn.*, 26, 48, 1969.
5. Altarelli, M. and Beck, P., *Solid State Commun.*, 22, 675, 1977.
6. Gerritsen, A. N. and Linde, J. O., *Physica*, 17, 573, 584, 1951; 18, 877, 1952.
7. Haen, P., Souletie, J., and Teixeira, J., *J. Low Temp. Phys.*, 23, 191, 1976.
8. Laborde, O. and Radhakrishna, P., *J. Phys. F*, 3, 1731, 1973.
9. Ford, P. J. and Mydosh, J. A., *Phys. Rev. B*, 14, 2057, 1976.
10. Beck, P. A. and Chakrabarti, D. J., *Amorphous Magnetism*, Hooper, H. O. and deGraaf, A. M., Eds., Plenum Press, New York, N.Y., 1973, 273.
11. Dean, R. S., Potter, E. V., Long, J. R., and Huber, R. W., *Trans. ASM*, 34, 465, 1945.
12. Schmitt, R. W. and Jacobs, I. S., *Can. J. Phys.*, 34, 1285, 1956.
13. Smith, C. S. and Palmer, E. W., *Trans AIME*, 117, 225, 1935.
14. Schmitt, R. W. and Jacobs, I. S., *J. Phys. Chem. Solids*, 3, 324, 1957.
15. Schmitt, R. W., *Phys. Rev.*, 103, 83, 1956.
16. Gerritsen, A. N. and Linde, J. O., *Physica*, 18, 877, 1952.
17. Gerritsen, A. N., *Handbook of Physics*, Springer-Verlag, Berlin, 1957.

Cu-Ni (Copper-Nickel)

Cu and Ni form a complete series of disordered solid solution alloys at all temperatures. The NiCu alloys are ferromagnetic where the Curie temperature decreases with increasing Cu concentration and becomes 0°K at the Ni - 40 a/o Cu composition.

The electrical resistivity of the CuNi system has been investigated exhaustively for both dilute[1-10] and concentrated[10-22] alloys. For the CuNi alloys, the residual resistivity at 4.2°K is a linear function of the Ni concentration to approximately 30 a/o Ni (see Figure 1), resulting in a coefficient of $d\varrho(\text{CuNi})/dc = 1.1 \ \mu\Omega\text{cm}/(\text{a/o})\text{Ni}$. This behavior is reproduced almost exactly at more elevated temperatures[21] to 500°C, as shown in Figure 2.

In contrast to the CuNi behavior, Figure 2 shows that the dependence of the residual resistivity of NiCu alloys on the Cu concentration changes significantly with increasing temperature. The linear response becomes less extensive and gives larger values for the concentration coefficient $d\varrho(\text{NiCu})/dc$ of resistivity as temperature increases. The coefficient is approximately 0.9 $\mu\Omega\text{cm}/(\text{a/o})\text{Cu}$ at 78°K (see Figure 3) and 1.5 $\mu\Omega\text{cm}/(\text{a/o})\text{Cu}$ at 25°C. The temperature dependence of the resistivity of concentrated CuNi alloys from 0 to 300 (see Figure 4) and from 200 to 800°K (see Figure 5) shows that the resistivity is almost temperature insensitive for the Cu-40 a/o Ni to Cu - 50 a/o Ni compositions, making these alloys attractive material candidates for high-quality resistors. The resistivity of the concentrated magnetic NiCu alloys as a function temperature (see Figures 5 and 6) display the characteristic discontinuity at the Curie temperature where magnetic ordering commences. The DMR for the concentrated NiCu alloys (see Figure 7) between 300 and 700°K shows a peak at the Curie temperature for each alloy.

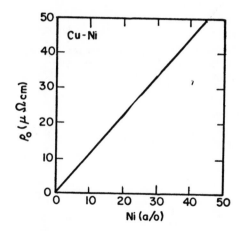

FIGURE 1. Residual resistivity at 4.2°K for CuNi alloys.[5]

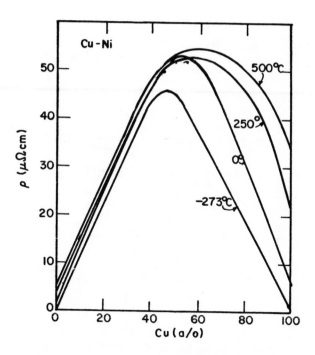

FIGURE 2. Resistivity of Cu-Ni alloys at 0, 250, and 500°C[21]; at 273°C.[20]

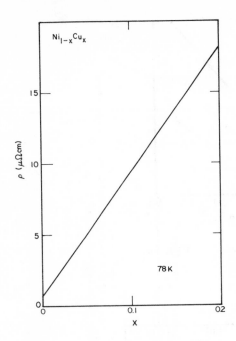

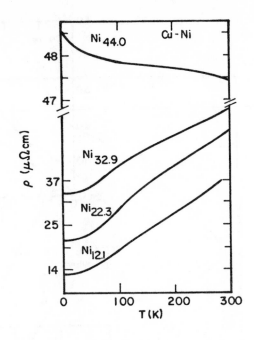

FIGURE 3. Electrical resistivity of Ni-Cu alloys as a function of copper concentration at 78°K.[10]

FIGURE 4. Electrical resistivity vs. temperature for Cu-Ni alloys.

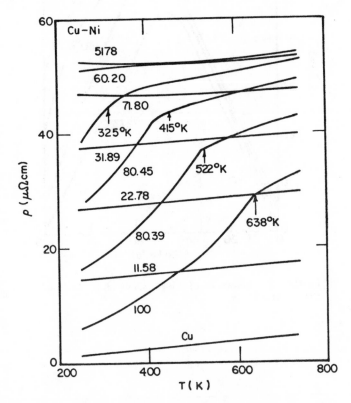

FIGURE 5. Resistivity as a function of temperature for several Cu-Ni alloys. The numbers indicate the Ni amount (in a/o).[23]

227

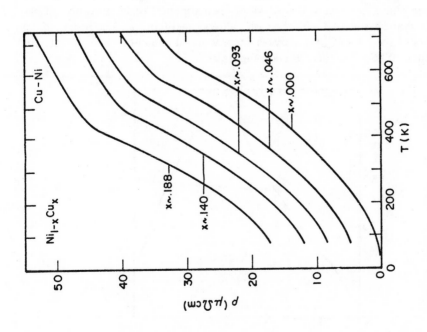

FIGURE 7. $\Delta \varrho$ (x,T) as a function of T for Ni-Cu alloys between 300 and 700°K. Numbers give Cu (a/o).[10]

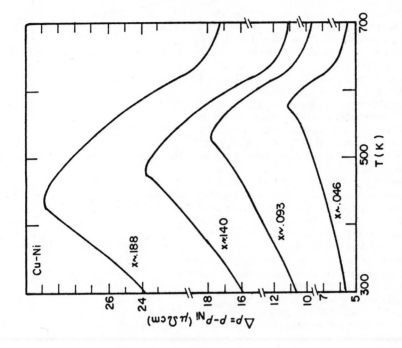

FIGURE 6. Electrical resistivity of Ni-Cu alloys as a function of absolute temperature between 78 and 700°K.[10]

REFERENCES

1. Linde, J. O., *Ann. Phys.,* 15, 219, 1932.
2. Lengeler, B., Schilling, W., and Wenzl, *J. Low Temp. Phys.,* 2, 59, 1970.
3. Domenicali, C. A. and Christenson, E. L., *J. of Appl. Phys.,* 32, 2450, 1961.
4. Norbury, A. L. and Kuwada, K., *Philos. Mag.,* 4, 1338, 1927.
5. Legvoid, J., Peterson, D. T., Burgardt, P., Hofer, R. J., Lundell, B., Vyrostek, T. A., and Gartner, H., *Phys. Rev. B,* 9, 2386, 1974.
6. MacDonald, D. K. C. and Pearson, W. B., *Acta Metall.,* 3, 392, 1955.
7. Ferrell, T. and Greig, D., *J. Phys. C,* 1, 1359, 1968.
8. Schwerer, F. C. and Cuddy, L. J., *Phys. Rev. B,* 2, 1575, 1970.
9. Dorleijn, J. W. F. and Miedema, A. R., *J. Phys. F,* 5, 487, 1975.
10. Yao, D., *Ann. Rep. Inst. Phys. Academic Sinica,* 8, 11, 1978.
11. Jackson, P. J. and Saunders, N. H., *Phys. Lett. A,* 28, 19, 1968.
12. Houghton, R. W., Sarachik, M. P., and Kouvel, J. S., *Solid State Commun.,* 8, 943, 1970; *Phys. Rev. Lett.,* 25, 238, 1970.
13. Coles, B. R., *Proc. R. Soc. London Ser. B,* 65, 221, 1952.
14. Skoskiewicz, T. and Baranowski, B., *Solid State Commun.,* 7, 647, 1969.
15. Crangle, J. and Butcher, R. J. L., *Phys. Lett. A,* 32, 80, 1970.
16. Gartner, H. and Auer, W., *Phys. Status Solidi A,* 49, 149, 1978.
17. Ahmad, H. M. and Greig, D., *Phys. Rev. Lett.,* 32, 833, 1974.
18. Chevenard, P., *J. Inst. Met.,* 36, 39, 1926.
19. Smith, C. S. and Palmer, E. W., *Trans. AIME,* 117, 225, 1935.
20. Krupkowski, A. and de Haas, W. J., *Comm. Leiden,* 194, 1, 1928.
21. Svensson, B., *Ann. Phys.,* 25, 263, 1936; 22, 97, 1935.
22. Iguchi, E. and Udagawa, K., *J. Phys. F.,* 5, 214, 1975.
23. Gerritsen, A. N., *Handbook of Physics,* Springer-Verlag, Berlin, 1957.
24. Schroeder, P. A., Wolf, R., and Woollam, J. A., *Phys. Rev. A.,* 138, 105, 1965.

Cu-Pd (Copper-Palladium)

The Cu-Pd phase diagram is a continuous series of solid solution alloys containing two compositions, Pd_1Cu_3 and Pd_1Cu_1, that order at room temperatures. The resistivity of the singe-phase disordered Cu-Pd alloys as a function of composition display the assymetric shape associated with the Coles alloys (see Figure 1). The room temperature resistivity of the dilute CuPd alloys is linear with Pd content and is characterized by a coefficient of 0.86 $\mu\Omega$cm/(a/o)Pd (see Figure 2).

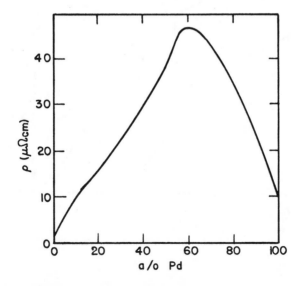

FIGURE 1.   Composition dependence of the resistivity of the disordered Cu-Pd alloys at room temperature.[1,2]

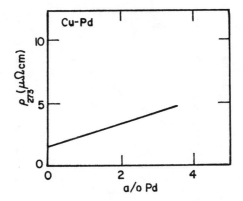

FIGURE 2. Resistivity of the dilute CuPd alloy vs. composition.[3]

REFERENCES

1. Svensson, B., *Ann. Phys.*, 14, 699, 1932.
2. Borelius, G., Johansson, C. H., and Linde, J. O., *Ann. Phys.*, 4, 86, 299, 1928.
3. Gerritsen, A. N., *Handbook of Physics*, Springer-Verlag, Berlin, 1957.

Cu-Pt (Copper-Platinum)

Cu and Pt form a continuous series of solid solution alloys at temperatures above 812°C. At lower temperatures, order-disorder transformations occur over a wide range of compositions, extending from about 5 to 92 a/o at 300°C.

As a result of these ordering reactions, a specification of the resistivity for the Cu-Pt alloys below 800°C is meaningless unless the state of order is known. For the disordered alloys, the resistivity is modestly consistent with Nordheim's rule as shown in Figure 1, where the influence of ordering at lower temperatures is also apparent over practically the entire range of Pt-Cu compositions. The resistivity of the CuPt alloys at 273° is linearly related to the Pt concentration to about 5 a/o Pt (see Figure 2) with a coefficient of 2 $\mu\Omega$cm/(a/o)Pt.

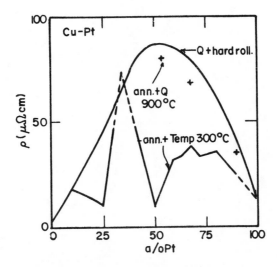

FIGURE 1. Composition dependence of the resistivity of Cu-Pt alloys.[1]

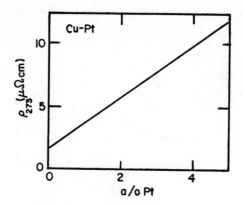

FIGURE 2.   Resistivity of several Cu-rich alloys
vs. impurity concentration.[4]

REFERENCES

1. Linde, J. O., *Ann. Phys.,* 30, 151, 1937.
2. Johansson, C. H. and Linde, J. O., *Ann. Phys.,* 82, 459, 1927.
3. Kurnakov, N. S. and Nemilow, V. A., *Z. Anorg. Chem.,* 210, 1, 1933.
4. Gerritsen, A. N., *Handbook of Physics,* Springer-Verlag, Berlin, 1957.

Cu-Rh (Copper-Rhodium)

The Cu-Rh phase diagram contains a peritectic reaction at 1100°C which suggests that the single-phase Cu-Rh alloys at 25°C exist as primary solid solutions of α-CuRh between 0 and 20 a/o Rh and α-RhCu between 90 and 100 a/o Rh. The alloys of intermediate composition are two-phase mixtures of α-CuRh and α-RhCu.

The resistivity of the CuRh primary phase alloys at 273°K is linear with concentration (see Figure 1), and the resulting coefficient is $d\varrho/dc = 5 \ \mu\Omega cm/(a/o)Rh$. Linde[3] gives for $\Delta\varrho/\Delta c$ for dilute alloys a value of 4.3 $\mu\Omega cm/(a/o)Rh$ for CuRh. The DMR for several CuRh alloys have been determined at low temperatures (see Figure 2).

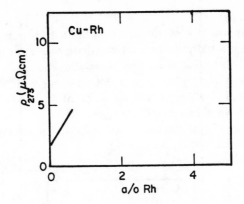

FIGURE 1. Resistivity at 273°K for C̲u̲Rh alloys as a function of Rh concentration.[2]

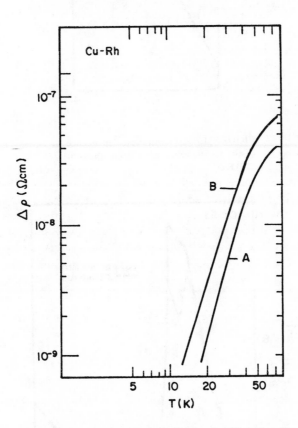

FIGURE 2. Deviations from MR, $\Delta\varrho$, plotted against T
for C̲u̲Rh alloys (a/o Rh). (A) 0.4; (B) 2.

REFERENCES

1. Loegel, B., *J. Phys. F.,* 3, L106, 1973.
2. Gerritsen, A. N., *Handbook of Physics,* Springer-Verlag, Berlin, 1957.
3. Linde, J. O., *Helv. Phys. Acta,* 41, 1013, 1968.

Cu-Sb (Copper-Antimony)

The Cu-Sb phase diagram at 25°C shows a very limited Cu-rich primary single-phase region and two compositionally narrow intermediate phases at approximately 19 Sb and 33% Sb. All other compositions are two-phase alloys.

The resistance of the dilute CuSb alloy at 0°C is linear with Sb concentration (see Figure 1). The associated coefficient is 5 $\mu\Omega$cm/(a/o)Sb. Linde[3] gives for $\Delta\varrho/\Delta c$ for dilute alloys a value of 5.4 $\mu\Omega$cm/(a/o)Sb for CuSb.

The composition dependence of the resistivity at 25°C for the entire range of Cu-Sb alloys is shown in Figure 2.

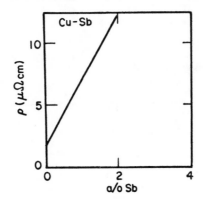

FIGURE 1.   Concentration dependence of the resistivity of dilute CuSb alloys at 273°K.[1]

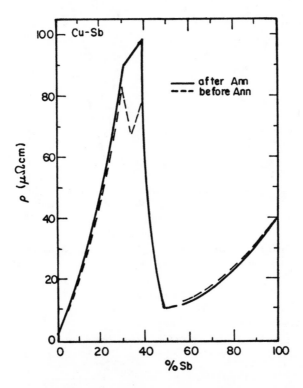

FIGURE 2.   Resistivity of Cu-Sb alloys at 25°C.[2]

REFERENCES

1. Gerritsen, A. N., *Handbook of Physics*, Springer-Verlag, Berlin, 1957.
2. Stephens, E. and Evans, E. J., *Philos. Mag.*, 7, 161, 1929.
3. Linde, J. O., *Helv. Phys. Acta*, 41, 1013, 1968.

Cu-Si (Copper-Silicon)

The single-phase regions of the Cu-Si phase diagram at 25°C consists of a Cu-rich primary solution region extending to 8% of Si and three intermediate phases, Cu-17 a/o Si, Cu-21 a/o Si, and Cu-24 a/o Si, whose ranges of compositional homogeneity are restricted to several atomic percent Si.

The resistivity of the primary-phase CuSi alloys have been determined as a function of temperature (see Figure 1). The resistivity of these alloys at 273°K is linearly related to the Si concentration (see Figure 2) with a coefficient of $d\rho/dc = 5\ \mu\Omega cm/(a/o)Si$.

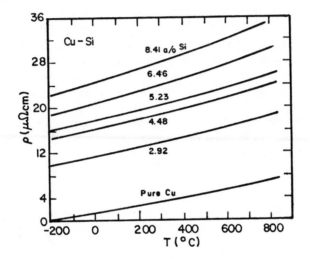

FIGURE 1.   Resistivity of Cu-Si alloys.[1]

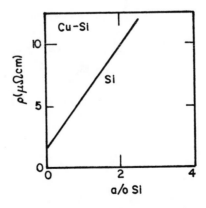

FIGURE 2.   Resistivity at 273°K of CuSi alloys for various silicon concentrations.[3]

REFERENCES

1. Domenicali, C. A. and Otter, F. A., *J. Appl. Phys.*, 26, 377, 1955.
2. Norbury, A. L. and Kuwada, K., *Philos. Mag.*, 4, 1338, 1927.
3. Gerritsen, A. N., *Handbook of Physics*, Springer-Verlag, Berlin, 1957.

Cu-Sn (Copper-Tin)

At 25°C, Cu is not soluble in solid Sn and Sn is not significantly soluble in solid Cu. Except for narrow intermediate phases at Cu-25 a/o Sn and Cu-45 a/o Sn, the Cu-Sn system exists as two-phase alloys at all compositions at 25°C.

The electrical resistivity of the whole spectrum of Cu-Sn alloys compositions at 25°C has been investigated and shows a rather complex behaviorial pattern (see Figure 1) that apparently reflects the uncertainties in the resistivities of the two-phase alloy systems.

The resistivity of several dilute primary-phase CuSi alloys has been measured as a function of temperature between −191 and 438°C (see Figure 2).

The resistivity of these dilute single-phase CuSn alloys at 25°C is a linear function of composition (see Figure 3) with a slope of 3.1 $\mu\Omega$-cm/(a/o)Sn.

Linde[8] gives for $\Delta\rho/\Delta c$ for dilute alloys a value of 2.88 $\mu\Omega$cm/(a/o)Sn for CuSn. Dukin and Aleksandrov[7] gave a critical account of pseudopotential calculations and the resistivity increase due to 1% impurity. They gave $[\Delta\rho/\Delta c]_{exp} = 2.0$ $\mu\Omega$cm/(a/o)Cu in SnCu.

The DMR has been evaluated for CuSn at various temperatures to 300°K (see Figure 4). The characteristic maximum at low temperatures (∼70°K) for nonmagnetic alloys is evident in these CuSn alloys.

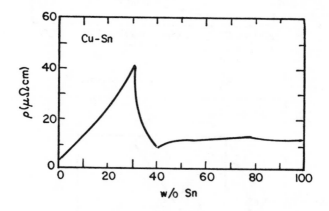

FIGURE 1.   Resistivity of Cu-Sn alloys at 25°C.[1]

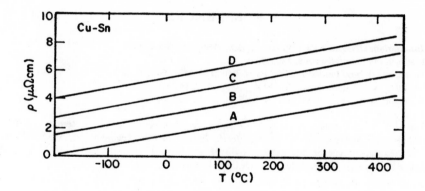

FIGURE 2. Electrical resistivities of certain annealed Cu α-solid solutions at temperatures between −191 and 438°. (A) Pure Cu, (B) 1.0 a/o Sn, (C) 2.03 a/o Sn, (D) 2.96 a/o Sn.[4]

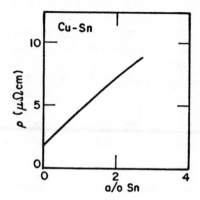

FIGURE 3. Residual resistivity of dilute CuSn alloys as a function of tin concentration.[5]

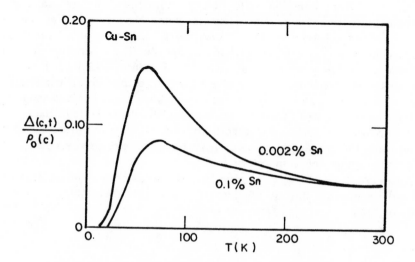

FIGURE 4. Deviations from Matthiessen's rule for CuSn alloys.[6]

REFERENCES

1. *International Critical Tables,* Vol. 6, McGraw-Hill, New York, 1928.
2. Linde, J. O., *Ann. Phys.,* 15, 239, 1932.
3. Dugdale, J. S. and Basinski, Z. S., *Phys. Rev.,* 157, 152, 1967.
4. Norbury, A. L. and Kuwada, K., *Philos. Mag.,* 4, 1338, 1927.
5. Gerritsen, A. N., *Handbook of Physics,* Springer-Verlag, Berlin, 1957.
6. Bass, J., *Adv. Phys.,* 21, 431, 1972.
7. Dukin, V. V. and Aleksandrov, B. N., *Pseudopotential Calculations of the Residual Resistivities of Dilute Solid Alloys Based on Normal Metals, Physics of Low Temperatures,* Akademia Nauk Uk SSR Academy of Science, USSR, Kharkov, 1978, 1.
8. Linde, J. O., *Helv. Phys. Acta,* 41, 1013, 1968.

Cu-Zn (Copper-Zinc)

The familiar Cu-Zn phase diagram shows extensive solid solubility ($\sim$35a/o) of Zn in the Cu lattice and some solubility of Cu ($\sim$0.3% Cu) in the Zn lattice at 25°C. Three prominent broad intermediate phases, β, γ, and ε, exist in the intermediate composition range between the terminal solid solutions. The β-phase orders below 460°C.

The resistivity at room temperature of the Cu-Zn alloys is a complex function of composition (see Figure 1) which results from the presence of the intermediate phases and associated two-phase regions.

From this data, the resistivity of the three intermediate phases is approximately:

Phase	Composition (a/o Zn)	Resistivity (at 25°C, $\mu\Omega$cm)
β	Cu-50	4
γ	Cu-60	12
ε	Cu-85	11

The anomalously low resistivity of the β-phase apparently results from the ordering reaction. The resistivity of the concentrate α-CuZn primary phase solid solution alloys at 25°C is not a linear function of the Zn concentration (see Figure 2) and linearity is only apparent when the Zn content is less than 2 a/o (see Figure 3). The concentration coefficient of the resistivity at 273°K is $d\varrho/dc = 0.3$ $\mu\Omega$cm/(a/o)Zn. Linde[14] gives for $\Delta\varrho/\Delta c$ for dilute alloys a value of 0.63 $\mu\Omega$cm/(a/o)Zn. The resistivities of the α-CuZn alloys as a function of composition at selected temperatures is shown in Figure 4.

Resistivity measurements on the β-phase alloys as a function of temperature show the influence of the ordering phenomena since a knee in the curve is apparent at the ordering temperature (see Figure 5). The composition dependence of the resistivity of these β-brass alloy at temperatures both above and below the ordering temperature also illustrates the ordering effect (see Figure 6) since the resistivity at 25°C is a minimum near the equiatomic composition.

Dukin and Aleksandrov[13] gave a critical account of pseudopotential calculations and the resistivity increase due to 1% impurity. They gave $[\Delta\varrho/\Delta c]_{exp} = 0.36$ $\mu\Omega$cm/(a/o)Cu in ZnCu.

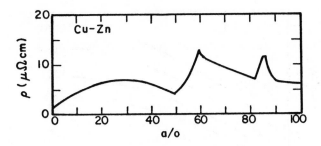

FIGURE 1.   Resistivity of the Cu-Zn alloys. Abscissa: a/o Zn.[1]

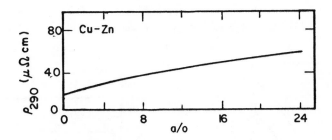

FIGURE 2.   Concentration dependence of the resistivities of the solid solutions of Zn in Cu at 290°K.[2]

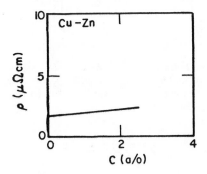

FIGURE 3.   Residual resistivity of CuZn alloys at 278°K.[7]

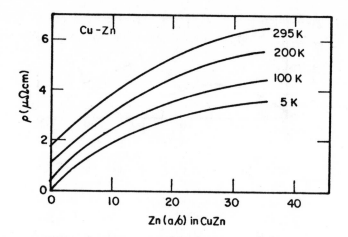

FIGURE 4.   Resistivity of C̲uZn alloys at various temperatures.[2]

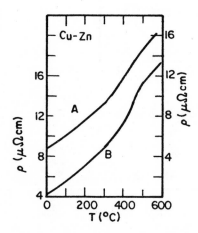

FIGURE 5.   Resistivity-temperature. Curve A for crystal of 55.50 a/o Cu. The ordinate scale for curve A is on the right. Curve B is for 51.25 a/o Cu (Scale on the left)[8].

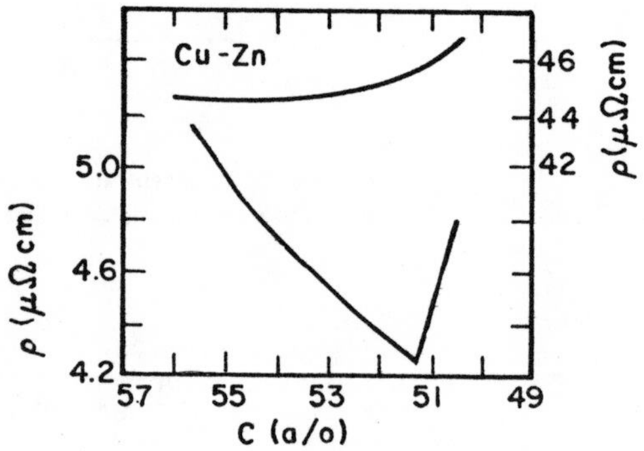

FIGURE 6.   Resistivity as a function of composition. Top curve at 25°C (scale at left). Bottom curve at 500°C (scale at right). Abscissa: a/o Cu.[8]

REFERENCES

1. *International Critical Tables,* Vol. 6, McGraw-Hill, New York, 1929.
2. Crisp, R. S., Henry, W. G., Schroeder, R. A., and Wilson, R. W., *J. Chem. Eng. Data,* 11, 556, 1966.
3. Linde, J. O., *Ann. Phys.,* 15, 219, 1932.
4. Alley, P. and Serin, B., *Phys. Rev.,* 116, 334, 1959.
5. Kierspe, W., *Z. Mettalkd.,* 58, 895, 1967.
6. Wilosewicz, D., Bartkowski, K., and Rofalowicz, J., *Acta Phys. Palon.,* A56, 779, 1979.
7. Gerritsen, A. N., *Handbook of Physics,* Springer-Verlag, Berlin, 1957.
8. Webb, W., *Phys. Rev.,* 55, 297, 1939.
9. Haughton, J. S. and Griffiths, W. T., *J. Inst. Met.,* 34, 245, 1924.
10. Matthiessen, A. and Vogt, C., *Ann. Phys.,* 2, 19, 1864.
11. Matthiessen, A. and Vogt, C., *Philos. Trans. R. Soc. London,* 154, 167, 1865.
12. Gerritsen, A. N. and Linde, J. O., *Physica,* 18, 877, 1952.
13. Dukin, V. V. and Aleksandrov, B. N., *Pseudopotential Calculations of the Residual Resistivities of Dilute Solid Alloys Based on Normal Metals, Physics of Low Temperatures,* Akademia Nauk Uk SSR Academy of Science, USSR, Kharkov, 1978, 1.
14. Linde, J. O., *Helv. Phys. Acta,* 41, 1013, 1968.

Cu-Zr (Copper-Zirconium)

Linde[1] gives for $\Delta\varrho/\Delta c$ for dilute alloys a value of $11.0 \, \mu\Omega cm/(a/o)Zr$ for C̲u̲Zr.

REFERENCES

1. Linde, J. O., *Helv. Phys. Acta*, 41, 1013, 1968.

IRON (Fe)

Fe-Ge (Iron-Germanium)

The Fe-Ge phase diagram indicates the presence of a primary α-FeGe phase field capable of dissolving up to 12 a/o Ge at 700°C. No Ge-rich primary phase appears to exist. Two intermediate phases with an appreciable range of homogeniety exist at the nominal compositions of Fe-40 a/o Ge and Fe-70 a/o Ge.

The residual resistivities of FeGe alloys (see Figure 1) give a concentration coefficient of K(Fe-Ge) = 6.8 μ/Ωcm/ (a/o) Ge. The DMR in FeGe at room temperature is small and concentration independent.

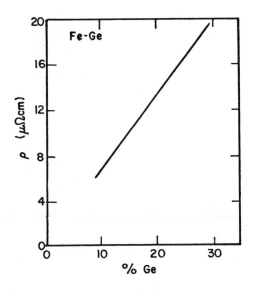

FIGURE 1.   Electrical resistivities of FeGe alloys at 4.2°K vs. solute concentration.[1]

REFERENCES

1. Arajs, S., Schwerer, F. C., and Fisher, R. M., *Phys. Status Solidi,* 33, 731, 1969.
2. Schwerer, F. C. and Cuddy, L. J., *J. Appl. Phys.,* 41, 1419, 1970.
3. Dorleijn, J. W. F., *Philips Res. Rep.,* 31, 287, 1976.

Fe-Mn (Iron-Manganese)

Fe and Mn tend to form a continuous series of disordered FCC solid-solution alloys at elevated temperatures that is modified at lower temperatures by the allotropic transformations occurring in both Fe and Mn. At 300°C, the α-Fe primary phase displays a limited Mn solubility of 2 a/o, while the α-Mn primary single-phase region extends to 32 a/o Fe. The high-temperature FCC phase is stable at 300°C in the composition range between 32 and 53 a/o Mn. The remaining compositions are two-phase mixtures of the FCC phase with either the α-FeMn or α-MnFe primary phases.

Unlike most dilute alloys, the resistivity of the dilute FeMn alloys at 25°C is not linearly related to the concentration of the solute (see Figure 1). The resistivity of the high-temperature FCC phase has been determined as a function of temperature for selected compositions in its room temperature compositional stability range. These alloys order antiferromagnetically and a knee in the resistivity curves occurs near the Neel temperature (see Figure 2).

The residual resistivity of dilute $\underline{Fe}Mn$ alloys is linearly related to the concentration (see Figure 3) through a coefficient of $d\rho/dc = 1.4~\mu\Omega cm/(a/o)Mn$, and the DMR at room temperature of these alloys is large and increases significantly with Mn concentration.

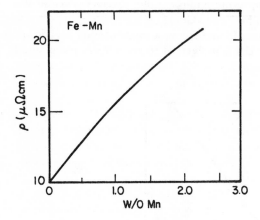

FIGURE 1.   Resistivity of $\underline{Fe}Mn$ alloys.[9]

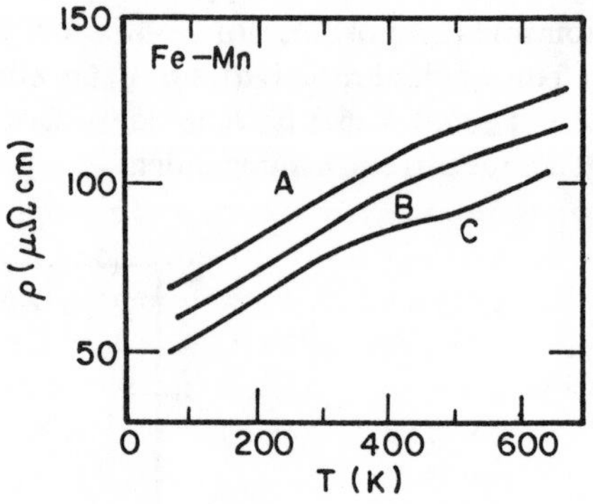

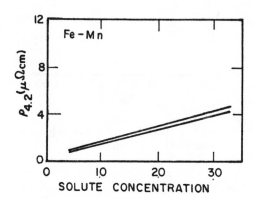

FIGURE 2.   Electrical resistivities of FCC Fe-Mn alloys as a function of temperature (a/o Mn). (A) 56.5; (B) 46.5; (C) 35.5.[8]

FIGURE 3.   Electrical resistivities of $\underline{Fe}Mn$ at 4.2°K vs. solute concentration.[2]

REFERENCES

1. Dorleijn, J. W. F., *Philips Res. Rep.*, 31, 287, 1976.
2. Arajas, S., Schwerer, F. C., and Fisher, R. M., *Phys. Status Solidi*, 33, 731, 1969.
3. Schwerer, F. C. and Cuddy, L. J., *J. Appl. Phys.*, 41, 1419, 1970.
4. Campbell I. A., Fert, A., and Pomeroy, A. R., *Philos. Mag.*, 15, 977, 1967.
5. Fert, A. and Campbell, I. A., *J. Phys. F*, 6, 849, 1976.
6. Dorleijn, J. W. F. and Miedema, A. R., *J. Phys. F*, 7, L23, 1977.
7. Bendick, B. and Pepperhoff, W., *J. Phys. F*, 8, 2535, 1978.
8. Shiga, M. and Nakamura, Y., *J. Phys. Soc. Jpn.*, 19, 1743, 1964.
9. Arrott, A. and Noakes, J. E., *Iron and Its Dilute Solid Solution*, Spencer R. and Werner, L., Eds., Interscience, New York, 1963, 112.

Fe-Mo (Iron-Molybdenum)

The Fe-Mo phase diagram at 600°C consists of two primary solid solution phases, each with a limited solute solubility of about 2 to 3 a/o. A narrow intermediate-phase ε exists at a composition of about Fe-40a/oMo, and the remaining compositions are two-phase mixtures of ε with the primary solid solution phases.

The resistivities of MoFe alloys have been determined at low temperatures for Fe concentrations well beyond the solubility limit. The very dilute MoFe alloys where the Fe concentration is less than 0.25a/o are Kondo systems, and the resistance as a function of temperature of these alloys is a minimum at the Kondo temperature (see Figure 1). For Fe concentrations between 0.25 and 3 a/o, the Fe impurities are no longer isolated, and their interaction results in an obvious change in the resistivity behavior of these Mo-Fe alloys (see Figure 1).

The MoFe alloys in the supersaturated composition range from 3 to 15 a/o are suggested to be mictomagnetic, and the resistivity at low temperatures of several of these alloys is shown in Figure 2. The residual resistivity of the MoFe alloy as a function of Fe concentration increases with the Fe concentration to a maximum in the micromagnetic composition range (see Figure 3).

The resistivities of the FeMo alloys have also been investigated. The residual resistivity is linear with Molybdenum concentration to 3 a/oMo (see Figure 4) with an associated coefficient of K(FeMo) = 1.8 $\mu\Omega$cm/(a/o)Mo. The deviation from Matthiessen's rule in these alloys at 298°K is strongly concentration dependent.[5]

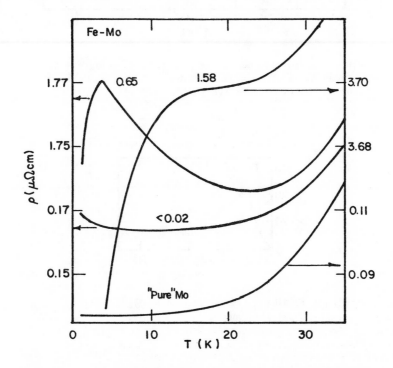

FIGURE 1. Low-temperature resistivities (in $\mu\Omega$cm) of dilute Mo-Fe alloys. The Fe contents in atomic percentages are given by the numbers attached to the curves. The marked < 0.02 is an arc-melted specimen of commercial purity Mo, containing some, but less than 0.02%, Fe. The purest Mo contained less than 0.001% of Fe or Mn.[2]

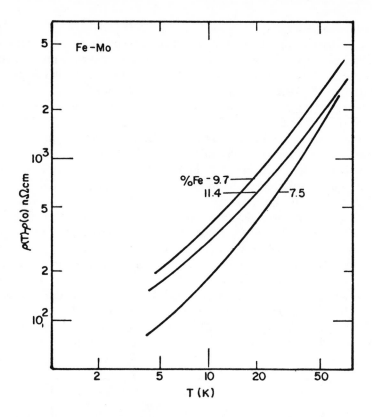

FIGURE 2.   Temperature dependence of the resistivity of MoFe alloys of
several compositions. The residual resistivities are 7.5 a/o Fe, 46.16 $\mu\Omega$cm;
9.7 a/o Fe, 38.215 $\mu\Omega$cm; 11.4 a/o Fe, 35.450 $\mu\Omega$cm.[3]

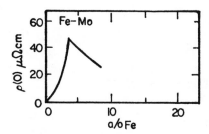

FIGURE 3.   Residual resistivity of MoFe
alloys vs Fe composition.[3]

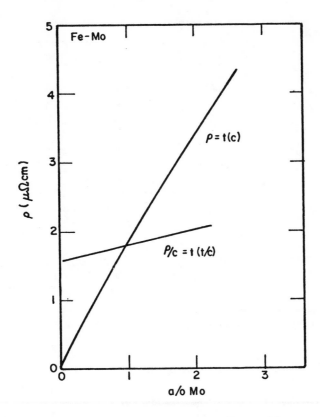

FIGURE 4. Residual resistivity of FeMo alloys vs. Mo concentration.[4]

REFERENCES

1. Hansen, M. and Anderko, K., *Constitution of Binary Alloys*, 2nd ed., McGraw-Hill, New York, 1958.
2. Coles, B. R., *Philos. Mag.*, 8, 335, 1963.
3. Amamou, A., Caudron, R., Costa, P., Friedt, J. M., Gautier, F., and Loegel, B., J. Phys. F., 6,
4. Biolluz, A., Thesis, Strasbourg, 1978.
5. Arajs, S., Schwerer, F. C., and Fisher, R. M., *Phys. Status Solidi*, 33, 731, 1969.
6. Dorleijn, J. W. F., *Philips Res. Rep.*, 31, 287, 1976.
7. Dorleijn, J. W. F. and Miedema, A. R., *J. Phys. F*, 7, L23, 1977.

Fe-Nb (Iron-Niobium)

The only Fe-Nb compositions that exist as single-phase alloys at 25°C occur for an intermediate phase near the Nb_1Fe_2 composition. No primary solid phases are evident at room temperature.

The electrical resistivity as a function of temperature for several alloys within the composition range of the intermediate phase is shown in Figure 1. Several of these alloys are ferromagnetic and the resistivity is seen to decrease relatively rapidly below the Curie temperature of each alloy. The residual resistivity of these alloys is compositionally dependent and is minimized at the Fe-33 a/o Nb composition (see Figure 2).

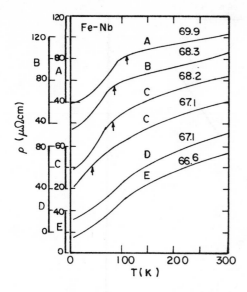

FIGURE 1.   Temperature variation of the electrical resistivity in Fe₂Nb₁. The arrow shows T_c. Numbers give (a/o)Fe.[1]

FIGURE 2.   Composition dependence (a/o Fe) of the electrical resistivity at 4.2°K in Fe-Nb

REFERENCES

1. Ikeda, K. and Nakamichi, T., *J. Phys. Soc. Jpn.*, 39, 963, 1975.

Fe-Ni (Iron-Nickel)

γ-Fe and Ni crystallize in a complete series of FCC solid solution alloys at temperatures above 910°C. At 910°C, a crystallographic transformation occurs in pure Fe and the low-temperature α-Fe BCC allotrope modifies the phase diagram. At lower temperatures, the compositions between Fe-10 a/o Ni and Fe-70 a/o Ni exist as a two-phase mixture of the primary phases α-Fe containing 10 a/o Ni and the Ni primary solid solution phase containing 30 a/o Fe. There is ample evidence that Fe₁Ni₃ is an order-disorder alloy.

Several resistivity studies[1-5] have been made on the concentrated primary solid solution alloys. The resistivity as a function of concentration for the concentrated Ni and Fe primary-phase alloys at several different temperatures is presented in Figure 1.

The Invar alloys containing approximately 30 to 50 a/o Ni are of commercial interest. Their residual resistivities are shown in Figure 2.

The resistivities of the dilute NiFe alloys have been adequately investigated.[7-13] The residual resistivity at 4.2°K as a function of Fe concentration (see Figure 3) yields a coefficient of $d\varrho/dc = 0.33$ $\mu\Omega$cm/(a/o)Fe. The residual resistivities of the dilute FeNi alloys at 4.2°K are concentration dependent through a coefficient of 2 $\mu\Omega$cm/(a/o)Ni. The resistivity at room temperature of these dilute FeNi and NiFe alloys is also linearly related to the solute concentration by a coefficient that is 3.2 and 6.8 $\mu\Omega$cm/(a/o), respectively.

The DMR for various dilute NiFe alloys is relatively independent of concentration and increases with temperature to a relatively large value at 300°K (see Figure 4). The DMR in the dilute FeNi alloys increases with temperature to 400°K and then decreases to zero as the temperature continues to increase to 1000°K. Figure 5 gives the impurity resistivity of Ni-1 a/o Fe. The DMR at 300°K in dilute FeNi is relatively independent of Ni concentration (see Figure 6).

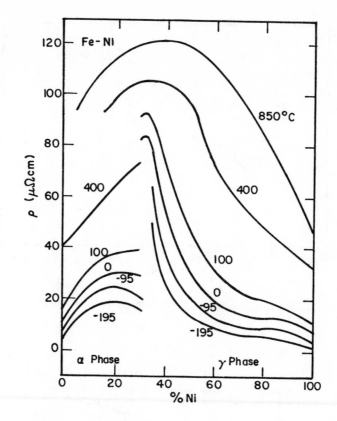

FIGURE 1. Electrical resistivities of Fe-Ni alloys at various temperatures.[20]

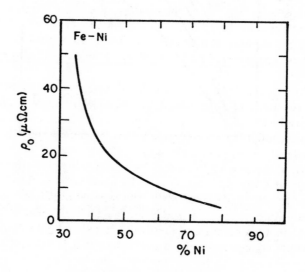

FIGURE 2. Residual resistivity.

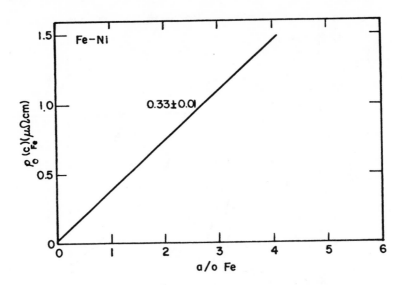

FIGURE 3.   Resistivty at 4.2°K; ϱ_0 (c) vs. solute concentration for N̲iFe alloys.[7]

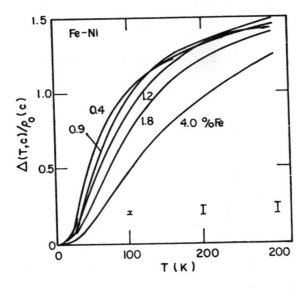

FIGURE 4.   Normalized deviations, $\Delta(T,c)/\varrho_0$ (c), for N̲iFe alloys. Numbers indicate approximate Fe concentrations and error bars are representative for the 0.4 a/o Fe alloy.[9]

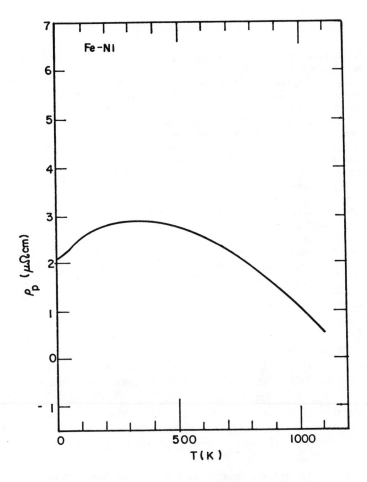

FIGURE 5.   Variation with temperature of the impurity resistivity for a
FeNi alloy with 1 a/o Ni. No corrections have been applied for changes
in Curie temperature upon alloying.[16]

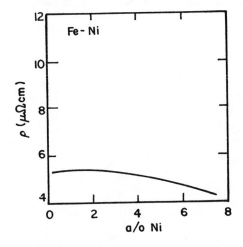

FIGURE 6.   DMR in FeNi alloys at 300°K vs.
nickel concentration.[17]

REFERENCES

1. Shirakawa, Y., *Sci. Rep. Tohoku Imp. Univ.*, 27, 485, 1939.
2. Bendick, W. and Pepperhoff, W., *J. Phys. F*, 8, 2535, 1978.
3. Eliezer, Z. and Hoggins, J. T., *Mater. Res. Bull.*, 12, 227, 1977.
4. Armstrong, B. E. and Fletcher, R., *Can. J. Phys.*, 50, 244, 1972.
5. Kondorskii, E. and Sedor, V. L., *J. Appl. Phys.*, 31, 3315, 1960.
6. Soffer, S., Dreesen, J. A., and Pugh, E. M., *Phys. Rev.*, 140, A668, 1965.
7. Schwerer, F. C. and Conway, J. W., *J. Phys. F*, 1, 877, 1971.
8. Fervell, T. and Grieg, D., *J. Phys. C*, 1, 1359, 1968.
9. Schwerer, F. C. and Cudy, L. J., *Phys. Rev. B*, 2, 1575, 1970.
10. Durand, J. and Gautier, F., *J. Phys. Chem. Solids*, 31, 2773, 1970.
11. Fert, A. and Campbell, I. A., *J. Phys. F*, 6, 849, 1976.
12. Dorleijn J. W. F., *Philips Res. Rep.*, 31 287, 1976.
13. Dorleijn, J. W. F. and Miedema, A. R., *J. Phys. F*, 5, 487, 1975.
14. Dorleijn, J. W. F. and Miedema, A. R., *J. Phys. F*, 7, L23, 1977.
15. Campbell, I. A., Fert, A., and Pomeroy, A. R., *Philos. Mag.*, 15, 977, 1967.
16. Schwerer, F. C. and Cuddy, L. J., *J. Appl. Phys.*, 41, 1419, 1970.
17. Arajs, S., Schwerer, F. C., and Fisher, R. M., *Phys. Status Solidi*, 33, 1731, 1969.
18. Dorleijn, J. W. F., *Philips Res. Rep.*, 31, 287, 1976.
19. Biolluz, A., Thesis, Strasbourg, 1978.
20. Bozorth, R., *Ferromagnetism*, Van Nostrand, Princeton, N.J., 1951.

Fe-Pd (Iron-Palladium)

At temperatures above 910°C, Fe and Pd combine in a continuous series of solid solution disordered FCC alloys. Below 910°C, the allotropic transformation in Fe and the presence of order-disorder reactions at FePd and Fe_1Pd_3 complicate the phase diagram considerably so that at room temperature only the Pd-rich compositions exist as disordered single-phase alloys. The remaining compositions are either ordered alloys or two-phase mixtures involving α-Fe with a limited Pd solubility as one of the constituents.

The resistivity of the PdFe primary phase alloys as a function of temperature up to 300°K for Fe concentrations to 12 a/o has been determined (see Figure 1). A relatively rapid decrease in the resistivity with decreasing temperature is evident for each of these alloys. This behavior is also apparent at very low temperatures for some of the very dilute PdFe alloys (see Figure 2). The residual resistivity of the dilute FePd alloys is linearly related to Pd composition where $K_{FePd} = 2 \ \mu\Omega\text{cm}/(\text{a/o})\text{Pd}$.

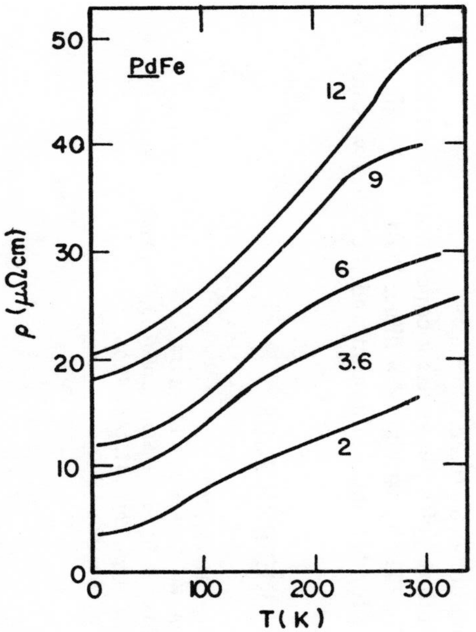

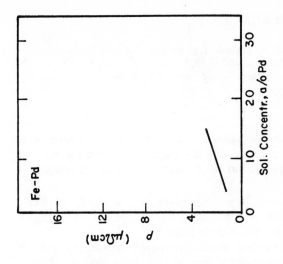

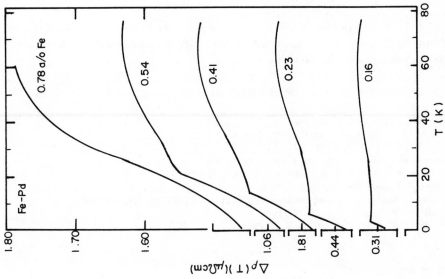

FIGURE 3. Resistance at 4.2°K for FePd alloys vs. Pd concentrations.[3]

FIGURE 2. Δρ(T) (in μΩcm) vs. T(in °K) for all five alloys examined. Note the changes of zero for the incremental resistivities.[2]

FIGURE 1. Electrical resistivity, ρ, of a series of PdFe alloys as a function of temperature. The numbers associated with each curve are the Fe concentrations (in a/o).[1]

REFERENCES

1. Skalski, S., Kowatra, M. P., Mydosh, J. A., and Budnick, J. I., *Phys. Rev. B*, 2, 3613, 1970.
2. Williams, G. and Loram, J. W., *J. Phys. Chem. Solids*, 30, 1827, 1969.
3. Arajs, S., Schwerer, F. C., and Fisher, R. M., *Phys. Status Solidi*, 33, 731, 1969.

Fe-Rh (Iron-Rhodium)

Rh appears to be extremely soluble in the primary α-Fe phase. The solubility limit is about 50 a/o Rh at room temperature. Phase stability for the Rh-rich alloys has not been completely established, but it appears that the solubility of Fe in Rh extends to at least 20 a/o.

Most of the experimental investigations have focused on the resistivity of dilute RhFe alloys[1-5] at low temperatures where these primary phase alloys have a resistance minimum, even when the Fe concentration is relatively large.[5]

The nearly equiatomic FeRh alloys are also of particular interest because they exhibit a first-order antiferromagnetic-to-ferromagnetic transition, and the resistivity of these alloys as a function of temperature shows an unusual maximum-minimum effect associated with the transition temperature.

The resistivity investigations of the Fe-rich compositions have been confined mostly to the dilute FeRh alloys where the residual resistivity is related to the concentration by the coefficient $K_{FeRh} = 1.0\ \mu\Omega\text{cm}/(\text{a/o})\text{Rh}$.

REFERENCES

1. Coles, B. R., *Phys. Lett.*, 8, 243, 1964.
2. Laborde, O. and Radhakrishina, P., *Phys. Lett. A*, 37, 209, 1971.
3. Rusby, R. L., *J. Phys. F*, 4, 1265, 1974.
4. Kaiser, A. B. and Doniach, S., *Int. J. Magn.*, 1, 11, 1970.
5. Murani, A. P. and Coles, B. R., *J. Phys. C*, Suppl. 2, 159, 1970.
6. Oliveira, N. F. and Foner, S., *Phys. Lett. A*, 34, 15, 1971.
7. Schinkel, C. J., Hartog, R., and Hochstenbach, F. H. A. M., *J. Phys. F*, 4, 1412, 1974.
8. Dorleijn, D. W. J., *Res. Rep.*, 31, 287, 1976.
9. Dorleijn, D. W. J. and Miedema, A. R., *J. Phys. F*, 7, L23, 1977.

Fe-Ru (Iron-Ruthenium)

Fe-Ru below 1400°C is a simple eutectoid system with an invarient temperature of about 500°C. The primary terminal solid solution phases are α-Fe containing 4.8 a/o Ru and a sizeable α-Ru phase containing 76 a/o Fe.

The residual resistivity of the dilute FeRu alloys has a concentration coefficient of $\Delta\varrho(4.2°\text{K})/\Delta c = 2\ \mu\Omega\text{cm}/(\text{a/o})\text{Ru}$. The resistivity of a dilute Ru-1 a/o Fe alloy was investigated at low temperatures for a Kondo minimum and the effect was not observed.

REFERENCES

1. Dorleijn, D. W. J., *Philips Res. Rep.*, 31, 287, 1976.
2. Dorleijn, D. W. J. and Miedema, A. R., *J. Phys. F*, 7, L23, 1977.
3. Arajs, S., Schwerer, F. C., and Fisher, R. M., *Phys. Status Solidi*, 33, 731, 1969.
4. Schwerer, F. C. and Cuddy, L. J., *J. Appl. Phys.*, 41, 1419, 1970.
5. Sarachik, M. D., *J. Appl. Phys.*, 39, 699, 1968.

Fe-Sb (Iron-Antimony)

At 25°C, Fe and Sb are insoluble in each other. An intermediate phase, ε, with a modest range of homogeneity near the equiatomic composition and an intermetallic compound, Fe_1Sb_2, both exist at room temperature.

The resistivity of dilute FeSb alloys has been measured at low temperatures. The DMR for a Fe-0.34 a/o Sb alloy increases with temperature to 300°K. The residual resistivity as a function of Sb concentration (see Figure 1) gives a coefficient of $d\varrho/dc = 7 \mu\Omega cm/(a/o)Sb$.

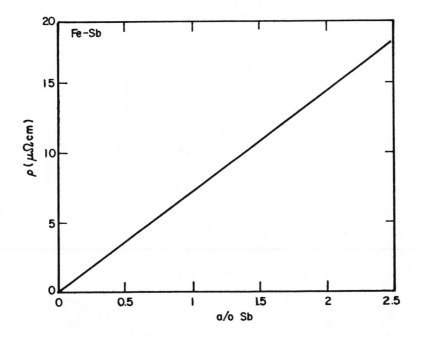

FIGURE 1.   Concentration dependence of the resistivity of FeSb alloys at 4.2°K.[2]

REFERENCES

1. Ross, R. N., Price, D. C., and Williams, G., *J. Mag. Magn. Mater.*, 10, 59, 1979.
2. Biolluz, A., Thesis, Strasbourg, 1978.

Fe-Si (Iron-Silicon)

The solubility limit of Si in the α-Fe lattice is approximately 26 a/o at 25°C, while Fe is not soluble in solid Si. The primary α-Fe phase is expected to order for compositions between 11 and 26 a/o of Si. Two intermediate phases, each with a narrow homogeneity range, exist at nominal compositions Fe-50 a/o Si and Fe-70 a/o Si.

The temperature dependence of the resistivity of various FeSi alloy compositions in the primary-phase field have been determined[1-3] and are shown in Figures 1 and 2. The resistivity at room temperature as a function of Si concentration in these alloys shows a peak at a composition of about 11 a/o Si (see Figure 3) where the Fe-Si alloys are first expected to order. The resistivity of Fe-Si alloys with compositions between 11 and 26 a/o must be specified with caution depending upon the state of order in these alloys.

The resistivity of these FeSi alloys at 4.2, 78, and 300°K is linearly related to the Si concentration to approximately 6 a/o Si (see Figures 3 and 4), and the resulting coefficient is approximately 6 $\mu\Omega$cm/(a/o)Si at all three temperatures. The DMR has been investigated for these primary-phase alloys.[3] The temperature dependence of the solute resistivity for a Fe-1 a/o Si alloy at temperatures to 1000°K is presented in Figure 5.

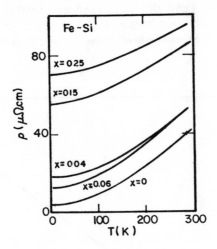

FIGURE 1.   Resistivity of Fe$_{3+x}$Si$_{1-x}$[1]

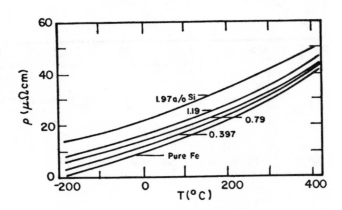

FIGURE 2.   Resistivity of Fe-Si alloys.

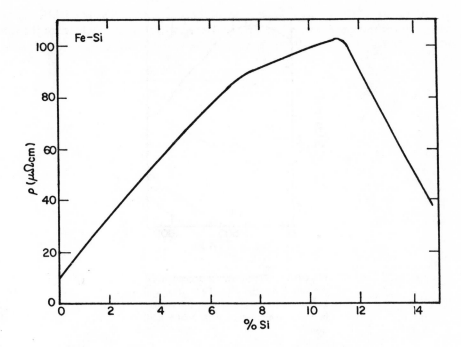

FIGURE 3.   Resistivity of Fe-rich Fe-Si alloys at room temperature.[10]

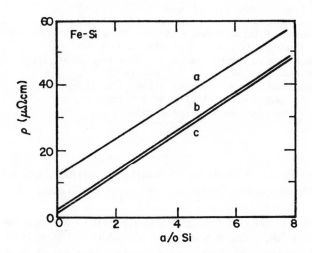

FIGURE 4.   Resistivity vs. (a/o) Si. (a) 300°K; (b) 78°K; (c) 4.2°K.[3]

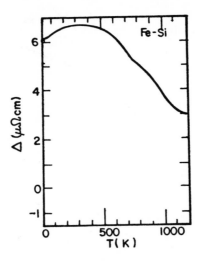

FIGURE 5. Temperature dependence of the solute resistivity for Fe + 1 a/o Si. No corrections were applied for changes in Curie temperature with solute additions.[5]

REFERENCES

1. Budnick, J. D., Muir, W. B., Niculescu, V., and Raj, K., *Transition Metals*, Institute of Physics, London, 1978, 196.
2. Domenicali, C. A. and Otter, F. A., *J. Appl. Phys.*, 26, 377, 1955.
3. Schwerer, F. C., Conroy, J. W., and Arajs, S., *J. Phys. Chem. Solids*, 30, 1513, 1969.
4. Arajs, S., Schwerer, F. C., and Fisher, R. M., *Phys. Status Solidi*, 33, 731, 1969.
5. Schwerer, F. C. and Cuddy, L. J., *J. Appl. Phys.*, 41, 1419, 1970.
6. Campbell, I. A., Fert, A., and Pomeroy, A. R., *Philos. Mag.*, 15, 977, 1967.
7. Glaser, F. W. and Ivanick, W., *J. Met.*, 206, 1290, 1956.
8. Yensen, T. O., *Trans. Am. Inst. Elec. Eng.*, 34, 2601, 1915; 43, 145, 1924.
9. Corson, H. G., *Trans. Am. Inst. Met. Eng.*, 80, 249, 1920.
10. Bozorth, R., *Ferromagnetism*, Van Nostrand, Princeton, N.J., 1951.

Fe-Sn (Iron-Tin)

Except for an Fe-rich primary phase containing approximately up to 5a/o Sn and two intermetallic compounds, Fe_1Sn_2 and Fe_1Sn_1, all Fe-Sn compositions are two-phase alloys at room temperature.

The electrical resistivity of the FeSn primary phase alloys has been determined at 4.2, 77, and 297°K for compositions up to 1.83a/oSn (see Figure 1).

The residual resistance of these FeSn alloys as a function of Sn concentration (see Figures 1 and 2) gives a coefficient of approximately 5.5 $\mu\Omega$cm/(a/o)Sn. The resistivity of these dilute FeSn primary phase alloys have been evaluated also for the DMR as a function of temperature and concentration. The impurity resistivity between 10 and 300°K for Fe-0.3a/oSn and Fe-0.95a/oSn is presented in Figure 3.

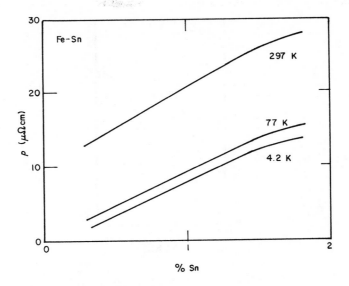

FIGURE 1. Resistivity of dilute FeSn alloys as a function of Sn concentration at several temperatures.[2,4]

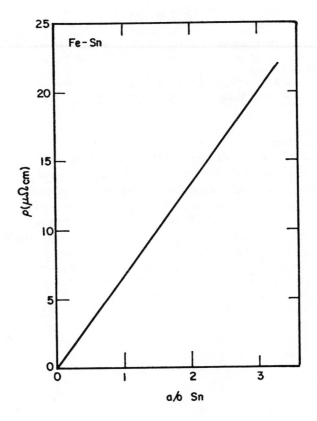

FIGURE 2. Resistivity of FeSn alloys at 4.2°K vs. composition.[3]

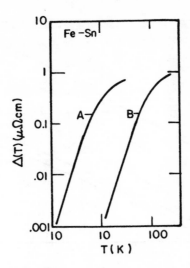

FIGURE 3. The details of the parameter $\Delta(T) = \varrho_{alloy}(T) - \varrho(T)$ (in $\mu\Omega$cm) below 100°K for the Fe$_3$Sn alloys. The solid lines represent T^3 dependences (a/o Sn). (A) 0.3; (B) 0.95.[1]

REFERENCES

1. Ross, R. N., Price, D. D., and Williams, G., *J. Magn. Mater.*, 10, 59, 1979.
2. Arajs, S., Schwerer, F. C., and Fisher, R. M., *Phys. Status Solidi*, 33, 731, 1969.
3. Biolluz, A., Thesis, Strasbourg 1978.
4. Arajs, S., Chessin, H., and Dunmyre, G. R., *J. Appl. Phys.*, 36, 1370, 1965.
5. Price, D. C. and Williams, G., *J. Phys. F*, 3, 810, 1973.

Fe-Ta (Iron-Tantallum)

Fe and Ta in the solid state are insoluble in each other at temperatures below 1000°C. One intermediate phase exists around the Ta$_1$Fe$_2$ composition. The remaining Fe-Ta compositions are mostly two-phase mixtures of TaFe$_2$ and either Fe or Ta.

The resistivity of this intermediate Fe$_2$Ta$_1$ Laves phase of various compositions has been measured at temperatures to 300°K (see Figure 1). The residual resistivity as a function of composition is minimum near the Fe$_2$Ta$_1$ stoichiometry (Figure 2). The concentration dependence of the residual resistivity has been determined for the very dilute FeTa alloys and the coefficient is $K_{FeTa} = 2.5 \ \mu\Omega$cm/(a/o)Ta.

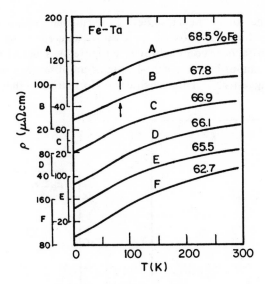

FIGURE 1. Temperature variation of the electrical resistivity in $Fe_{2+x}Ta_{1-x}$.[1]

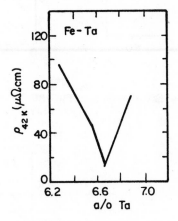

FIGURE 2. Compositions dependence of the residual resistivity.[2]

REFERENCES

1. Ikeda, K. and Nakamichi, T., *J. Phys. Soc. Jpn.*, 39, 963, 1975.
2. Arajs, S., Schwerer, F. C., and Fisher, R. M., *Phys. Status Solidi*, 33, 731, 1969.

Fe-Ti (Iron-Titanium)

The α-Fe BCC lattice can accommodate only a limited amount of Ti ($\sim$2a/o) at 25°C, while Fe is relatively insoluble in α-Ti. Two intermediate phases exist at nominal compositions, Ti_1Fe_2 and Ti_1Fe_1. The Fe_1Ti_1 phase orders at low temperatures.

The resistivity of the primary phase FeTi alloy has been investigated. The concentration dependence of the residual resistivity is shown in Figure 1, while the DMR at 300°K as a function of Ti concentration is presented in Figure 2.

The resistivity behavior of the $Ti_{1-x}Fe_{2+x}$ intermediate phase alloys has also been determined. The residual resistivity of these alloys as a function of composition is a minimum at the Ti-67 a/o Fe composition (see Figure 3). The resistivity of these alloys at any temperature cannot be specified because of an observed thermal hysteresis (see Figure 4) which is suggestive of a martensitic transformation in these alloys.

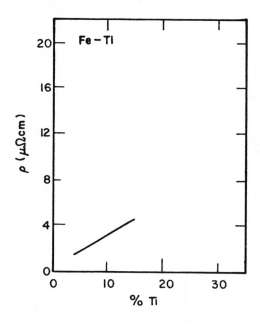

FIGURE 1.   Electrical resistivities of FeTi alloys at 4.2°K.[1]

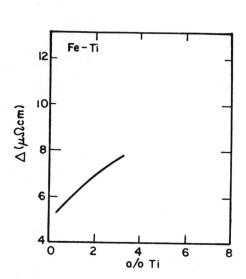

FIGURE 2.   Apparent deviations from Matthiessen's rule at room temperature for FeTi.[1]

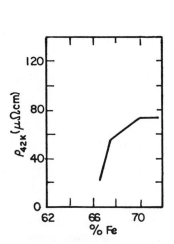

FIGURE 3.   Composition dependence of the electrical resistivity at 4.2°K in $Fe_{2+x}Ti_{1-x}$.[8]

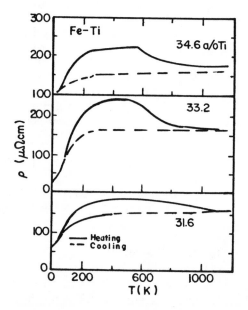

FIGURE 4.   Thermo-hysteresis curves of the electrical resistivity in three Laves phase compounds in the Fe-Ti alloys system with compositions of 31.6, 33.2, and 34.6 a/o Ti.[8]

REFERENCES

1. Arajs, S., Schwerer, F. C., and Fisher, R. M., *Phys. Status Solidi,* 33, 731, 1969.
2. Fert, A. and Campbell, I. A., *J. Phys. F,* 6, 849, 1976.
3. Dorleijn, J. W. F. and Miedema, A. R., *J. Phys. F,* 7, L23, 1977.
4. Campbell, I. A., Fert, A., and Pomeroy, A. R., *Philos. Mag.,* 15, 977, 1967.
5. Dorleijn, J. W. F., *Philips Res. Rep.,* 31, 287, 1976.
6. Farrel, T. and Grieg, D., *J. Phys. C,* 1, 1359, 1968.
7. Ikeda, K. and Nakamichi, T., *J. Phys. Soc. Jpn.,* 39, 963, 1975.
8. Ikeda, K., Nakamichi, T., and Yamamoto, M., *Phys. Status Solidi A,* 12, 595, 1972.

Fe-V (Iron-Vanadium)

At elevated temperatures, Fe and V tend to form a continuous series of disordered BCC solid solution alloys that are stable to room temperature for only the Fe-rich and V-rich compositions. For compositions around FeV_1, the disordered BCC high-temperature phase transforms below 1200°C to the σ-phase.

The resistivity of several FeV alloys at temperatures to 1200°K is shown in Figure 1. No clearly evident anomalous behavior is obvious around the Curie temperature of these alloys. The residual resistivity of the more dilute FeV alloys as a function of concentration (see Figure 2) is characterized by a coefficient of approximately dρ/dc = 1.0 μΩcm/(a/o)V.

The DMR has also been determined for these Fe-V alloys at 300°K, the deviation increases with increasing V concentration (see Figure 3). At higher temperatures, the deviation is employed to determine the Curie temperature (see Figure 4).

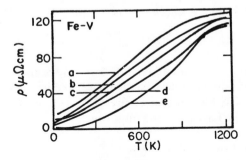

FIGURE 1. Resistivity of Fe-V alloys as a function of temperature. (a) 21.7 a/o V; (b) 16.2 a/o V; (c) 10.8 a/o V; (d) 5.5 a/o V; (e) Fe.[1]

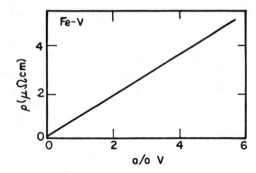

FIGURE 2. Residual resistivity of dilute FeV alloys vs. V concentrations. [2]

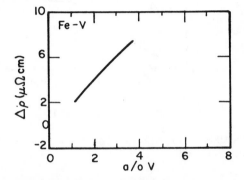

FIGURE 3. Apparent deviations from Matthiessen's rule at room temperature for Fe-V.[3]

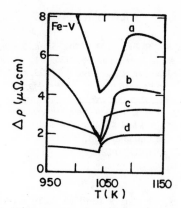

FIGURE 4. $\Delta\varrho(c,T)$ as a function of T for Fe-V alloys between 950 and 1150°K (a/o V). (a) 6.1 (b) 2.7; (c) 0.9; (d) 0.5.[7]

REFERENCES

1. Fujiwara, H., Sueda, N., and Fujiwara, Y., *J. Phys. Soc. Jpn.*, 28, 527, 1970.
2. Dorleijn, J. W. F., *Philips Res. Rep.*, 31, 287, 1976.
3. Campbell, I. A., Fert, A., and Pomeroy, A. R., *Philos. Mag.*, 15, 977, 1967.
4. Arajs, S., Schwerer, F. C., and Fisher, R. M., *Phys. Status Solidi*, 33, 731, 1969.
5. Fert, A. and Campbell, I. A., *J. Phys. F*, 6, 849, 1976.
6. Dorleijn, J. W. F. and Miedema, A. R., *J. Phys. F*, 7, L23, 1977.
7. Yao, Y. D. and Arajs, S., *Phys. Status Solidi B*, 89, K201, 1978.
8. Teoh, W., Arajs, S., Abukay, D., and Anderson, E. E., *J. Magn. Magn. Mater.*, 3, 260, 1976.

Fe-Zr (Iron-Zirconium)

Fe and Zr form several intermediate phases, but do not apparently form any primary solid solution alloys. Fe_2Zr appears to be a prominent intermediate phase, and its electrical resistivity has been determined from 0 to 1000°K (see Figure 1). A knee in the curve is evident at 600°K and is attributed to the ferromagnetic transition.

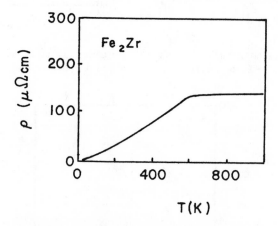

FIGURE 1.   Temperature variation of the electrical resistivity in Fe_2Zr.

REFERENCES

1. Ikeda, K. and Nakamichi, T., *J. Phys. Soc. Jpn.*, 39, 963, 1975.

GALLIUM (Ga)

Ga-Ge (Gallium-Germanium)

Dukin and Aleksandrov[1] gave a critical account of pseudopotential calculations and the resistivity increase due to 1% impurity. They gave $[\Delta\varrho/\Delta c]_{exp} = 3.33$ $\mu\Omega$cm/(a/o)Ge in GaGe.

REFERENCES

1. Dukin, V. V. and Aleksandrov, B. N., *Pseudopotential Calculations of the Residual Resistivities of Dilute Solid Alloys Based on Normal Metals, Physics of Low Temperatures*, Akademia Nauk Uk SSR Academy of Science, USSR, Kharkov, 1978, 1.

Ga-Hg (Gallium-Mercury)

Dukin and Aleksandrov[1] gave a critical account of pseudopotential calculations and the resistivity increase due to 1% impurity. They gave $[\Delta\varrho/\Delta c]_{exp} = 4.7$ $\mu\Omega$cm/(a/o)Hg in GaHg and $[\Delta\varrho/\Delta c]_{exp} = 1.21$ $\mu\Omega$cm/(a/o)Ga in HgGa.

REFERENCES

1. Dukin, V. V. and Aleksandrov, B. N., *Pseudopotential Calculations of the Residual Resistivities of Dilute Solid Alloys Based on Normal Metals, Physics of Low Temperatures*, Akademia Nauk Uk SSR Academy of Science, USSR, Kharkov, 1978, 1.

Ga-In (Gallium-Indium)

Pashayev et al.[1] measured the resistivity of Ga-In alloys between −20 and +800°C. The resistivity of pure Ga in the liquid state is given by $\varrho = 25 + 2.17 \times 10^{-2}$ ($\mu\Omega$cm) (T in °C), and for pure In in the liquid state, resistivity is given by $\varrho = 30.27 + 2.1 \times 10^{-2}$($\mu\Omega$cm) (T in °C).

Boughton et al.[2] determined deviations from Matthiessen's rule on single crystal GaIn alloys. Ga was of 99.9999% purity. They obtained the following values for the residual resistivity.

In (impurity concentration)

ppm	ϱ_o (nΩcm)
5	0.0521
15	0.0865
60	0.1256

Dukin and Aleksandrov[3] gave a critical account of pseudopotential calculations and the resistivity increase due to 1% impurity. They gave $[\Delta\varrho/\Delta c]_{exp} = 0.045$ $\mu\Omega$cm/(a/o)In in GaIn and $[\Delta\varrho/\Delta c]_{exp} = 0.14$ $\mu\Omega$cm/(a/o)Ga in InGa.

REFERENCES

1. Pashayev, B. P., Palchayev, D. K., Chalabov, R. I., and Revels, V. G., *Phys. Met. Metallogr.*, 37 (3), 525, 1974.
2. Boughton, R. I., Pollick, J. J., and Morelli, L., *Phys. Rev. B*, 17, 1611, 1978.
3. Dukin, V. V. and Aleksandrov, B. N., *Pseudopotential Calculations of the Residual Resistivities of Dilute Solid Alloys Based on Normal Metals, Physics of Low Temperatures*, Akademia Nauk Uk SSR Academy of Science, USSR, Kharkov, 1978, 1.

Ga-Li (Gallium-Lithium)

Dukin and Aleksandrov[1] gave a critical account of pseudopotential calculations and the resistivity increase due to 1% impurity. They gave $[\Delta\varrho/\Delta c]_{exp} = 7.4 \ \mu\Omega cm/(a/o)Ga$ in L̲iGa.

REFERENCES

1. Dukin, V. V. and Aleksandrov, B. N., *Pseudopotential Calculations of the Residual Resistivities of Dilute Solid Alloys Based on Normal Metals, Physics of Low Temperatures,* Akademia Nauk Uk SSR Academy of Science, USSR, Kharkov, 1978, 1.

Ga-Mg (Gallium-Magnesium)

Poga et al.[1] measured the resistivity of Ga_2Mg_5. The compound crystallizes in the orthorhombic system. The space group is D_{2k}^{26}. They melted elements of 6N purity in stochiometric proportions under a protective layer of flux in a graphite crucible at 700°C. The alloy cooled then rapidly (30 min). Figure 1 shows $\varrho(T)$. The alloy showed no superconductive behavior in the temperature range under investigation (7.4 to 300°K). The residual resistivity ϱ_o is 2.000 $\mu\Omega cm$ (RRR $\approx$ 11) $\varrho_i = \varrho - \varrho_o \propto T^\alpha$. $\alpha = 3.3$ for 15°K < T < 23°K. The authors calculate the Debye temperature as 200°K.

Dukin and Aleksandrov[2] gave a critical account of pseudopotential calculations and the resistivity increase due to 1% impurity. They gave $[\Delta\varrho/\Delta c]_{exp} = 0.1 \ \mu\Omega cm/(a/o)Mg$ in G̲aMg.

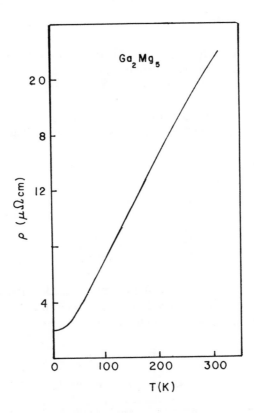

FIGURE 1.   Resistivity of Ga_2Mg_5.

REFERENCES

1. Poga, M., Bradea, I., Ivanciu, O., Niculescu, D., and Cruceanu, E., *Mater. Res. Bull.*, 10, 1349, 1975.
2. Dukin, V. V. and Aleksandrov, B. N., *Pseudopotential Calculations of the Residual Resistivities of Dilute Solid Alloys Based on Normal Metals, Physics of Low Temperatures,* Akademia Nauk Uk SSR Academy of Science, USSR, Kharkov, 1978, 1.

Ga-Ni (Gallium-Nickel)

The Ga-Ni system has a complex phase diagram that shows a large primary solid solution Ni phase field that can dissolve up to 27a/o Ga at 25°C and shows also the existence of five intermediate phases whose ranges of homogeneity at 25°C are all fairly narrow except for the Ni_1Ga_1 phase. The primary phase field apparently contains an ordered Ni_3Ga_1 alloy.

The resistivity of dilute NiGa alloys has been investigated[1] for the concentration dependence of the residual resistivity and the temperature dependence of the DMR (see Figure 1). $\Delta\varrho$ for alloys with 1 to 3 a/o Ga are similar. The concentration coefficient of the residual resistivity is $K(NiGa) = 1.9 \ \mu\Omega cm/(a/o)Ga$ and the DMR increases with temperature in the range 10 to 40 °K.

Resistivity measurements on the more concentrated NiGa primary phase alloys indicate that Ni_3Ga_1 exists as an ordered compound. The residual resistivity as a function of composition for concentrated primary phase NiGa alloys has a minimum at the Ni_3Ga_1 composition (see Figure 2). The ratio of the resistivity at room temperature to the residual value for these alloys shows a dramatic increase at the Ni_3Ga_1 composition suggesting that this stoichiometric alloy has a large temperature coefficient of resistivity and that the off-stoichiometric alloys have a relatively small temperature coefficient of resistivity.

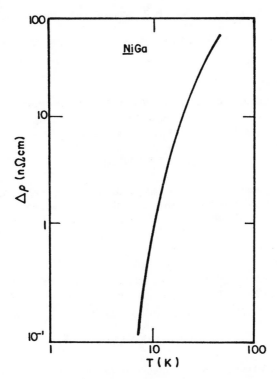

FIGURE 1. Deviations from Matthiessen's rule vs. temperature in a NiGa alloy with 1 a/o Ga.[1]

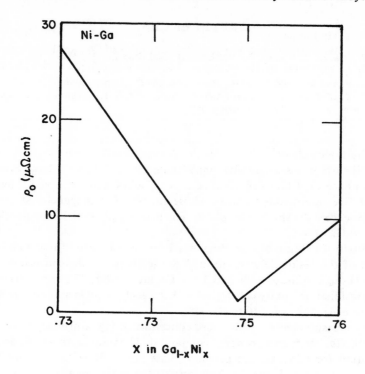

FIGURE 2.   Residual resistivity in $Ga_{1-x} Ni_x$ alloys.[2]

REFERENCES

1. Hugel, J., *J. Phys. F,* 3, 1723, 1973.
2. Fluitman, J. H. J., Boom, R., DeChatel, P. F., Schinkel, C. J., Tilanus, J. L. L., and De Vries, B. R., *J. Phys. F,* 3, 109, 1973.
3. Yamaguchi, Y., Kiewit, D. A., Aoki, T., and Brittian, J. O., *J. Appl. Phys.,* 39, 231, 1968.

Ga-Pb (Gallium-Lead)

Dukin and Aleksandrov[1] gave a critical account of pseudopotential calculations and the resistivity increase due to 1% impurity. They gave $[\Delta \varrho / \Delta c]_{exp} = 14.1 \ \mu\Omega cm/(a/o)Pb$ in GaPb and $[\Delta \varrho / \Delta c]_{exp} = 0.55 \ \mu\Omega cm/(a/o)Ga$ in PbGa.

REFERENCES

1. Dukin, V. V. and Aleksandrov, B. N., *Pseudopotential Calculations of the Residual Resistivities of Dilute Solid Alloys Based on Normal Metals, Physics of Low Temperatures,* Akademia Nauk Uk SSR Academy of Science, USSR, Kharkov, 1978, 1.

Ga-Sn (Gallium-Tin)

Dukin and Aleksandrov[1] gave a critical account of pseudopotential calculations and the resistivity increase due to 1% impurity. They gave $[\Delta \varrho / \Delta c]_{exp} = 0.80 \ \mu\Omega cm/(a/o)Sn$ in GaSn and $[\Delta \varrho / \Delta c]_{exp} = 0.28 \ \mu\Omega cm/(a/o)Ga$ in SnGa.

REFERENCES

1. Dukin, V. V. and Aleksandrov, B. N., *Pseudopotential Calculations of the Residual Resistivities of Dilute Solid Alloys Based on Normal Metals, Physics of Low Temperatures,* Akademia Nauk Uk SSR Academy of Science, USSR, Kharkov, 1978, 1.

Ga-Tl (Gallium-Thallium)

Dukin and Aleksandrov[1] gave a critical account of pseudopotential calculations and the resistivity increase due to 1% impurity. They gave $[\Delta\varrho/\Delta c]_{exp} = 4.8\ \mu\Omega cm/(a/o)Tl$ in $\underline{Ga}Tl$.

REFERENCES

1. Dukin, V. V. and Aleksandrov, B. N., *Pseudopotential Calculations of the Residual Resistivities of Dilute Solid Alloys Based on Normal Metals, Physics of Low Temperatures,* Akademia Nauk Uk SSR Academy of Science, USSR, Kharkov, 1978, 1.

Ga-U (Gallium-Uranium)

Levitin et al.[1] measured the temperature dependence of the resistivity of UGa_2 both for cooling and heating. Figure 1 gives experimental results. $d\varrho/dT$ shows a maximum at 123°K. This should be close to the Curie temperature T_c. Magnetic measurements give $T_c = 129°K$.

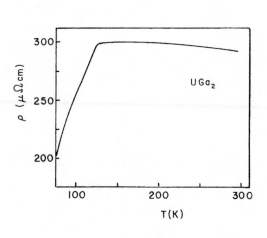

FIGURE 1A.   Resistivity of UGa_2.

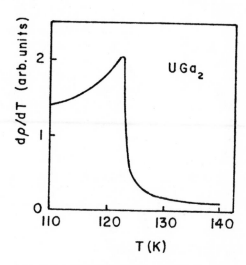

FIGURE 1B. $d\varrho/dT$ of Ga_2U.

REFERENCES

1. Levitin, R. Z., Dmitrievskii, A. S., Henkie, Z., and Misivk, A., *Phys. Status Solidi A,* 27, K109, 1975.

Ga-Zn (Gallium-Zinc)

Boughton et al.[1] studied deviations from Matthiessen's rule on $\underline{Ga}Zn$ single crystals. Ga was of 99.9999% purity. They found the following values for the residual resistivity:

Zn (ppm)	ϱ_o (nΩcm)
50	0.1401
75	0.1697
100	0.5568
200	2.700
500	11.47
1000	51.80
2000	57.05

Dukin and Aleksandrov[2] gave a critical account of pseudopotential calculations and the resistivity increase due to 1% impurity. They gave $[\Delta\varrho/\Delta c]_{exp} = 0.63\ \mu\Omega\text{cm}/(\text{a/o})$Ga in ZnGa and $[\Delta\varrho/\Delta c]_{exp} = 2.45\ \mu\Omega\text{cm}/(\text{a/o})$Zn in GaZn.

REFERENCES

1. Boughton, R. I., Pollick, R. I., and Morelli, J. J., *Phys. Rev. B,* 17(14), 1611, 1978.
2. Dukin, V. V. and Aleksandrov, B. N., *Pseudopotential Calculations of the Residual Resistivities of Dilute Solid Alloys Based on Normal Metals, Physics of Low Temperatures,* Akademia Nauk Uk SSR Academy of Science, USSR, Kharkov, 1978, 1.

GADOLINIUM (Gd)

Gd-Mg (Gadolinium-Magnesium)

Bijvoet et al.[1] added up to 0.25 a/o to Mg. They melted the sample by induction heating. Their pure Mg sample showed the following resistivity values: ϱ (4.7°K) = 0.025 $\mu\Omega$ cm; ϱ (77°K) = 0.55 $\mu\Omega$ cm; ϱ (293°K) = 4.55 $\mu\Omega$ cm. The increase to ϱ with alloying, $\Delta\varrho$, is essentially proportional to c. This shows that the samples should be single phase: $\Delta\varrho/c$ = 8.2 $\mu\Omega$cm/(a/o) Nd. Salkovitz et al.[2] obtained $\Delta\varrho/c$ = 9.5 $\mu\Omega$cm/(a/o)Nd.

REFERENCES

1. Bijvoet, J., de Hon B., Dekker, J. A., and Ratheneu, G. W., *Solid State Commun.*, 1, 237, 1963.
2. Salkovitz, E. J., Schindler, A. I., and Kammer, E. W., *Phys. Rev.*, 105, 887, 1957.

Gd-Nb (Gadolinium-Niobium)

Koch and Kroeger[1] prepared Nb-alloys with small amounts of Gd by arc-melting. Nb was of about 99.9% purity. Cold worked and annealed samples were studied. The normal state at 4.2°K was obtained by applying a magnetic field of 6 kOe. Table 1 gives experimental results.

Table 1

	Condition	ϱ(298°K) ($\mu\Omega$cm)	ϱ(4.2°K) ($\mu\Omega$cm)
Nb	Cold work, 92% RA	14.78	0.51
Nb	1 hr at 1500°C	14.69	0.44
Nb-0.1 a/o Gd	1 hr at 1500°C	14.59	0.19
Nb-1.0 a/o Gd	1 hr at 1500°C	14.87	0.21

REFERENCES

1. Koch, C. C. and Kroeger, D. M., *J. Less Common Met.*, 40, 29, 1975.

Gd-Pd (Gadolinium-Palladium)

Cannella et al.[1] studied the resistivity of PdGd alloys with 2, 3, and 5 a/o Gd. Samples were annealed for over 24 hr at 950°C. Figure 1 shows $\Delta\varrho$(T) of the alloy with 5 a/o Gd. $\Delta\varrho$(T) = ϱ_{alloy}(T) − ϱ_{Pd}(T). The sharp change in the slope of ϱ is associated with the decrease in spin disorder in the ferromagnetic state. The Curie temperature for this alloy, T_c, obtained from susceptibility measurements is 3.0 ± 0.2°K. T_c = 1.5 ± 0.1°K of the alloy with 2 a/o Gd and 2.2 ± 0.1°K for the alloy with 3 a/o Gd. On the other hand, Sarachik and Shaltiel[2] and Crangle[3] found no magnetic contribution to the resistivity near the Curie temperatures.

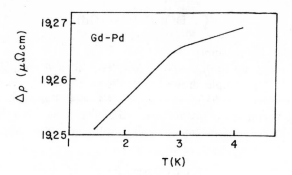

FIGURE 1.   Pd + 5 a/o Gd; $\Delta \varrho = \varrho_{alloy} - \varrho_{Pd}$.

REFERENCES

1. Cannella, V., Burch, T. J., and Budnick, J. I., *AIP Conf. Proc.*, 24, 464, 1974.
2. Sarachik, M. and Shaltiel, D., *J. Appl. Phys.*, 38, 1155, 1967.
3. Crangle, J., *Phys. Rev. Lett.*, 13, 569, 1964.

Gd-Ti (Gadolinium-Titanium)

Shamashov et al.[1] measured the thermal and the electrical conductivity of a Ti alloy with 6.5a/o Gd. Alloys were prepared from 99.9% pure Ti and 99.8% pure Gd by arc-melting. Samples were obtained by forging ingots 55 to 60%, followed by an anneal at 700°C for 2 hr at 10^{-4} Torr. Figure 1 gives $\varrho(T)$ of pure Ti and Ti + 6.5a/o Gd. The drop in ϱ at $\sim$860°C is due to the phase transformation. The sample contains two phases under under equilibrium condition.[2]

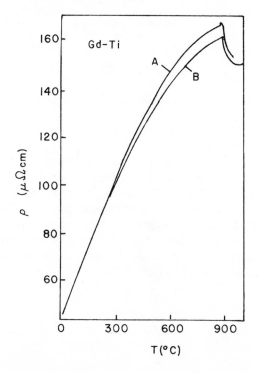

FIGURE 1.   Resistivity of (A) Ti + 6.5 a/o Ga; (B) pure Ti.

REFERENCES

1. Shamashov, F. P., Neimark, B. E., Grebennikov, R. V., Merkulev, A. N., and Poselskii, V. B., *High Temp.*, 9, 1208, 1971.
2. Hansen, M. and Anderko, K., *Constitution of Binary Alloys*, 2nd ed., McGraw-Hill, New York, 1958.

Gd-Y (Gadolinium-Yttrium)

Koch and Kroeger[1] prepared Nb-alloys with small amounts of Y by arc-melting. Nb was of about 99.9% purity. Cold-worked and annealed samples were studied. The normal state at 4.2°K was obtained by applying a magnetic field of 6 KOe. Table 1 gives experimental results.

The cold-worked state may be poorly defined. That may explain why $\varrho(4.2°K)$ of the alloy with 0.1 a/o Y is the same as that of the alloy with 2 a/o Y. However, it is difficult to understand why the pure Nb alloy in the annealed state has a larger residual resistivity than all annealed Nb alloys.

Table 1

	Condition	$\varrho(298°K)$ $(\mu\Omega cm)$	$\varrho(4.2°K)$ $(\mu\Omega cm)$
Nb	Cold work, 92% RA	14.78	0.51
Nb-0.1 a/o Y	Cold work, 92% RA	15.53	0.87
Nb-2.0 a/o Y	Cold work, 92% RA	15.55	0.87
Nb	1 hr at 1500°C	14.69	0.44
Nb-0.1 a/o Y	1 hr at 1500°C	14.59	0.24
Nb-2.0 a/o Y	1 hr at 1500°C	15.21	0.23

REFERENCES

1. Koch, C. C. and Kroeger, D. M., *J. Less Common Met.*, 40, 29, 1975.

GERMANIUM (Ge)

Ge-Nb (Germanium-Niobium)

Müller et al.[1] determined the transition temperature, T_c, and the residual resistivity, ϱ_o, of irradiated Nb_3Ge as a function of the annealing temperature, T_A.

REFERENCES

1. Müller, P., Adrian, H., Ischenko, G., and Braun, H., *J. Phys. Coolqu.*, 39 (C-6, Pt. 1), 387, 1978.

Ge-Ni (Germanium-Nickel)

Solid Ni at room temperature will dissolve a maximum of 11 a/o Ge, while Ni is not at all soluble in the Ge lattice. Three intermediate compounds exist at 25°C with narrow ranges of homogeneity so the majority of the Ni-Ge compositions exist as two-phase alloys at 25°C.

The resistivity of several dilute primary solid solution NiGe alloys have been determined as a function of temperature to evaluate both the residual resistivities at selected Ga concentrations (see Figure 1) and DMR at various temperatures (see Figure 2).

The concentration coefficient of the residual resistivity is $d\varrho/dc = 2.8$ $\mu\Omega$cm/(a/o)Ge and the DMR increases with temperature in the range 10 to 60°K.

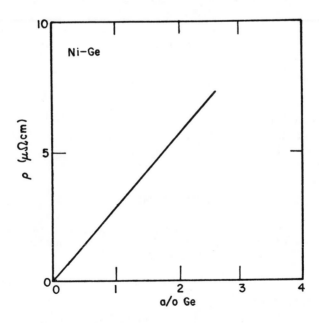

FIGURE 1.   Residual resistivity of NiGe alloys as a function of Ge concentration.[1]

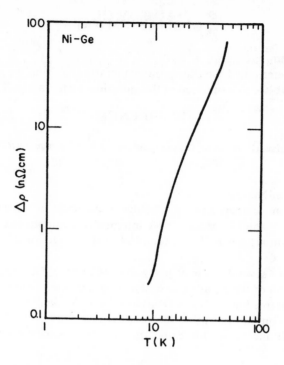

FIGURE 2. Deviations from Matthiessen's rule at various temperatures for Ni + 1 a/o Ge.[1]

REFERENCES

1. Hugel, J., *J. Phys. F,* 3, 1723, 1973.

Ge-Sn (Germanium-Tin)

Dukin and Aleksandrov[1] gave a critical account of pseudopotential calculations and the resistivity increase due to 1% impurity. They gave $[\Delta \varrho / \Delta c]_{exp} = 0.62 \ \mu\Omega cm/$ (a/o)Ge in SnGe.

REFERENCES

1. Dukin, V. V. and Aleksandrov, B. N., *Pseudopotential Calculations of the Residual Resistivities of Dilute Solid Alloys Based on Normal Metals, Physics of Low Temperatures,* Akademia Nauk Uk SSR Academy of Science, USSR, Kharkov, 1978, 1.

Ge-Zn (Germanium-Zinc)

Dukin and Aleksandrov[1] gave a critical account of pseudopotential calculations and the resistivity increase due to 1% impurity. They gave $[\Delta \varrho / \Delta c]_{exp} = 3.5 \ \mu\Omega cm/$(a/o)Ge in ZnGe.

REFERENCES

1. Dukin, V. V. and Aleksandrov, B. N., *Pseudopotential Calculations of the Residual Resistivities of Dilute Solid Alloys Based on Normal Metals, Physics of Low Temperatures,* Akademia Nauk Uk SSR Academy of Science, USSR, Kharkov, 1978, 1.

HAFNIUM (Hf)

Hf-Nb (Hafnium-Niobium)

Haen and Teixeira[1] measured the resistivity of Nb-Hf alloys with 6 to 56 a/o Hf between the superconducting state and room temperature. The residual resistivity is given in Figure 1.[2,3]

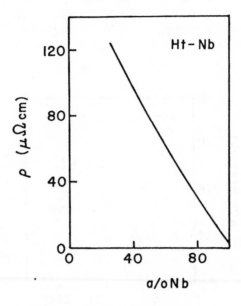

FIGURE 1.   Resistivity of Hf-Nb alloys at 1.2°K.[2]

REFERENCES

1. Haen, P. and Teixeira, J., *Rev. Phys. Appl.,* 9, 879, 1944.
2. Berlincourt, T. G. and Hake, R. R., *Phys. Rev.,* 131, 140, 1963.
3. Siemens, R. E., Oden, L. L., and Deardorff, D. K., *U.S. Bureau of Mines* Report of Investigation, 7258, Washington, D.C., 1969.

Hf-Sc (Hafnium-Scandium)

Prudnikov et al.[1] studied the Hall effect, the magnetic susceptibility, and the resistivity of Sc-Hf alloys from 77 to 300°K. $d\rho/dT$ was constant for all alloys. Figure 1 gives the room temperature resistivities.

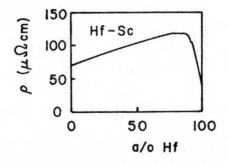

FIGURE 1.   Resistivity of Hf-Sc at room temperature.

REFERENCES

1. Prudnikov, V. N., Volkov, D. I., and Kozlova, T. M., *Phys. Met. Metallogr.*, 30(1), 32, 1970.

Hf-V (Hafnium-Vanadium)

Kimura and Umeda[1] prepared V_2Hf compounds by arc-melting. The purity of the raw material was 99.8%. Figure 1 gives the $\varrho(T)/\varrho$ (300°K) of the alloy. V_2Hf undergoes a cubic to hexagonal transformation. The data between 300°K and the phase transition can be given by (T in °K): $\varrho = 72.4 + 0.044 \cdot T + 22.9 \exp(-120/T)$ $\mu\Omega$cm.

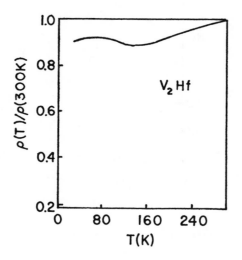

FIGURE 1.   Relative resistivity of V_2Hf.

REFERENCES

1. Kimura, Y. and Umeda, M., *Phys. Status Solidi*, 47, K111, 1978.

MERCURY (Hg)

Hg-In (Mercury-Indium)

Dukin and Aleksandrov[1] gave a critical account of pseudopotential calculations and the resistivity increase due to 1% impurity. They gave $[\Delta\varrho/\Delta c]_{exp} = 0.15$ $\mu\Omega$cm/(a/o)Hg in InHg and $[\Delta\varrho/\Delta c]_{exp} = 0.44$ $\mu\Omega$cm/(a/o)In in HgIn.

REFERENCES

1. Dukin, V. V. and Aleksandrov, B. N., *Pseudopotential Calculations of the Residual Resistivities of Dilute Solid Alloys Based on Normal Metals, Physics of Low Temperatures,* Akademia Nauk Uk SSR Academy of Science, USSR, Kharkov, 1978, 1.

Hg-Li (Mercury-Lithium)

Krill and Lapierre[1] determined the resistivity of LiHg alloys with up to 6.4 a/o Hg from liquid He temperatures to about 65°K to study deviations from Matthiessen's rule. They found that $d\varrho_o/dc \simeq 3.63$ $\mu\Omega$cm/(a/o)Hg.

REFERENCES

1. Krill, G. and Lapierre, E. M., *Solid State Commun.,* 9, 835, 1971.

Hg-Mn (Mercury-Manganese)

Nakawaga and Hori[1] studied $\varrho(T)$ of MnHg, an intermetallic compound of the CsCl structure. Samples were prepared by mixing Hg and Mn powder and annealing it at 400°C for 35 hr in an evacuated and sealed quartz tube. The resistivity at 0°C is $\varrho_o \simeq 5 \times 10^{-5}\Omega$cm.

REFERENCES

1. Nakagawa, Y. and Hori, T., *J. Phys. Soc. Jpn.,* 16, 1470, 1961.

Hg-Pb (Mercury-Lead)

Pb is nearly insoluble in solid Hg, but more than 20 a/o Hg will dissolve in Pb.[1] The intermediate phase, $HgPb_2$, is stable from $\sim$65 to $\sim$68 a/o Pb. Cohen et al.[2] measured ϱ of PbHg alloys from −196 to 240°C and found that $\Delta\varrho/c = 2.0$ $\mu\Omega$cm/(a/o)Hg, independent of temperature.

Dukin and Aleksandrov[3] gave a critical account of pseudopotential calculations and the resistivity increase due to 1% impurity. They gave $[\Delta\varrho/\Delta c]_{exp} = 2.63$ $\mu\Omega$cm/(a/o)Pb in HgPb.

REFERENCES

1. Hansen, M. and Anderko, K., *Constitution of Binary Alloys,* 2nd ed., McGraw-Hill, New York, 1958.
2. Cohen, B. M., Turnbull, D., and Warburton, W. K., *Phys. Rev. B,* 16, 2491, 1977.
3. Dukin, V. V. and Aleksandrov, B. N., *Pseudopotential Calculations of the Residual Resistivities of Dilute Solid Alloys Based on Normal Metals, Physics of Low Temperatures,* Akademia Nauk Uk SSR Academy of Science, USSR, Kharkov, 1978, 1.

Hg-Tl (Mercury-Thallium)

Dukin and Aleksandrov[1] gave a critical account of pseudopotential calculations and the resistivity increase due to 1% impurity. They gave $[\Delta\varrho/\Delta c]_{exp} = 0.9 \; \mu\Omega cm/(a/o)Hg$ in TlHg and $[\Delta\varrho/\Delta c]_{exp} = 0.32 \; \mu\Omega cm/(a/o)Tl$ in HgTl.

REFERENCES

1. Dukin, V. V. and Aleksandrov, B. N., *Pseudopotential Calculations of the Residual Resistivities of Dilute Solid Alloys Based on Normal Metals, Physics of Low Temperatures,* Akademia Nauk Uk SSR Academy of Science, USSR, Kharkov, 1978, 1.

Hg-Zn (Mercury-Zinc)

Dukin and Aleksandrov[1] gave a critical account of pseudopotential calculations and the resistivity increase due to 1% impurity. They gave $[\Delta\varrho/\Delta c]_{exp} = 0.34 \; \mu\Omega cm/(a/o)Zn$ in HgZn and $[\Delta\varrho/\Delta c]_{exp} = 0.67 \; \mu\Omega cm/(a/o)Hg$ in ZnHg.

REFERENCES

1. Dukin, V. V. and Aleksandrov, B. N., *Pseudopotential Calculations of the Residual Resistivities of Dilute Solid Alloys Based on Normal Metals, Physics of Low Temperatures,* Akademia Nauk Uk SSR Academy of Science, USSR, Kharkov, 1978, 1.

INDIUM (In)

In-Li (Indium-Lithium)

Yahagi and Iwamura[1] measured the resistivity of the InLi compound between 30 and 400°C. Data agree well with a previous investigation.[2]

Dukin and Aleksandrov[3] gave a critical account of pseudopotential calculations and the resistivity increase due to 1% impurity. They gave $[\Delta\varrho/\Delta c]_{exp} = 5.8$ $\mu\Omega$cm/(a/o)In in LiIn.

REFERENCES

1. Yahagi, M. and Iwamura, K., *Phys. Status Solidi A,* 39, 189, 1977.
2. Junod, P. and Mooser, E., *Helv. Phys. Acta,* 29, 194, 1956.
3. Dukin, V. V. and Aleksandrov, B. N., *Pseudopotential Calculations of the Residual Resistivities of Dilute Solid Alloys Based on Normal Metals, Physics of Low Temperatures,* Akademia Nauk Uk SSR Academy of Science, USSR, Kharkov, 1978, 1.

In-Mg (Indium-Magnesium)

In and Mg elements form extensive solid solutions with each other.[1] The existence of several intermediate phases has been reported. Salkovitz et al.[2] find $\Delta\varrho/c = 2.0$ $\mu\Omega$cm/(a/o)In in MgIn alloys.

Dukin and Aleksandrov[3] gave a critical account of pseudopotential calculations and the resistivity increase due to 1% impurity. They gave $[\Delta\varrho/\Delta c]_{exp} = 0.45$ $\mu\Omega$cm/(a/o)Mg in InMg.

REFERENCES

1. Hansen, M. and Anderko, K., *Constitution of Binary Alloys,* 2nd ed., McGraw-Hill, New York, 1958.
2. Salkovitz, E. I., Schindler, A. I., and Kammer, E. W., *Phys. Rev.,* 105, 887, 1959.
3. Dukin, V. V. and Aleksandrov, B. N., *Pseudopotential Calculations of the Residual Resistivities of Dilute Solid Alloys Based on Normal Metals, Physics of Low Temperatures,* Akademia Nauk Uk SSR Academy of Science, USSR, Kharkov, 1978, 1.

In-Pb (Indium-Lead)

In and Pb form extensive solid solution ranges.[1] The room temperature resistivity is given in Table 1.[2]

Dukin and Aleksandrov[1] gave a critical account of pseudopotential calculations and the resistivity increase due to 1% impurity. They gave $[\Delta\varrho/\Delta c]_{exp} = 0.73$ $\mu\Omega$cm/(a/o)In in PbIn.

Table 1
ϱ (25°C) of In-Pb

Pb (w/o)	ϱ ($\mu\Omega$cm)	Pb (w/o)	ϱ ($\mu\Omega$cm)
0	9.913	86.4	30.98
16.8	16.52	94.2	28.27
31.1	24.28	97.1	25.41
54.6	27.79	99.2	22.83
64.4	31.92	100.00	21.50
73.0	32.66		

REFERENCES

1. Hansen, M. and Anderko, K., *Constitution of Binary Alloys,* McGraw-Hill, New York, 1958.
2. *International Critical Tables,* Vol. 6, McGraw-Hill, New York, 1929, 156.
3. Dukin, V. V. and Aleksandrov, B. N., *Pseudopotential Calculations of the Residual Resistivities of Dilute Solid Alloys Based on Normal Metals, Physics of Low Temperatures,* Akademia Nauk Uk SSR Academy of Science, USSR, Kharkov, 1978, 1.

In-Sb (Indium-Antimony)

Dukin and Aleksandrov[1] gave a critical account of pseudopotential calculations and the resistivity increase due to 1% impurity. They gave $[\Delta\varrho/\Delta c]_{exp} = 2.0$ $\mu\Omega$cm/(a/o)Sb in InSb.

REFERENCES

1. Dukin, V. V. and Aleksandrov, B. N., *Pseudopotential Calculations of the Residual Resistivities of Dilute Solid Alloys Based on Normal Metals, Physics of Low Temperatures,* Akademia Nauk Uk SSR Academy of Science, USSR, Kharkov, 1978, 1.

In-Sc (Indium-Scandium)

Masuda et al.[1] published $\varrho(T)$ of Sc_3In. Figure 1 shows their results.

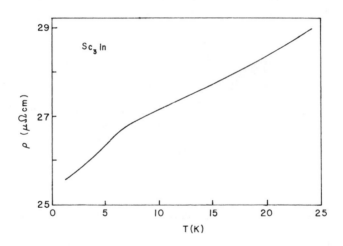

FIGURE 1.   Resistivity of Sc_3In.

REFERENCES

1. Masuda, Y., Hioki, T., and Oota, A., *Physica,* 91(B,C), 291, 1977.

In-Sn (Indium-Tin)

Buckbuchler and Reynolds[1] studied deviations from Matthiessen's rule and the orientation dependence of the electrical resistivity of Sn with Sb impurities. Single-crystal samples were grown with the Bridgman method from 99.999% pure Sn. The orientation dependence in a tetragonal crystal like Sn can be written as $\varrho(\Theta) = p\perp(1 - [\{\varrho(o)/\varrho(90°C)] - 1\}\cos^2\Theta)$, with Θ being the angle between current direction and tetragonal axis. $\varrho\perp = \varrho(90°)$. The residual resistivity is given as $\varrho_o\perp/c = 0.54 \pm 0.02$ $\mu\Omega$cm/(a/o)In for alloys with up to 2 a/o In.

Mukerjee and Chaudhuri[2] prepared SnIn simple crystals from 99.999% pure In and Sn with the Bridgeman method. Measurements were taken along the (001) axis. They found that Nordheim's rule describes the residual resistivity adequately.

$$\rho_o = c(1 - c) \times 4.4 \times 10^{-6} \ \Omega cm$$

The actual measurements gave:

Impurity concentration (a/o)	Residual resistence (Ωcm)
0.0016	5.5×10^{-9}
0.0031	8.93×10^{-9}
0.0093	4.09×10^{-8}

These values are much larger than expected from Reference 1. ϱ of dilute alloys and deviations from Matthiessen's rule were studied by Bressau,[3] and the transition from the normal to the superconducting state was studied by Toxen et al.[4]

Dukin and Aleksandrov[5] gave a critical account of pseudopotential calculations and the resistivity increase due to 1% impurity. They gave $[\Delta\varrho/\Delta c])_{exp} = 0.37 \ \mu\Omega cm/(a/o)Sn$ in InSn.

REFERENCES

1. Buckbuchler, F. V. and Reynolds, C. R., *Phys. Rev.*, 175, 550, 1968.
2. Mukerjee, A. K. and Chaudhuri, K. D., *Solid State Commun.*, 15, 1085, 1974.
3. Bressau, O. J., *J. Phys. F*, 5, 481, 1975.
4. Toxen, A. M., Burns, M. J., and Quin, D. J., *Phys. Rev. A*, 138, 1145, 1965.
5. Dukin, V. V. and Aleksandrov, B. N., *Pseudopotential Calculations of the Residual Resistivities of Dilute Solid Alloys Based on Normal Metals, Physics of Low Temperatues,* Akademia Nauk Uk SSR Academy of Science, USSR, Kharkov, 1978, 1.

In-Tl (Indium-Thallium)

Dukin and Aleksandrov[1] gave a critical account of pseudopotential calculations and the resistivity increase due to 1% impurity. They gave $[\Delta\varrho/\Delta c]_{exp} = 0.2 \ \mu\Omega cm/(a/o)Tl$ in InTl and $[\Delta\varrho/\Delta c]_{exp} = 0.34 \ \mu\Omega cm/(a/o)In$ in TlIn.

REFERENCES

1. Dukin, V. V. and Aleksandrov, B. N., *Pseudopotential Calculations of the Residual Resistivities of Dilute Solid Alloys Based on Normal Metals, Physics of Low Temperatures,* Akademia Nauk Uk SSR Academy of Science, USSR, Kharkov, 1978, 1.

In-Zn (Indium-Zinc)

Dukin and Aleksandrov[1] gave a critical account of pseudopotential calculations and the resistivity increase due to 1% impurity. They gave $[\Delta\varrho/\Delta c]_{exp} = 0.03 \ \mu\Omega cm/(a/o)Zn$ in InZn and $[\Delta\varrho/\Delta c]_{exp} = 0.42 \ \mu\Omega cm/(a/o)In$ in ZnIn.

REFERENCES

1. Dukin, V. V. and Aleksandrov, B. N., *Pseudopotential Calculations of the Residual Resistivities of Dilute Solid Alloys Based on Normal Metals, Physics of Low Temperatures,* Akademia Nauk Uk SSR Academy of Science, USSR, Kharkov, 1978, 1.

IRIDIUM (Ir)

Ir-Pt (Iridium-Platinum)

Pt and Ir form a complete series of solid solutions.[1] There is some evidence of ordering below 1000°C. The room temperature resistivities are given in Table 1.[2]

Table 1

Ir (w/o)	ϱ ($\mu\Omega$cm)	Temperature coefficient[a] (10^{-3} 1°C)
0	10.07	3.48
1	12.51	3.23
2.5	14.66	2.61
5	18.23	2.03
10	24.39	1.38
15	26.9	1.02
20	29.5	0.81
25	31.55	0.66
30	32.9	0.58
35	36.9	0.58

[a]   25 to 100°C.

REFERENCES

1. Hansen, M. and Anderko, K., *Constitution of Binary Alloys,* McGraw-Hill, New York, 1958.
2. *International Critical Tables,* 6, McGraw-Hill, New York, 1929, 156.

Ir-U (Iridium-Uranium)*

Brodsky et al.[1] studied the cubic intermetallic compounds, UIr_2 (MgCu$_2$ type) and UIr_3 (ordered AuCu$_3$ type), between 2 and 300°K. $\varrho(T)$ follows at low temperatures a temperature dependence of the form $\varrho - \varrho_o = AT^n$, with n = 1.9 for UIr_2 and n = 3.7 for UIr_3. This suggests the UIr_2 may be a spin fluctuation compound. UIr_3 behaves similarly as simple transition metal compounds. The experimental results, obtained on arc-melted single-phase samples is shown in Figure 1.

[1]   URh$_3$ data probably from Nellis, W. J., Harvey, A. R., and Brodsky, M. B., *AIP Conf. Proc. USA,* 10, 1076, 1973; and Harvey, A. R., Brodsky, M. B., and Nellis, W. J., *Phys. Rev. B,* 7, 4137, 1973.

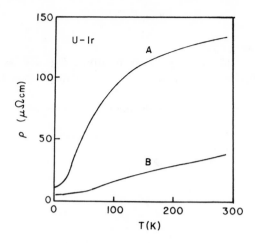

FIGURE 1.   Resistivity of (A) UIr$_2$; (B) UIr$_3$.

REFERENCES

1. Brodsky, M. B., Tainor, R. J., Arko, A. J., and Culbert, H. V., *AIP Conf. Proc. USA*, 29, 317, 1976.

LITHIUM (Li)

Li-Mg (Lithium-Magnesium)

Dugdale and Guyan[1] studied the effect of the martensitic transformation on the resistivity of Li and LiMg alloys. The residual resistivities of their samples ϱ_o (4.2°K) and ϱ (295°K) are given in Table 1.

Dukin and Alksandrov[2] gave a critical account of pseudopotential calculations and the resistivity increase due to 1% impurity. They gave $[\Delta\varrho/\Delta c]_{exp} = 0.50 \ \mu\Omega cm/(a/o)Li$ in MgLi.

Table 1

Mg (a/o)	ϱ_o (4.2°K) (Ωcm)	ϱ (295°K) (Ωcm)
0	13.8×10^{-9}	9.52×10^{-6}
0.05	59.5	9.58
0.07	78	—
0.21	250	9.78
0.47	672	10.27
0.84	1036	10.68
0.94	1300	11.07

REFERENCES

1. Dugdale, J. S. and Guyan, D., *Cryogenics*, 2, 103, 1961.
2. Dukin, V. V. and Aleksandrov, B. N., *Pseudopotential Calculations of the Residual Resistivities of Dilute Solid Alloys Based on Normal Metals, Physics of Low Temperatures*, Akademia Nauk Uk SSR Academy of Science, USSR, Kharkov, 1978, 1.

Li-Pb (Lithium-Lead)

Grube and Klaiber[1] studied the phase diagram of Li-Pb. They prepared samples from 99% Li and Pb of Kahlbaum purity. Table 1 gives the electrical conductivity.

Dukin and Aleksandrov[2] gave a critical account of pseudopotential calculations and the resistivity increase due to 1% impurity. They gave $[\Delta\varrho/\Delta c]_{exp} = 12.3 \ \mu\Omega cm/(a/o)Pb$ in LiPb.

Table 1
ELECTRICAL CONDUCTIVITY ($10^4 \ \Omega^{-1} cm^{-1}$)

Li (a/o)	Temperature (°C) 50	100	150	200	Li (a/o)	Temperature (°C) 50	100	150	200
Pb	4.315	3.676	3.166	2.756	40.0	1.613	1.562	1.575	1.667
1.0	4.123	3.597	3.154	2.702	45.0	1.261	1.277	1.371	1.567
2.0	4.014	3.448	2.937	2.481	47.5	1.108	1.162	1.295	1.568
3.0	3.802	3.289	2.841	2.434	49.0	0.986	1.051	1.190	1.438
4.0	3.868	3.333	2.899	2.499	50.0	1.155	1.228	1.37	1.655
5.0	3.939	3.313	2.814	2.379	51.0	1.236	1.344	1.529	1.880
10.0	3.707	3.153	2.711	2.290	52.0	1.250	1.445	1.771	2.045
20.0	3.168	2.796	2.455	2.241	53.0	1.567	1.711	1.957	2.066
25.0	2.703	2.448	2.283	2.203	54.0	1.402	1.501	1.669	1.934
30.0	2.527	2.344	2.240	2.220	55.0	1.396	1.501	1.696	1.991
35.0	1.995	1.878	1.804	1.832	60.0	1.685	1.661	1.802	1.667

REFERENCES

1. Grube, G. and Klaiber, H., *Z. Elektrochem.,* 40(11), 745, 1934.
2. Dukin, V. V. and Aleksandrov, B. N., *Pseudopotential Calculations of the Residual Resistivities of Dilute Solid Alloys Based on Normal Metals, Physics of Low Temperatures,* Akademia Nauk Uk SSR Academy of Science, USSR, Kharkov, 1978, 1.

Li-Sn (Lithium-Tin)

Grube and Meyer[1] studied the phase diagram of Li-Sn by thermal analysis and found the following compounds: Li_4Sn, Li_3Sn, $LiSn_4$. Samples were prepared from high purity Sn (Kahlbaum) and 99% Li. The isotherms of the electrical conductivities are shown in Table 1.

Table 1
ELECTRICAL CONDUCTIVITY $\varkappa \cdot 10^{-4} \Omega^{-1} cm^{-1}$

Li a/o	Temperature (°C)						
	50	100	150	200	250	300	350
0	7.40	6.28	5.32	4.63	—	—	—
1	7.29	6.22	5.25	4.60	—	—	—
2	7.27	6.12	5.26	4.56	—	—	—
5	7.15	6.12	5.25	4.53	—	—	—
7.5	7.12	6.03	5.16	4.46	—	—	—
10	7.08	6.07	5.17	4.45	—	—	—
15	6.82	5.90	5.04	4.40	—	—	—
20	6.66	5.72	4.97	4.33	—	—	—
25	6.57	5.65	4.87	4.28	3.51	2.87	—
30	6.26	5.44	4.97	4.15	3.23	2.88	—
33.3	6.15	5.34	4.62	4.07	3.19	2.83	2.01
35	5.78	5.07	4.44	3.91	3.05	2.80	2.07
40	5.02	4.53	3.99	3.58	2.88	2.61	2.22
45	4.40	3.95	3.62	3.20	2.79	2.52	2.18
49	3.88	3.58	3.22	2.93	2.63	2.33	2.07
50	3.32	3.08	2.85	2.67	2.34	2.09	1.89
51	3.23	2.99	2.67	2.40	2.18	2.08	1.85
52	3.05	2.78	2.53	2.32	2.12	1.95	1.78
55	2.84	2.55	2.33	2.13	1.93	1.75	1.59
57	2.56	2.32	2.09	1.87	1.74	1.60	1.47
60	2.34	2.12	1.92	1.76	1.64	1.51	1.40
63	2.10	1.93	1.75	1.63	1.50	1.40	1.29
66	1.94	1.83	1.66	1.55	1.43	1.33	1.24

REFERENCES

1. Grube, G. and Meyer, E., *Z. Elektrochem.,* 40, 771, 1934.

Li-Th (Lithium-Thorium)

Grube and Schaufler[1] studied the phase diagram of Li-Th by thermal analysis. Impurity levels in Th were about 0.05%; Li was 99% pure. These authors found the following compounds: $LiTh$, Li_2Ti, Li_5Tl_2, Li_3Tl, and Li_4Tl. The other samples were two-phase alloys. These authors determined the isotherms of the electrical conductivities.

REFERENCES

1. Grube, G. and Schaufler, G., *Z. Elektrochem.,* 40, 593, 1934.

Li-Tl (Lithium-Thallium)

Dukin and Aleksandrov[1] gave a critical account of pseudopotential calculations and the resistivity increase due to 1% impurity. They gave $[\Delta \varrho / \Delta c]_{exp} = 7.26 \; \mu\Omega$cm/(a/o)Tl in LiTl.

REFERENCES

1. Dukin, V. V. and Aleksandrov, B. N., *Pseudopotential Calculations of the Residual Resistivities of Dilute Solid Alloys Based on Normal Metals, Physics of Low Temperatures*, Akademia Nauk Uk SSR Academy of Science, USSR Kharkov, 1978, 1.

Li-Zn (Lithium-Zinc)

Dukin and Aleksandrov[1] gave a critical account of pseudopotential calculations and the resistivity increase due to 1% impurity. They gave $[\Delta \varrho / \Delta c]_{exp} = 2.8 \; \mu\Omega$cm/(a/o)Zn in LiZn.

REFERENCES

1. Dukin, V. V. and Aleksandrov, B. N., *Pseudopotential Calculations of the Residual Resistivities of Dilute Solid Alloys Based on Normal Metals, Physics of Low Temperatures*, Akademia Nauk Uk SSR Academy of Science, USSR, Kharkov, 1978, 1.

MAGNESIUM (Mg)

Mg-Nd (Magnesium-Neodymium)

Bijvoet et al.[1] added up to 0.25 a/o Nd to Mg. Samples were prepared by induction heating. Pure Mg showed the following resistivity values: $\varrho(4.2 \text{ K}) = 0.025 \ \mu\Omega\text{cm}$, $\varrho(77 \text{ K}) = 0.55 \ \mu\Omega\text{cm}$, and $\varrho(293 \text{ K}) = 4.55 \ \mu\Omega\text{cm}$. The increase in ϱ with alloying, $\Delta\varrho$, is essentially proportional to c. This shows that the samples should be single phase. $\Delta\varrho/c = 9.5 \ \mu\Omega\text{cm}/(\text{a/o})\text{Nd}$ at 77 K and $\Delta\varrho/c = 8.2 \ \mu\Omega\text{cm}/(\text{a/o})\text{Nd}$ at room temperature. Das and Gerritsen[2] found $\Delta\varrho/c = 1.47 \ \mu\Omega\text{cm}/(\text{a/o})\text{Nd}$ at room temperature. These samples were cold rolled, and may have exceeded the solubility limit of the impurities.

REFERENCES

1. Bijvoet, J., de Hon, B., Dekker, J. A., and Rathenau, G. W., *Solid State Commun.*, 1, 237, 1963.
2. Das, K. B. and Gerritsen, N. A., *J. Appl. Phys.*, 33, 3301, 1962.

Mg-Pb (Magnesium-Lead)

Up to 7.75 a/o Pb will dissolve in Mg at 466°C, and up to 5.9 a/o Mg will dissolve in Pb at 253°C.[1] The intermetallic compound, Mg_2Pb, melts at 550°C.

Knappwost[2] measured $\varrho(T)$ of Mg_2Pb in the liquid and solid state. Figure 1 shows his results. The difference to earlier measurements by Pietenpol and Miley[3] is attributed to differences in sample composition. Small deviations from the stochiometric composition Mg_2Pb changes $\varrho(T)$ markedly.

Dukin and Aleksandrov[4] gave a critical account of pseudopotential calculations and the resistivity increase due to 1% impurity. They gave $[\Delta\varrho/\Delta c]_{exp} = 2.3 \ \mu\Omega\text{cm}/(\text{a/o})\text{Mg}$ in PbMg and $[\Delta\varrho/\Delta c]_{exp} = 5.8 \ \mu\Omega\text{cm}/(\text{a/o})\text{Pb}$ in MgPb.

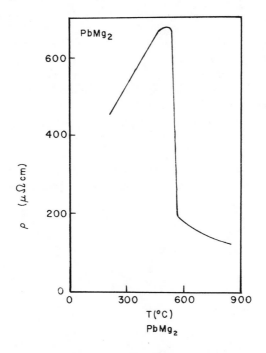

FIGURE 1.   Resistivity of Mg_2Pb.

REFERENCES

1. Hansen, M., and Anderko, K., *Constitution of Binary Alloys*, 2nd ed., McGraw-Hill, New York, 1958.
2. Knappwost, A., *Z. Elektrochem.*, 57(7), 618, 1953.
3. Pietenpol, W. B. and Miley, H. A., *Phys. Rev.*, 34, 158, 1929.
4. Dukin, V. V. and Aleksandrov, B. N., *Pseudopotential Calculations of the Residual Resistivities of Dilute Solid Alloys Based on Normal Metals, Physics of Low Temperatures*, Akademia Nauk Uk SSR Academy of Science, USSR, Kharkov, 1978, 1.

Mg-Sn (Magnesium-Tin)

Dukin and Aleksandrov[1] gave a critical account of pseudopotential calculations and the resistivity increase due to 1% impurity. They gave $[\Delta\varrho/\Delta c]_{exp} = 5.25 \ \mu\Omega cm/(a/o)Sn$ in M̲gSn.

REFERENCES

1. Dukin, V. V. and Aleksandrov, B. N., *Pseudopotential Calculations of the Residual Resistivities of Dilute Solid Alloys Based on Normal Metals, Physics of Low Temperatures*, Akademia Nauk Uk SSR Academy of Science, USSR, Kharkov, 1978, 1.

Mg-Tl (Magnesium-Thallium)

Grube and Hille[1] studied the phase diagram of Mg-Tl with resistivity measurements. They found three intermetallic compounds: Tl_3Mg_8, $TlMg_2$, and Tl_2Mg_3. Hansen and Anderko[2] list Mg_5Tl_2, Mg_2Tl, and $MgTl$, using more recent investigations and corrections.

The electrical conductivity for alloy samples prepared from 99.93% Mg and Tl with traces of sesquioxydes and Pb are shown in Table 1. Only samples with about 8 to 15 a/o Tl may be single-phase alloys. Salkovitz et al.[3] find that $\delta\varrho/c = 3.2 \ \mu\Omega cm/(a/o)Tl$ for M̲gTl alloys.

Poga et al.[4] measured the resistivity of Tl_2Mg_5. This compound recrystallizes in the orthorhombic system. The space group is D_{2h}^{26}. The authors melted elements of 6N purity in stochiometric proportions under a protective layer of flux in a graphite crucible at 700°C. The alloy cooled then rapidly (30 min). Figure 1 gives $\varrho(T)$. There is no evidence of superconductivity even at 2.4°K. The residual resistivity is $\varrho_o = 0.234 \ \mu\Omega cm$ (RRR $\simeq$ 110). This is one of the highest values found for intermetallic compounds. $\varrho_i = \varrho - \varrho_o \propto T^\alpha$, with $\alpha = 3.4$ for $15°K < T < 23°K$. The authors calculate as Debye temperature: $\Theta_D = 125°K$.

Dukin and Aleksandrov[5] gave a critical account of pseudopotential calculations and the resistivity increase due to 1% impurity. They gave $[\Delta\varrho/\Delta c]_{exp} = 2.5 \ \mu\Omega cm/(a/o)Mg$ in T̲lMg.

Table 1
ELECTRICAL CONDUCTIVITY OF Mg-Tl
ALLOYS IN $10^4 \, \Omega^{-1} \, cm^{-1}$. TEMPERATURE IN °C

Tl (a/o)	Temperature (°C)					
	40	80	120	160	200	280
0.00	20.04	—	15.70	—	12.94	11.06
3.00	7.50	—	6.91	—	6.39	5.97
4.00	6.05	—	5.71	—	5.41	5.15
5.82	4.66	—	4.48	—	4.31	4.16
8.00	3.91	—	3.79	—	3.68	3.55
9.82	4.01	—	3.82	—	3.64	3.35
16.00	3.94	—	3.54	—	3.27	2.97
20.00	4.07	—	3.49	—	3.09	2.68
24.00	3.96	—	3.28	—	2.81	2.41
28.57	3.88	—	3.18	—	2.55	2.21
42.50	6.17	5.41	4.81	4.30	3.83	—
45.00	5.41	4.75	4.24	3.79	3.40	—
47.50	6.19	5.43	4.85	4.31	3.83	—
50.00	6.87	5.98	5.29	4.75	4.23	—
60.00	6.41	5.46	4.70	4.28	3.68	—
70.00	5.89	5.04	4.33	3.70	2.98	—
80.00	5.65	4.86	4.07	3.45	2.79	—
90.00	5.24	4.55	3.86	3.12	2.34	—
95.17	5.18	4.46	3.69	3.00	2.33	—
96.73	5.01	4.26	3.56	2.93	2.65	—
97.53	4.79	4.07	3.52	2.96	2.79	—
98.34	4.74	4.02	3.43	3.00	2.97	—
99.17	4.98	4.38	3.84	3.43	3.15	—
100.00	5.53	4.70	4.1i	3.66	3.33	—

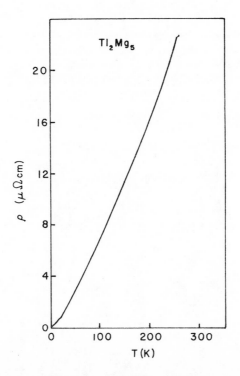

FIGURE 1.   Resistivity of Tl_2Mg_5.

REFERENCES

1. Grube, G. and Hille, J., *Z. Elektrochem.*, 40, 101, 1934.
2. Hansen, M. and Anderko, K., *Constitution of Binary Alloys,* 2nd ed., McGraw-Hill, New York, 1958.
3. Salkovitz, E. I., Schindler, A. I., and Kammer, E. W., *Phys. Rev.*, 105, 887, 1957.
4. Poga, M., Bradea, I., Ivanciu, O., Niculescu, D., and Cruceanu, E., *Mater. Res. Bull.*, 10, 1349, 1975.
5. Dukin, V. V. and Aleksandrov, B. N., *Pseudopotential Calculations of the Residual Resistivities of Dilute Solid Alloys Based on Normal Metals, Physics of Low Temperatures,* Akademia Nauk Uk SSR Academy of Science, USSR, Kharkov, 1978, 1.

Mg-Zn (Magnesium-Zinc)

Dukin and Aleksandrov[1] gave a critical account of pseudopotential calculations and the resistivity increase due to 1% impurity. They gave $[\Delta\varrho/\Delta c]_{exp} = 0.73$ $\mu\Omega$cm/(a/o)Mg in $\underline{Zn}$Mg.

REFERENCES

1. Dukin, V. V. and Aleksandrov, B. N., *Pseudopotential Calculations of the Residual Resistivities of Dilute Solid Alloys Based on Normal Metals, Physics of Low Temperatures,* Akademia Nauk Uk SSR Academy of Science, USSR, Kharkov, 1978, 1.

MANGANESE (Mn)

Mn-Ni (Manganese-Nickel)

A primary solid solution phase field of Ni that dissolves up to 38 a/o of Mn exists at 25°C. The alloys in this range of compositions are magnetic, and the Mn_1Ni_3 composition is suspected to be an order-disorder alloy. An intermediate phase with a large homogeneity range exists around the equiatomic composition. The high-temperature disordered solid solution phase is stable at low temperatures in the 20 to 30 a/o Ni composition range. The Mn-rich primary solid solution phase has a limited Ni solubility ($\sim$2 a/o Ni).

The resistivity work on the Ni-Mn system has been mostly confined to the NiMn primary solid solution alloys. The resistivity of the concentrated NiMn primary solid solution alloys at room temperature is shown in Figure 1 for alloys in the quenched and annealed states. The difference in the quenched and annealed resistivities for the alloys around 25 a/o Mn composition results from the ordering associated with the Ni_3Mn_1, alloy. The influence of ordering on the temperature dependence of the resistivity of this alloy is illustrated in Figure 2. From these data, it is apparent that the room temperature resistivity of Ni-Mn alloys in the composition range 20 to 30 a/o Mn will be highly dependent on the degree of order in these alloys. The temperature dependence of the resistivity of several Ni-Mn alloys whose managanese concentration is less than 20 a/o is shown in Figure 3.

The resistivity of the dilute NiMn alloy has been investigated, and the concentration coefficient of the residual resistivity is found to be approximately 0.7 $\mu\Omega$cm/(a/o) Mn. The DMR for the NiMn is appreciable at 297°K.[8]

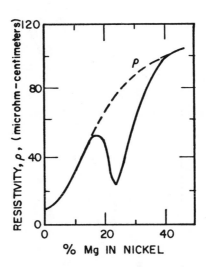

FIGURE 1.   Electrical resistivities of Ni-Mn alloys: quenched (upper curve, disordered structure) or annealed (lower curve, ordered structure).[9]

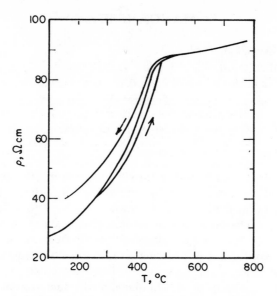

FIGURE 2.   Resistivity of Ni_3Mn_1 (23.8 a/o Mn) on heating and cooling, showing the effect of ordering. Equilibrium is represented by the middle curve.[9]

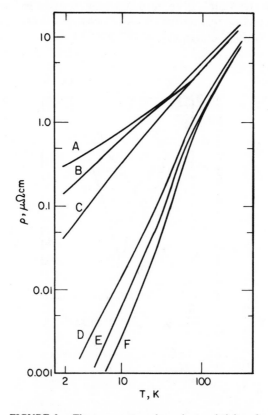

FIGURE 3.   The temperature dependent resistivity of some Ni-Mn alloys. The Mn concentration was determined by electron microprobe, $\varrho_o(\mu\Omega cm)$. (A) 8.37, 7.4 a/o Mn; (B) 23.7, 13.1 a/o Mn; (C) 35.1, 17.5 a/o Mn; (D) 1.58, 2.1 a/o Mn; (E) 1.12, 1.0 a/o Mn; (F) 0.28, Ni.[7]

REFERENCES

1. Valentiner, S. and Becker, G., *Z. Phys.*, 93, 795, 1935.
2. Kaya, S. and Nakayama, N., *Proc. Phys. Math. Soc. Jpn.*, 22, 126, 1940.
3. Fert, A. and Campbell, I. A., *Phys. Rev. Lett.*, 21, 1190, 1968; *J. Phys.*, 32, C1-40-50, 1971; *J. Phys. F.*, 6, 849, 1976.
4. Farrell, T. and Grieg, D., *J. Phys. C*, 1, 1359, 1968.
5. Dorleijn, J. W. F., *Philips Res. Rep.*, 31, 287, 1976.
6. Dorleijn, J. W. F. and Miedema, A. R., *J. Phys. F*, 5, 487, 1975.
7. Rowlands, J. A., *J. Phys. F*, 3, L149, 1973.
8. Durand, J. and Gautier, F., *J. Phys. Chem. Solids*, 31, 2773, 1970.
9. Bozorth, R. M., *Ferromagnetism*, Van Nostrand, Princeton, N.J., 1951.

Mn-Pt (Manganese-Platinum)

The phase diagram of Mn-Pt shows[1] that more than 35 a/o Mn will dissolve in Pt. The solubility of Pt in Mn is of the order of a few percent in α-, β-, and δ-Mn; it is quite extensive in γ-Mn. An intermediate phase is centered around MnPt.

Sarkissian and Taylor[2] prepared Pt-Mn alloys with up to 18 a/o Mn from 4N pure Mn and spectroscopically pure Pt by arc-melting, followed by an anneal at 750°C for about 1/2 hr in a vacuum. $\Delta\varrho = \varrho_{alloy} - \varrho_{host}$ for alloys with 1 to 12 a/o Mn is given in Figure 1. $\Delta\varrho$ for alloys with 1 a/o Mn or less is given in Figure 2. The data were explained on the basis of spin glass behavior.

Kästner et al.[3] prepared PtMn alloy in a levitation melting arc furnace. These investigations at lower temperatures than Reference 2 shows a pronounced minimum of ϱ, associated with the transition to the spin glass sate. Masumoto et al.[4] studied the thermal expansion, Youngs modulus, the magnetic susceptibility, and the electrical resistivity from −150 to 500°C.

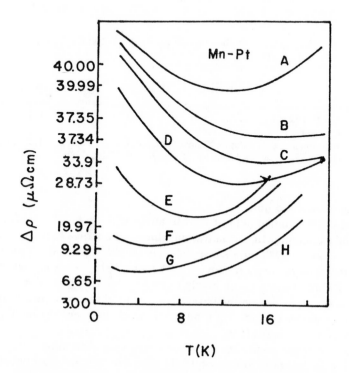

FIGURE 1. $\Delta\varrho = \varrho_{alloy} - \varrho_{host}$ of Mn-Pt alloys, (in a/o Mn). (A) 12; (B) 11; (C) 10; (D) 8; (E) 5.4; (F) 3; (G) 2; (H) 1.

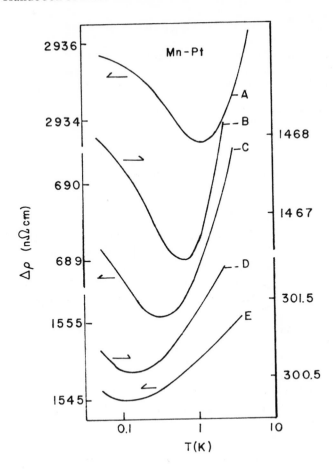

FIGURE 2.   Δϱ of PtMn alloys (in a/o Mn). (A) 1.0; (B) 0.60; (C) 0.25; (D) 0.10; (E) 0.05

REFERENCES

1. Hansen, M. and Anderko, K., *Constitution of Binary Alloys*, McGraw-Hill, New York, 1958.
2. Sarkissian, B. V., and Taylor, R. H., *J. Phys. F*, 4, L243, 1974.
3. Kästner, J., Wassermann, E. F., Matho, K., and Tholence, J. L., *J. Phys. F*, 8, 1, 103, 1978.
4. Masumoto, H., Sawaya, S., and Kikuchi, M., *Trans. Jpn. Inst. Met.*, 19(7), 390, 1978.

Mn-Rh (Manganese-Rhodium)

Murani and Coles[1] studied $\varrho(T)$ of Rh-Mn alloys with up to 14 a/o Mn; 14 a/o Mn seems to be the maximum Mn concentration which will dissolve in Rh at 800°C. Alloys were prepared from 99.99% pure Rh sponge and 99.995% pure Mn. Powder mixtures were arc-melted. The impurity resistivity of the alloys are shown in Figure 1. The behavior of the RhMn alloys is qualitatively similar to that of RhFe alloys. The data can therefore be again explained on the basis of the spin fluctuations model, of the simple phase shift analysis, by considering modifications of the exchange interaction, or by using a combination of these effects.

Coles et al.[3] reported earlier on the resistivity of a RhMn alloy wire with 0.5 a/o Mn. No T^2 region has been found. $d\varrho^2/dT^2$ does not become positive at the lowest temperature.

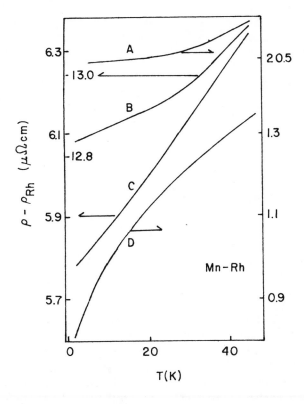

FIGURE 1. $\varrho - \varrho_{Rh}$ of Rh-Mn alloys (in a/o Mn). (A) 14.2; (B) 9.4; (C) 4.6; (D) 1.0.

REFERENCES

1. Murani, A. P. and Coles, B. R., *J. Phys. F,* 2, 1137, 1972.
2. Raub, E. and Mahler, W., *Z. Metallkd.,* 46, 282, 1955.
3. Coles, B. R., Mozumder, S., and Rusby, R., *12th Int. Conf. on Low Temp. Physics,* Keigaku, Publ. Co., Tokyo, 1971, 737.

Mn-Ru (Manganese-Ruthenium)

The solubility of Mn in Sb is negligible.[1] A few percent Sb may dissolve in various allotropic modifications of Mn. Intermediate phases of composition Mn_2Sb and MnSb have been found.

Williams and Stanford[2] measured the resistivity of MnRu alloys between 4.2°K and room temperature. Samples were arc-melted and then sliced by spark-erosion. A vacuum anneal at 620°C for 6 hr followed. Figure 1 shows $\varrho(T)$ of alloys with 0.89 and 2.40 a/o Ru. These samples should be single-phase alloys, since α-Mn can absorb more than 10 a/o Ru.[1] The minimum of ϱ between 110 and 140°K is associated with the onset of antiferromagnetic ordering. $T_N = 111°K$ for the alloy with 0.89 a/o Ru and about 135°K for the alloy with 2.40 a/o Ru.

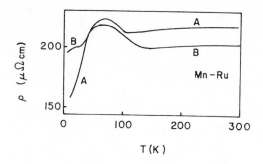

FIGURE 1.   Resistivity of <u>Mn</u>Ru (in a/o Ru). (A)
0.89; (B) 2.40.

REFERENCES

1. Hansen, M. and Anderko, *Constitution of Binary Alloys,* 2nd ed., McGraw-Hill, New York, 1958.
2. Williams, W., Jr. and Stanford, J. L., *Phys. Rev., B,* 7, 3244, 1973.

Mn-Sb (Manganese-Antimony)

The solubility of Mn in Sb is negligible. A few percent Sb may dissolve in the various allotropic modifications of Mn. The intermediate phases, Mn_2Sb and MnSb, have been found. Tu Chen et al.[2,3] prepared Mn_xSb single crystals by pulling them from the melt. Samples were cut and the composition was determined by wet chemical analysis. The samples are ferromagnetic. Figure 1 gives $\varrho(T)$. $\varrho(T)$ is nearly constant in the paramagnetic range above T_c. $\delta\varrho(T)/\varrho(T_c)\delta T$ is even negative for the alloy with x = 1.15. $\varrho(T)$ decreases for all alloys below the Curie temperature. The authors propose a simple band model to explain their data. ϱ of MnSb was also measured by Fischer and Pearson[6] and by Manneve-Tassy,[4] who found hysteresis effect above 450°C, whereas Fakidov and Grazhdankiva[5] studied the influence of a magnetic field on the electrical resistance of ferromagnetic alloys of Mn and Sb.

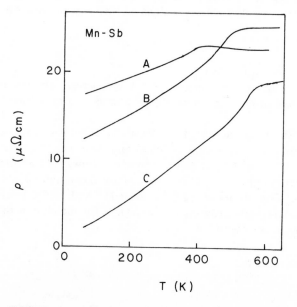

FIGURE 1.   Resistivity of Mn_xSb. (A) X = 1.15; (B)
X = 1.075; (C) X = 1.01.

REFERENCES

1. Hansen, M. and Anderko, K., *Constitution of Binary Alloys,* 2nd ed., McGraw-Hill, New York, 1958.
2. Chen, Tu, Stutius, W., Allen, J. W., and Stewart, G. R., *AIP Conf. Proc. USA,* 29, 532, 1976.
3. Chen, Tu, Rodowski, D., and White, R. M., *J. Appl. Phys.,* 49, 1425, 1978.
4. Manneve-Tassy, G., *C. R. Acad. Sci. Paris,* 226, 1992, 1948.
5. Fakidov, I. G. and Grazhdankiva, N. P., *Dokl. Adad. Nauk. SSSR,* 66, 847, 1949.
6. Fischer, G. and Pearson, W. B., *Can. J. Phys.,* 36, 1010, 1958.

Mn-Sn (Manganese-Tin)

Hansen and Anderko[1] report that three compounds exist. They are $MnSn_2$, Mn_2Sn, and Mn_3Sn. The last one has a homogeneity range from 23 to 24.5 a/o Sn.

Kouvel and Hartelius[2] melted Mn and Sn in an induction furnace. The chemical analysis showed 68.5 a/o Sn. The sample was annealed for 4 days at 490°C and furnace cooled. X-ray measurements showed mostly lines from $MnSn_2$, with a few very weak lines due to β-Sn. The sample contained about 0.2 a/o of ferromagnetic ordered Mn_2Sn. Figure 1 gives $\varrho(T)$ of the alloy. The authors suggest on the basis of magnetic measurements that the sample orders antiferromagnetically at the sharp drop of $\varrho(T)$ at 73°K. However, antiferromagnetic ordering is not completely destroyed above 73°K.

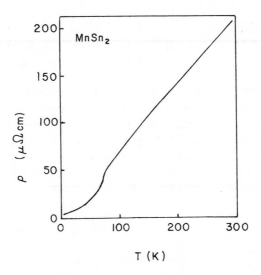

FIGURE 1.   Resistivity of $MnSn_2$.

REFERENCES

1. Hansen, M. and Anderko, K., *Constitution of Binary Alloys,* 2nd ed., McGraw-Hill, New York, 1958.
2. Kouvel, J. S. and Hartelius, C. C., *Phys. Rev.,* 123, 124, 1961.

Mn-Ti (Manganese-Titanium)

Hake et al.[1] measured $\varrho(T)$ for a number of dilute superconducting hcp Ti alloys between 1.1 and 35°K. Samples were prepared from "iodide process" Ti, typically of 99.92% w/o purity. The second component in the alloy was usually better than 99.4

w/o pure. The alloys were melted under Ar in an arc furnace. No heat treatment followed. Resistivity minima were found for 99.92 w/o pure Ti and Ti-Mn alloys with 0.02 to 2 a/o Mn. Table 1 gives $\varrho(4.7°K)$ and $\varrho(273°K)$ and the temperature of the minima. The authors associate these minima with localized moments.

Table 1

Mn (a/o)	$\varrho(273°K)$ ($\mu\Omega$cm)	$\varrho(4.2°K)$ ($\mu\Omega$cm)	T_{min}
≈0	42.2	1.44	14.1
0.018	42.1	1.71	14.3
0.101	43.5	3.66	—
0.114	42.8	2.88	16.2
0.212	46.5	5.47	18.3
0.407	47.8	6.85	—
1.00	49.5	7.81	—
106	50.0	8.22	17.0
2.01	54.1	10.7	17.9

REFERENCES

1. Hake, R. R., Leslie, D. H., and Berlincourt, T. G., *Phys. Rev.*, 127, 170, 1962.

Mn-V (Manganese-Vanadium)

There is evidence for four intermediate phases with compositions near 11, 24, 50, and 85 a/o V. The 50 a/o V sample is bcc. The 24 a/o V sample should be the σ-phase.[1]

Williams and Stanford[2] measured the resistivity of MnV alloys with 1, 2.1, and 2.4 a/o V between 4.2°K and room temperature. Samples were arc-melted and then sliced by spark-erosion. A vaccum anneal at 620°C for 6 hr followed. Figure 1 gives experimental results. The minimum in $\varrho(T)$ is associated with the onset of antiferromagnetism. T_N decreases with increasing V concentrations as in CrV alloys. Nagasawaba and Senba[3] found that $\varrho(T) - \varrho(0) = -0.01 \times T^2$ $\mu\Omega$cm for an alloy with 1 a/o V and attributed this temperature dependence to spin fluctuations.

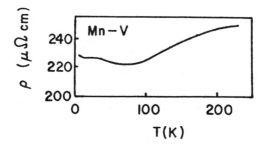

FIGURE 1A.   Resistivity of Mn- 1 a/o V.

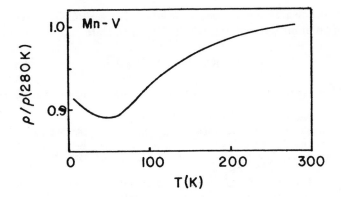

FIGURE 1B.   Resistivity of Mn- 2.1 a/o V.

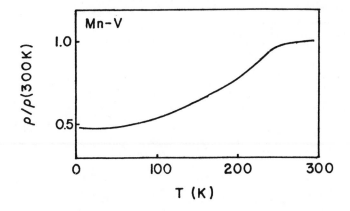

FIGURE 1C.   Resistivity of Mn- 2.4 a/o V.

REFERENCES

1. Hansen, M. and Anderko, K., *Constitution of Binary Alloys*, 2nd ed., McGraw-Hill, New York, 1958.
2. Williams, W., Jr. and Stanford, J. L., *Phys. Rev. B*, 7, 3244, 1973.
3. Nagasawaba, H. and Senba, M., *J. Phys. Soc. Jpn.*, 39, 70, 1975.

Mn-Zn (Manganese-Zinc)

The Mn-Zn phase diagram shows several intermediate phases.[1] More than 5 a/o Mn will dissolve in Zn. Most resistivity measurements[2-13] are concerned with effects of very small amount of Mn in Zn-Mn alloys.

Bell and Caplin[12] and Pilot et al.[5] studied the effect of isolated Mn impurity atoms on the resistivity of Zn. Hedgcock and Rizzuto[2] prepared ZnMn alloys from 99.999% Zn. The R(T)/R(273) show the characteristic minimum for Kondo alloys. Magnetic ordering becomes significant at temperatures higher than the resistance minimum.

Falke et al.[3] prepared ZnMn alloys usually from pure Zn and a master alloys 6N Zn and 3N Mn). The resistivity shows the characteristic maximum and minimum of the Kondo effect. The slope of the straight line is 1/(5 ± 0.5). This means that the Kondo result holds fairly will.

Ford et al.[4] prepared Zn alloys by mixing pure Zn with a master alloy under pure Ar in a pyrex container. In all alloys the resistivity has a $\log_{10} T$ dependence, but at very low temperature a T^2 dependence is observed. The residual resistivities is explained on the basis of an unmagnetized Friedel virtual bound state.

REFERENCES

1. Hansen, M. and Anderko, K., *Constitution of Binary Alloys,* 2nd ed., McGraw-Hill, New York, 1958.
2. Hedgcock, F. T. and Rizzuto, C., *Phys. Rev.,* 163, 517, 1967.
3. Falke, H. P., Jablonski, H. P., and Wassermann, E. F., *Z. Phys.,* 269, 285, 1974.
4. Ford, P. J., Rizzuto, C., and Salamoni, E., *Phys, Rev. B,* 6, 1851, 1972.
5. Pilot, A., Vaccarone, R., and Rizzuto, C., *Phys. Lett. A,* 40, 405, 1972.
6. Korn, D., *Z. Phys.,* 238, 275, 1976.
7. Stroink, G. W. R. and Muir, W. B., *Phys. Can.,* 29, 25, 1973.
8. Newrock, R. S., The Kondo Effect, Thesis, Rutgers State University, New Brunswick, N.J., 1970.
9. Collings, E. W., Hedgcock, F. T., and Muir, W. M., *Proc. 8th Int. Conf. Low Temperature Physics,* Butterworths, London, 1963, 253.
10. Wassermann, E. and Falke, H., *Z. Metallkd.,* 60, 623, 1969.
11. Muto, Y., Tawara, Y., Shibuya, Y., and Fukuroi, T., *J. Phys. Soc. Jpn.,* 14, 380, 1959.
12. Bell, A. E. and Caplin, A. D., *J. Phys. F,* 5, 143, 1974.
13. Wassermann, E., *Z. Phys.,* 234, 347, 1970.

MOLYBDENUM (Mo)

Mo-Nb (Molybdenum-Niobium)

Mo and Nb form a complete series of solid solutions. Cox et al.[1] prepared single-crystal metals and alloys by a floating zone electron beam melting method. Figure 1 gives the resistivities of the alloys used. The authors determined also the Hall coefficients of these alloys. They assumed a rigid band model and showed that a two-band model could explain the anisotropy of both the resistivity and the Hall coefficient of Mo-rich alloys.

Savitskii et al.[2] studied the resistivity of Mo-Nb single crystals. High-temperature measurements have been made by Inoue and Shimizu.[3] The normal state resistivity of $Nb_{0.8}Mo_{0.2}$ together with other properties was determined by Muto et al.[4] The effect of gas content and heat treatment on Mo-Nb alloys was studied by Sokolova.[5]

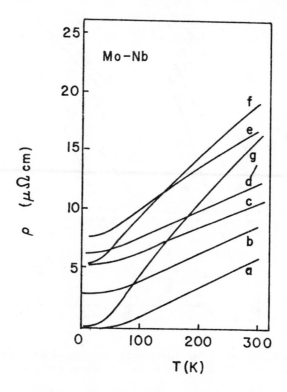

FIGURE 1. Resistivity of Mo, Nb, and Mo-Nb alloys.
(a) Mo; (b) 7.9 a/o Nb; (c) 24.8 a/o Nb; (d) 41.9 a/o Nb; (e) 64.8 a/o Nb; (f) 83.2 a/o Nb; (g) Nb.

REFERENCES

1. Cox, W. R., Hayes, D. J., and Brotzen, F. R., *Phys. Rev. B*, 7, 3580, 1973.
2. Savitskii, E. M., Burov, I. V., and Pirogova, S. V., *Sov. Phys. Dokl.*, 13, 951, 1969.
3. Inoue, J. and Shimizu, M., *J. Phys. Soc. Jpn.*, 41, 1211, 1976.
4. Muto, Y., Mori, K., and Noto, K., Physica, Proc. Int. Conf. Science Superconductivity, Stanford, 55, 362.
5. Sokolova, G. V., *Fiz. Met. Metalloved.*, 47(1), 118, 1979.

Mo-Ni (Molybdenum-Nickel)

The solubility limit of Mo in the Ni lattice at 25°C is 13 a/o, while Ni does not dissolve in the Mo lattice. Three intermediate compounds with narrow homogeneity ranges exist around the 20, 25, and 50 a/o Mo compositions.

Resistivity information on the N̲i̲Mo alloy system is meager. The resistivity of alloys in the N̲i̲Mo primary solid solution phase region have been determined at temperatures below 100°K (see Figure 1). The dilute NiMo alloys are characterized by a large concentration coefficient of residual resistance of $K_{N̲i̲Mo} = 5.8 \ \mu\Omega cm/(a/o)Mo$ and a relatively small DMR at 25°C.

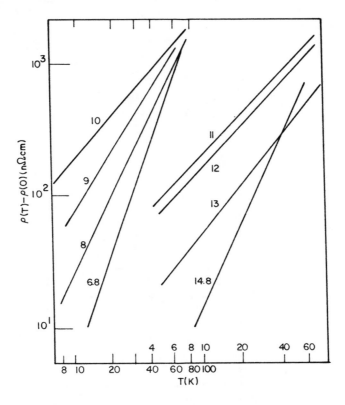

FIGURE 1.   Temperature-dependent part of the electrical resistivity (nΩcm) for Ni-Mo alloys. The number of each curve gives the Mo concentration (in a/o).[1]

REFERENCES

1. Amamou, A., Gautier, F., and Loegel, B., *J. Phys. F,* 5, 1342, 1975.
2. Durand, J. and Gautier, F., *J. Phys. J. Chem. Solids,* 31, 2773, 1970.
3. Durand, J., Thesis, Strasbourg, 1973.
4. Smit, J., *Physica,* 21, 877, 1955.

Mo-Pd (Molybdenum-Palladium)

Volkenshtein et al.[1] measured the resistivity of Pd-Mo alloys from about 20 to 273°K. Figure 1 gives the temperature dependence of $\varrho(T)$ of Mo-rich alloys. The authors also studied alloys with 5 to 20 a/o Mo.

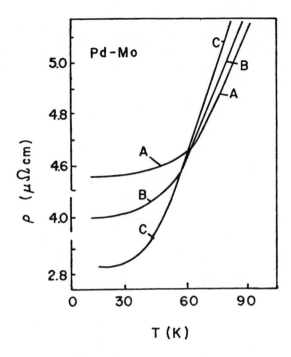

FIGURE 1.   Resistivity of Pd-Mo (in a/o Mo). (A) 80; (B) 90; (C) 95.

REFERENCES

1. Volkenshtein, N. V., Vgodnikova, L. A., and Tsiovkin, Yu N., *Sov. Phys. Solid State,* 12, 1160, 1970.

Mo-Re (Molybdenum-Rhenium)

The solubility of Re in solid Mo is 27 a/o at 1000°C.[1] Two intermediate phases exist. The maximum solubility of Mo in solid Re is 14 a/o. The resistivity of Mo-Re alloys[2] prepared by a floating zone electron beam technique is shown in Figure 1.

Peletskii and Druzhinin[3] prepared Mo-Re crystals by electron beam zone melting. Impurity levels are of the order of 3×10^{-3} a/o. Table 1 gives experimental results. Data are 5 to 6% lower than the data by Gaines and Sims[4] on polycrystalline samples. The data by Arutyunov and Filippov[5] on single crystals agree at low temperature with those of Reference 2 and at high temperature with those of Reference 3. The difference in results should be attributed to impurity variations.

Table 1
SPECIFIC ELECTRICAL RESISTIVITY OF Re AND Mo-Re ALLOYS

T(°K)	8% Re	20% Re	47% Re	Re
293	8.16	12.25	19.24	18.2
400	11.15	15.2	—	25.25
500	13.83	18.0	—	32.2
600	16.6	20.75	—	38.7
700	19.3	23.55	—	44.85
800	22.05	26.3	—	50.6
900	24.73	29.1	—	55.9
1000	27.5	31.85	—	60.9
1100	30.25	34.65	—	65.5
1200	33.0	37.45	48.35	69.6
1300	35.8	40.25	51.4	73.5
1400	38.75	43.2	54.3	77.4
1500	41.75	46.15	57.2	81.0
1600	44.8	49.1	60.05	84.3
1700	47.95	52.05	62.85	87.4
1800	51.2	55.0	65.65	90.3
1900	54.6	58.0	68.4	92.75
2000	58.05	61.0	71.0	94.85
2100	61.5	64.0	73.65	96.65
2200	64.95	66.9	—	98.4
2300	68.4	69.9	—	100.1
2400	—	72.9	—	101.8
2500	—	—	—	103.4
2600	—	—	—	105.0

Note: Data are not corrected for thermal expansion of the specimens.

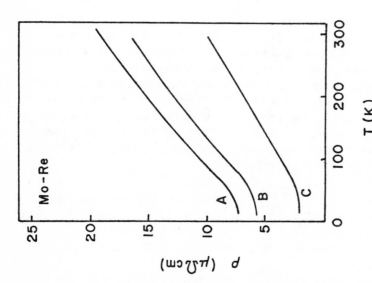

FIGURE 1.   Resistivity of Mo-Re alloy (in a/o Re). (A) 7.4; (B) 14.4; (C) 26.9.

REFERENCES

1. Hansen, M. and Anderko, K., *Constitution of Binary Alloys,* 2nd ed., McGraw-Hill, New York, 1958.
2. Cox, W. R., Hayes, D. J., and Brotzen, F. R., *Phys. Rev.,* 7, 3580, 1973.
3. Peletskii, V. E. and Druzhinin, V. P., *High Temp. USA,* 10, 584, 1972.
4. Gaines, G. B. and Sims, C. T., *Trans. ASTM,* 57, 759, 1957.
5. Arutyunov, A. V. and Filippov, L. P., *Teplofiz. Vys. Temp.,* 8(5), 655, 1970.

Mo-Te (Molybdenum-Tellurium)

Hughes and Friend[1] reported the resistivity of β-MoTe$_2$ between 10 and 300°K. In this material, the metallic atoms form a zig-zag chain. Figure 1 gives their experimental results.

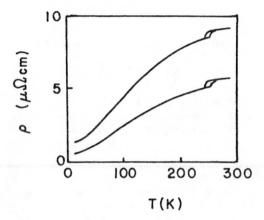

FIGURE 1. Temperature dependence of the two in-plane components of β-MoTe$_2$.

REFERENCES

1. Hughes, H. P. and Friend, R. H., *J. Phys. C,* 11, L103, 1978.

Mo-Ti (Molybdenum-Titanium)

β-Ti forms a continuous series of solid solutions with Mo, but the maximum solubility of Mo in α-Ti is of the order of a percent. Ho and Collings[1] prepared Ti-Mo alloys with 5 to 19.4 a/o Mo. These alloys were arc-melted and quenched. Most of the samples are multiphased ($\beta + \omega_1$, possibly even $\alpha' + \beta + \omega$); only the alloy with 19.4 a/o Mo should be a (metastable) single-phase bcc sample. The authors give a detailed analysis for the resistivity of the multiphased sample (see the section entitled "Introduction"). ϱ of the alloy with 19.4 a/o Mo after a 2-hr anneal at 1200°C and quenched in iced brine is 110 $\mu\Omega$cm at 4°K, 112 $\mu\Omega$cm at 78°K, and 115 to 118 $\mu\Omega$cm at 300°K. Hake et al.[2] obtained 97 $\mu\Omega$cm at 300°K and 88 $\mu\Omega$cm at 4°K for an alloy with 23.4 a/o Mo.

Berlincourt and Hake[3] prepared alloys from elements of usually better than 99.9% purity in an arc furnace. Figure 1 gives the normal state resistivity of the sample at 1.2°K, obtained by using very high current densities. The decrease in $\varrho(1.2°K)$, with increasing impurity concentration is typical for Group IV transition elements with impurities from Group V transition elements.

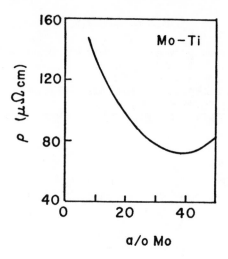

FIGURE 1.   Resistivity of Mo-Ti at 1.2°K.

REFERENCES

1. Ho, J. C. and Collings, E. W., *Phys. Rev. B*, 6, 3727, 1972.
2. Hake, R. R., Leslie, D. H., and Berlincourt, T. G., *Phys. Chem. Solids*, 90, 177, 1961.
3. Berlincourt, T. G. and Hake, R. R., *Phys. Rev.*, 131, 190, 1963.

Mo-W (Molybdenum-Tungsten)

Mo and W form a complete series of solid solutions.[1] Vertogradskii and Chekhvskoi[2] prepared W-Mo alloys with 34.4 and 86.5 a/o W by powder metallurgical methods from ≥99.95% W and ≥99.95% Mo. Their results, together with older data,[3-5] are shown in Figure 1. They found that $\varrho(34.4 \text{ a/o W}) = 0.02516 + 0.2094 \times 10^{-3}T + 0.0328 \times 10^{-6}T^2$ $\mu\Omega$m and $\varrho(86.5 \text{ a/o W}) = 0.01537 + 0.2240 \times 10^{-3}T + 0.0306 \times 10^{-6}T^2$ $\mu\Omega$m in the temperature range from 1200 to 2800°K. T is in °K. Reference 2 also measured the temperature coefficient of Mo-W alloys.

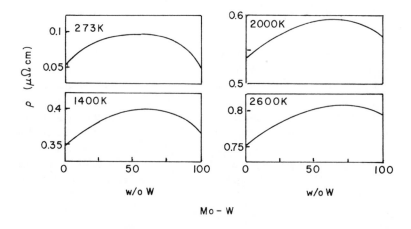

FIGURE 1.   Resistivity isotherms of Mo-W alloys.

REFERENCES

1. Geiss, W. and van Liempt, J. A. M., *Z. Anorg. Chem.*, 128, 335, 1923.
2. Vertogradskii, V. A. and Chekhovskoi, V. Ya, *High Temp. High Pres.*, 4, 1972; *High Temp. High Pres.*, 3, 6, 1971.
3. Kieffer, R. and Hotop, W., *Pulvermetallurgie and Sinterwerskstoffe,* Springer-Verlag, Berlin, 1943.
4. Gladkov, A. S., Amosa, V. M., Kopetskii, Ch., and Levin, H. V., *Metals and Alloys for Vacuum Tube Devices,* Energiya, Moscow, 1969.
5. Bossart, P. N., *Physics,* 7, 50, 1936.

SODIUM (Na)

Na-Pb (Sodium-Lead)

Hansen and Anderko[1] report practically no solubility of Pb in solid Na. There exist four intermediate phases ($\sim$20 to 22 a/o Pb near $Na_{15}Pb_4$, $\sim$27 to 30 a/o Pb near Na_5Pb_2, 47 to 50 a/o Pb near NaPb, and a β-phase with $\sim$67 to 73 a/o Pb) and possibly a Na_2Pb compound. Up to 12 a/o Na will dissolve in solid Pb at 307°C; 1.8 a/o Na will dissolve in Pb at 0°C.

Klaiber[2] used "Kahlbaum" "pro analysis" Pb and "Pro analysis" Na for his alloys. Samples were melted under Ar and homogenized in an evacuated glass tube for 5 days at 270°C. Table 1 and Figure 1 give the composition dependence of the electrical conductivity.

Durkin and Aleksandrov[3] gave a critical account of pseudopotential calculations and the resistivity increase due to 1% impurity. They gave $[\Delta\varrho/\Delta c]_{exp} = 3.88$ $\mu\Omega$cm/(a/o)Na in PbNa.

Table 1
ELECTRICAL CONDUCTIVITY IN UNITS OF 10^4 Ω^{-1} cm^{-1} OF Na-Pb ALLOYS

Na (w/o)	Na (a/o)	50	100	150	200	250	300
0.0	0.0	4.32	3.68	3.17	2.76	2.43	2.13
0.22	2.0	3.12	2.79	2.47	2.19	1.99	1.80
0.5	3.9	2.45	2.15	1.97	1.80	1.66	1.54
0.67	5.7	2.709	2.14	1.70	1.58	1.49	1.40
0.88	7.4	2.48	2.11	1.67	1.40	1.34	1.29
0.98	8.2	2.38	2.05	1.64	1.36	1.29	1.23
1.24	10.2	2.44	1.96	1.50	1.21	1.16	1.11
1.63	12.8	2.35	1.95	1.62	1.28	1.08	1.06
1.82	14.3	2.27	1.92	1.57	1.29	1.09	1.02
2.27	17.3	2.19	1.86	1.57	1.32	1.12	0.93
2.68	19.9	2.10	1.79	1.53	1.29	1.11	0.94
3.35	23.7	2.06	1.80	1.56	1.38	1.22	1.05
3.48	24.5	2.03	1.76	1.56	1.36	1.21	1.03
3.68	25.6	1.98	1.74	1.52	1.33	1.18	1.00
3.91	26.8	1.67	1.51	1.38	1.26	1.14	1.00
4.04	27.5	1.36	1.27	1.19	1.12	1.05	0.96
4.20	28.3	1.26	1.19	1.13	1.07	1.01	0.93
4.46	29.6	1.10	1.04	1.00	0.96	0.916	0.86
4.64	30.5	1.02	0.98	0.95	0.91	0.87	0.82
4.84	31.4	0.94	0.91	0.87	0.84	0.81	0.79
5.08	32.5	0.79	0.75	0.72	0.69	0.67	0.70
5.27	33.4	0.77	0.74	0.71	0.68	0.66	0.68
6.24	37.5	0.595	0.57	0.54	0.51	0.50	0.52

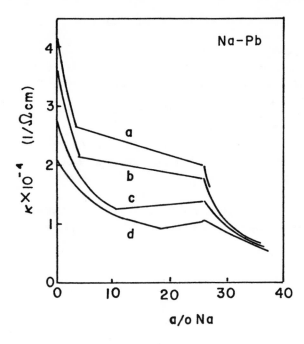

FIGURE 1.   Conductivity of Na-Pb alloys (in °C). (a) 50; (b) 100; (c) 200; (d) 300.[2]

REFERENCES

1. Hansen, M. and Anderko, K., *Constitution of Binary Alloys,* 2nd ed., McGraw-Hill, New York, 1958.
2. Klaiber, H., *Z. Elektrochem.*, 42, 258, 1936.
3. Dukin, V. V. and Aleksandrov, B. N., *Pseudopotential Calculations of the Residual Resistivities of Dilute Solid Alloys Based on Normal Metals, Physics of Low Temperatures,* Akademia Nauk Uk SSR Academy of Science, USSR, Kharkov, 1978, 1.

Na-Sn (Sodium-Tin)

Dukin and Aleksandrov[1] gave a critical account of pseudopotential calculations and the resistivity increase due to 1% impurity. They gave $[\Delta\varrho/\Delta c]_{exp} = 3.18$ $\mu\Omega cm/(a/o)$Na in $\underline{Sn}$Na.

REFERENCES

1. Dukin, V. V. and Aleksandrov, B. N., *Pseudopotential Calculations of the Residual Resistivities of Dilute Solid Alloys Based on Normal Metals, Physics of Low Temperatures,* Akademia Nauk Uk SSR Academy of Science, USSR, Kharkov, 1978, 1.

Na-Tl (Sodium-Thallium)

Grube and Schmidt[1] studied the phase diagram and recrystallization of Na-Tl. They used "pro analysis" Na and Tl of 99.95% purity. Figure 1 gives the concentration dependence of the conductivity for temperatures from 60 to 220°C. The single-phase regions extend from pure Tl to about 15 a/o Na at room temperature and to 30 a/o at 200°C (α-Tl). The γ-phase is stable from 50 to 56 a/o Na at room temperature and 47 to 53 a/o Na at 200°C. Solid Na can absorb only a small amount of Tl.

More recently Hansen and Anderko[2] report that up to 1.1 a/o Tl may dissolve in solid Na at 63.9°C. They list as stable the compounds Na_6Tl, Na_2Tl_1, and $NaTl_2$. An intermediate phase extends from 41.7 a/o Tl at 154°C to 53.8 a/o Tl at 238°C around NaTl.

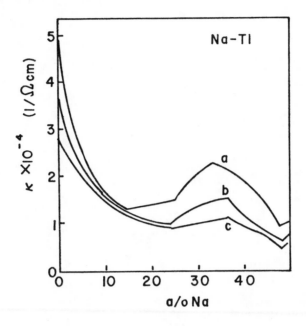

FIGURE 1. K of Na-Tl (in °C). (a) 60; (b) 160; (c) 220.

REFERENCES

1. Grube, G. and Schmidt, A., *Z. Elektrochem.*, 42, 201, 1936.
2. Hansen, M. and Anderko, K., *Constitution of Binary Alloys,* 2nd ed., McGraw-Hill, New York, 1958.

NIOBIUM (Nb)

Nb-Ni (Niobium-Nickel)

The Nb-rich end of the Ni-Nb phase diagram has not been investigated in any great detail and remains relatively undetermined. The Ni-rich end is known at 900°C and above. At 900°C a primary phase field exists which is capable of dissolving 7 a/o Nb in Ni. Two intermediate phases are apparently stable in a limited homogeneity range around the compositions Ni-25 a/o Nb and Ni-50 a/o Nb.

The electrical resistivity of the NiNb primary-phase alloys as a function of temperature for alloys containing up to 5.8 a/o Nb is shown in Figure 1. The bend in the curve is associated with the Curie temperature and becomes less apparent as the Nb concentration increases.

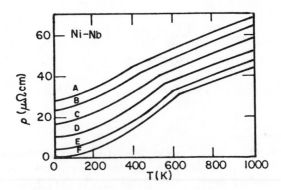

FIGURE 1.   Measured electrical resistivities of Ni and Ni-Nb alloys as a functon of temperature. (A) 5.8 a/o Nb; (B) 4.32 a/o Nb; (C) 3.20 a/o Nb; (D) 1.92 a/o Nb; (E) 0.63 a/o Nb; (F) pure Ni.[3]

REFERENCES

1. Durand, J. and Gautier, G., *J. Phys. Chem. Solids,* 31, 2773, 1970.
2. Durand, J., Thesis, Strausbourg, 1973.
3. Arajs, S., *J. Appl. Phys.* 32, 97, 1961.

Nb-Sn (Niobium-Tin)

Woodard and Cody[1] measured the resistivity of several Nb_3Sn samples from 18 to 850°K. Samples were obtained by chemical reduction from chlorides of Nb and Sn. Impurities are estimated to be less than 0.2%. This high purity leads to a narrow transition width to the superconducting state at 0.03°K. This is a factor of ten better than with samples prepared by the unsual metallurgical techniques.

Figure 1 shows the temperature dependence of the resistivity. It can be described with the equation $\varrho = \varrho_o + \varrho_1 T + \varrho_2 \exp(-T_o/T)$ to better than 1%.

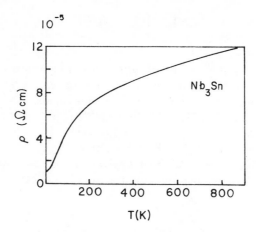

FIGURE 1.   Resistivity of Nb₃Sn.

REFERENCES

1. Woodard, D. W. and Cody, G. D., *Phys. Rev. A*, 136, 166, 1964.

Nb-Ta (Niobium-Tantalum)

Bückel reports that Nb-Ta forms a complete series of solid solutions.[1] Ul'yanov and Tarasov[2] prepared Nb-Ta samples by arc-melting. Nb (98.7% pure) was vacuum refined before arc-melting the alloys. The castings were rolled into bars at 110 to 1350°C and then vacuum annealed for 40 hr at 1400°C. Figure 1 gives $\varrho(T)$ values.

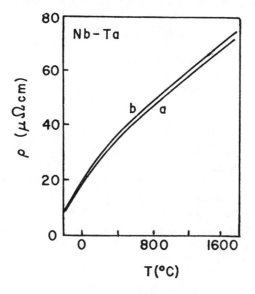

FIGURE 1.   Resistivity of N̲b̲Ta with (a) 2.63 and (b) 5.4 a/o Ta.

REFERENCES

1. Bückel, H., *Z. Metallkd.*, 37, 53, 1946.
2. Ul'yanov, R. A. and Tarasov, N. D., *Phys. Met. Metallogr.*, 17(2), 60, 1964.

Nb-Ti (Niobium-Titanium)

β-phase Ti and Nb form a complete series of solid solution.[1] The solubility of Nb in α-Ti is about 2 a/o Nb at 600°C. The two-phase region extends from about 1.5 a/o Nb to about 18 a/o Nb at 700°C. β can transform at high cooling rates in a martensitic transformation to α'.

Berlincourt and Hake[2] prepared alloys from elements of usually better than 99.9% purity in an arc furnace. Figure 1 gives the normal state resistivity of the sample at 1.2°K. The decrease in $\varrho(1.2°K)$ with increasing impurity concentration is typical for transition group IV elements with impurity transition elements of group V.

The resistivity of Nb-Ti alloys at $T < 0.1\ T_p$ is given by the equation:[3-5]

$$\rho = \rho_0 + \alpha\,(T/T_p)^3 - \beta\rho_0^2[\ln(T/T_c)]^{-1/2}$$

ϱ_0 is the residual resistivity. the second term is due to s-d scattering. The last term is due to fluctuation effects of the superconducting order parameter above the bulk critical temperature, T_c. β depends on the superconducting coherence length: $\beta = 0.037\ e^2\hbar^1\,(\ell\xi)^{-1/2}$; ℓ is the mean free path for electrons.

Nb-Ti alloys with a composition close to 45 a/o Ti are produced commercially for superconducting magnets. Annealed alloys (they recrystallize by annealing at 800°C) are subjected to deformation. Typically, a 4-cm-diameter rod is extended to 1.55-cm diameter, heat treated for 48 hr at 375°C, and then drawn to a diameter of about 1 cm. This material is then referred to as processed.

Figure 2 gives the resistivity of these alloys.[6] A metallographic study indicated that an annealed Nb-55 w/o Ti alloy strained at 4 and 76°K revealed some lattice transformations, possibly a martensitic tye.

Morton et al.[3] studied the thermal and electrical resistivity of $Nb_{35}Ti_{65}$ alloys which are superconductors. One sample was tested in the as-received state (as-rolled). The other annealed in vacuum at 550°C. Figure 3 gives the experimental results. Novikov et al.[7] determined $\varrho(T)$ of Ti-rich Ti-Nb alloys below 10°K.

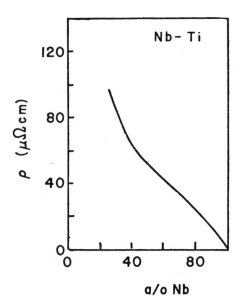

FIGURE 1.   Resistivity of Nb-Ti at 1.2°K.

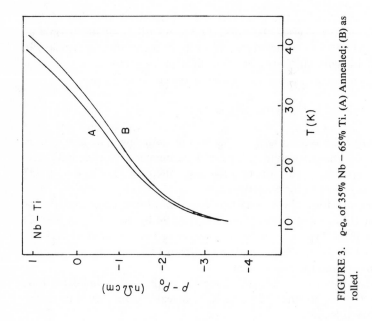

FIGURE 3. $\varrho-\varrho_o$ of 35% Nb – 65% Ti. (A) Annealed; (B) as rolled.

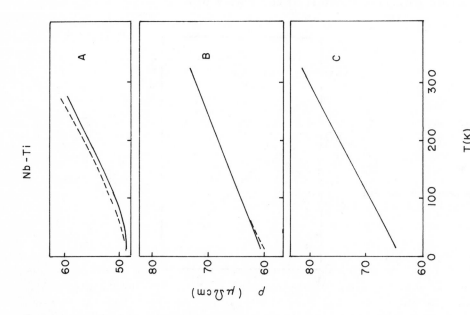

FIGURE 2.   Resistivity of Nb-Ti alloys; ———, processed; ———, annealed (in w/o Ti). (A) 34; (B) 45; (C) 55.

REFERENCES

1. Hansen, M. and Anderko, K., *Constitution of Binary Alloys,* 2nd ed., McGraw-Hill, New York, 1958.
2. Berlincourt, T. G., and Hake, R. R., *Phys. Rev.,* 13(1), 140, 1963.
3. Morton, N., James, B. W., Wostenholm, G. H., and Nichols, R. J., *J. Phys. F,* 5, 85, 1975.
4. Morton, N., James, B. W., Wostenholm, G. H., and Nattal, S., *J. Phys. F,* 5, 2098, 1951.
5. Feher, A., Reiffers, M., Petrovic, P., and Janos, S., *Phys. Status Solidi A,* 39(1), K67, 1977.
6. Read, D. T., *Cryogenics,* 18, 580, 1978.
7. Novikov, I. I., Borzyak, A. N., Boyarskii, S. V., Pepeshkin, Yu. D., Mikhailov, I. B., and Pishunov, A. N., *Sov. Phys. Dokl.,* 21(7), 404, 1976.

Nb-W (Niobium-Tungsten)

Bückel[1] reports that Nb and W form a continuous series of solid solutions. Ul'yanov and Tarasov[2] prepared Nb-W samples by arc-melting. Nb (98.7% pure) was vacuum refined before melting the alloy. The castings were rolled into bars at 1100 to 1350°C and then vacuum annealed for 40 hr at 1400°C. Figure 1 gives ϱ(T) values.

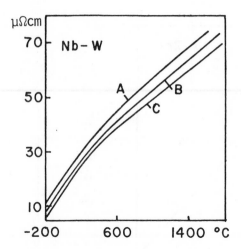

FIGURE 1.   Resistivity of Nb-W alloy. (A) 8.18 a/o W; (B) 5.3 a/o W; (C) pure Nb.

REFERENCES

1. Bückel, H., *Z. Metallkd.,* 37, 53, 1946.
2. Ul'yanov, R. A. and Tarasov, N. D., *Phys. Met. Metallogr.,* 17(2), 60, 1964.

Nb-Zr (Niobium-Zirconium)

β-Zr and Nb form a complete series of solid solutions.[1] Up to 6.4 a/o Nb will dissolve in α-Zr at 610°C. The two-phase regime, $\alpha + \beta$, extends at this temperature from 6.4 to 87 a/o Nb.

Berlincourt and Hake[2] prepared Zr-Nb alloys from elements of usually better than 99.9% purity in an arc furnace. Figure 1 gives the normal state resistivity at 1.2°K. ϱ(1.2°K) is a nearly linear function of composition for single-phase alloys.

Corsan et al.[3] studied the superconducting properties of single-phase and multiphase Zr-Nb alloys over the complete composition range. The samples were prepared from electron beam prepared Nb and iodized Zr by melting components in an Ar furnace. The buttons were homogenized at 1600°C in vacuo and forged. This followed an anneal at 585°C in a vacuum of 2×10^{-5} Torr for 5 weeks. Figure 2 gives the electrical resistivity in the normal state above T_c, where T_c is the transition temperature to the superconducting state. The exact range for the two-phase region is not completely clear. Zr may absorb 0.5 to 2 a/o Nb, not 6 a/o Nb as originally reported. The two-phase region extends to 88 a/o Nb. The high ϱ-value near 75 a/o Nb is difficult to explain, since in two-phase regions ϱ depends on the shape, size, and orientation of the phases.

Evans and Erickson[4] measured the resistivity of Nb-Zr alloys from the superconducting state to 273°K. Their resistivity curves are shown in Figure 3.

Morton et al.[5] measured the resistivity of Nb and a $Nb_{75}Zr_{25}$ alloy between the superconducting critical temperature and room temperature. The data were analyzed with phenomenological equations, and it was concluded that phonon induced s-d transitions are largely responsible for the observed resistivity. Figure 4 gives experimental results which show that $\varrho(T)$ depends markedly on the sample state.

Rogers and Atkins[6] measured ϱ (room temperature) of Nb-Zr alloys.

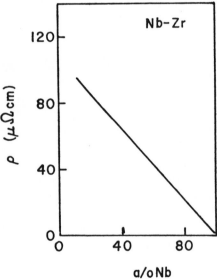

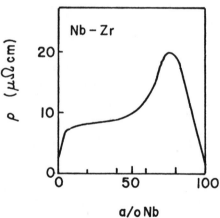

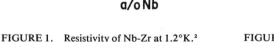

FIGURE 1.   Resistivity of Nb-Zr at 1.2°K.[2]

FIGURE 2.   Resistivity of Nb-Zr. Heat treatment; 585°C for 5 weeks.[3]

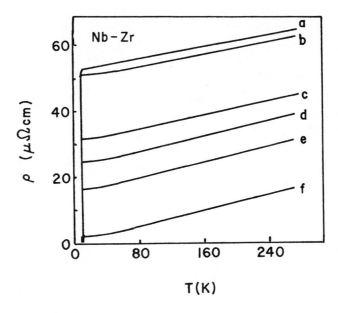

FIGURE 3. Resistivity of Nb-Zr. (a) 50 a/o Zr; (b) 50 a/o Zr; (c) 33 a/o Zr; (d) 25 a/o Zr; (e) 16 a/o Zr; (f) pure Nb.[5]

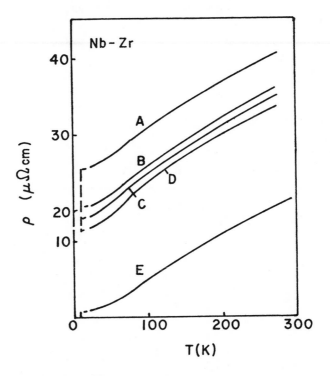

FIGURE 4. Resistivity of $Nb_{75}Zr_{25}$. (A) As rolled; (B) aged for 10 min at 800°C; (C) aged for 1 hr at 800°C; (D) aged for 5 hr at 800°C; (E) pure Nb.[5]

REFERENCES

1. Hansen, M. and Anderko, K., *Constitution of Binary Alloys,* 2nd ed., McGraw-Hill, New York, 1958.
2. Berlincourt, T. G. and Hake, R. R., *Phys. Rev.,* 131, 140, 1963.
3. Corsan, J. M., Williams, I., Cotterall, J. A., and Cook, A. J., *J. Less Common Met.,* 15, 437, 1968.
4. Evans, D. J. and Erickson, R. A., *J. Appl. Phys.,* 36, 11, 3517, 1965.
5. Morton, N., James, B. W., Wostenholm, G. H., and Nichols, R. J., *J. Phys. F,* 5, 85, 1975.
6. Rogers, B. A. and Atkins, D. F., *J. Met.,* 7, 1034, 1955.

NICKEL (Ni)

Ni-Pd (Nickel-Palladium)

Ni and Pd crystallize in a complete series of solid solution alloys with no evidence of ordering effects. Except for compositions near pure Pd, the alloys are ferromagnetic with a Curie temperature that is above 250°C for all alloys in the composition range between 0 and 80 a/o Pd.

The composition dependence of the resistivity of the Pd-Ni alloys at various temperatures is mildly reminiscent of Nordheim's rule (see Figure 1), while the temperature dependence of the resistivity below 350°K for various Ni-Pd alloys is moderately well behaved and relatively consistent with each other except for the Ni-82 (a/o) Pd alloy (see Figure 2). This alloy is near the critical composition at which ferromagnetic behavior occurs at low temperatures.

The resistivity of the dilute N̲iPd alloys has also been determined. The residual resistivity is concentration dependent, though the coefficient $d\varrho$ (NiPd)/dc = 2 $\mu\Omega$cm/(a/o)Pd and the DMR is practically zero at room temperature. The residual resistivity of the dilute P̲dNi alloys as a function of concentration gives $d\varrho/dc$ = 0.35 $\mu\Omega$cm(a/o)Ni. The temperature dependence of the resistivity of the dilute P̲dNi alloys whose compositions are just above and just below the critical composition (where the Curie temperature is 0°K) is also shown in Figure 3.

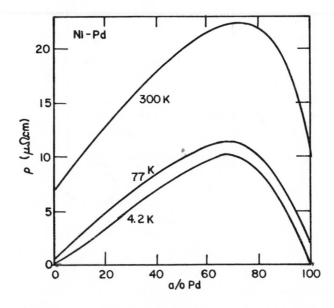

FIGURE 1.   The electrical resistivity as a function of composition for the Ni-Pd alloys at 4.2, 77, and 300°K.[1]

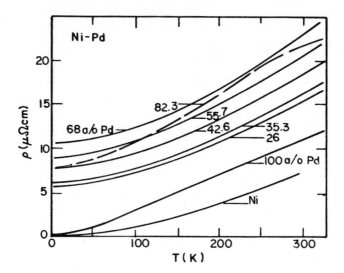

FIGURE 2.   The electrical resistivity as a function of temperature and composition for the Ni-Pd alloys. Numbrs give Pd(a/o).[1]

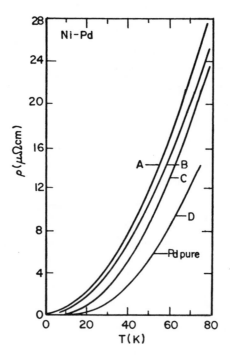

FIGURE 3.   The temperature dependence of the electrical resistivity, ϱ_T, which is the measured resistivity minus the residual resistivity. (A) 2.9 a/o Ni; (B) 3.1 a/o Ni; (C) 4.3 a/o Ni; (D) pure Pd.

REFERENCES

1. Schindler, A. I., Smith, R. J., and Salkovitz, E. I., *J. Phys. Chem. Solids,* 1, 39, 1956.
2. Dressen, J. A. and Pugh, E. M., *Phys. Rev.,* 120, 1218, 1960.
3. Grieg, D. and Rowlands, J. A., *J. Phys. F.,* 4, 232, 1974.

4. Farrell, T. and Grieg, D., *J. Phys. C*, 1, 1359, 1968.
5. Durand, J. and Gautier, F., *J. Phys. Chem. Solids*, 31, 2773, 1970.
6. Schwerer, F. C. and Cuddy, L. J., *Phys. Rev. B*, 2, 1575, 1970.
7. Schriempf, T. T., Schindler, A. I., and Mills, D. L., *Phys. Rev.*, 187, 959, 1969.
8. Kaiser, A. B. and Doniach, J., *Int. J. Magn.*, 1, 11, 1970.
9. Schindler, A. I. and Rice, M. J., *Phys. Rev.*, 164, 759, 1967.
10. Overhauser, A. W. and Schindler, A. I., *J. Appl. Phys.*, 28, 544, 1957.
11. Leger, H. G. and Muir, W. R., *Phys. Status Solidi B*, 49, 659, 1972.

Ni-Pt (Nickel-Platinum)

Ni and Pt mix completely in the solid state to form a continuous series of solid solutions. The Ni_1, Ni_1Pt_1 and Ni_3Pt_1 compositions apparently form ordered structures below 600°C. The Ni-rich alloys containing up to 35 a/o Pt are ferromagnetic at room temperature.

The dilute NiPt alloys are characterized by a concentration coefficient of residual resistivity of $d\varrho(NiPt)/dc = 1.0 \ \mu\Omega cm/(a/o)Pt$ and a large DMR at room temperature.[4]

REFERENCES

1. Oriani, R. A. and Jones, T. S., *Acta Metall.*, 1, 243, 153.
2. Dorleijn, J. W. F. and Meidema, A. R., *J. Phys. F*, 5, 487, 1975.
3. Dorleijn, J. W. F., *Philips Res. Rep.*, 31, 287, 1976.
4. Durand, J. and Gautier, *J. Phys. Chem. Solids*, 31, 2773, 1970.
5. Farrell, T. and Grieg, D., *J. Phys. C*, 1, 1359, 1968.

Ni-Rh (Nickel-Rhodium)

The Ni-Rh phase diagram is not completely established, but it appears that the solid Ni lattice can accomodate large amounts of Rh before the solubility limit is exceeded.

The electrical resistivity as a function of temperature was determined for NiRh alloys whose compositions are near the critical value (63 a/o Ni) which can sustain ferromagnetic behavior (Figure 1). The resistivity is well behaved and shows no analogous behavior associated with the Curie temperature or polarization clouds that characterize these particular alloys.

The resistivity of concentrated NiRh alloys at low temperatures (<100°K) is found[2] to obey a power law dependency on temperature of the form $\varrho(T) - \varrho_o = AT^n$, where the values of A and n are dependent on the composition of the alloy. The dilute NiRh alloys are characterized by both a concentration-residual resistivity coefficient[4] of K NiRh = $1.8 \ \mu\Omega cm/(a/o)$ Rh and a DMR that is appreciable at room temperature.[5]

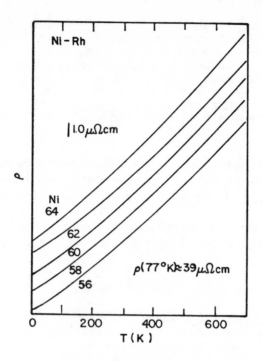

FIGURE 1.   Resistivity of Ni-Rh alloys as a function of temperature from 2 to 700°K. The five curves
have been separated vertically for clarity. The resistivity of all alloys is ~39 μΩcm at 77°K. The numbers give Ni-concentration in a/o.[1]

REFERENCES

1. Houghton, R. W., Sarachik, M. P., and Kouvel, J. S., *Solid State Commun.*, 10, 369, 1972.
2. Amamou, A., Gautier, F., and Loegel, B., *J. Phys. F*, 5, 1342, 1975.
3. Dorleijn, J. W. F. and Miedema, A. R., *J. Phys. F*, 5, 487, 1975.
4. Dorleijn, J. W. F., *Philips Res. Rep.*, 31, 287, 1976.
5. Durand, J. and Gautier, *J. Phys. Chem. Solids*, 31, 2773, 1970.
6. Durand, J., Thesis, Strasbourg, 1973.
7. Cadeville, M. C. and Durand, J., *Solid State Commun.*, 6, 399, 1968.

Ni-Ru (Nickel-Ruthenium)

The Ni-Ru phase diagram is a simple peritectic characterized by primary solid solution alloys that have large solute solubilities at elevated temperatures. At lower temperatures, the solubility of Ni in Ru and Ru in Ni is for both about 5%. The alloys
with compositions between N-5%Ru and Ni-95%Ru are two-phase mixtures of these
two primary phases.

The temperature dependence of the resistivity of Ni-Ru alloys in the vicinity of the
20 a/o Ru composition have been investigated at temperatures below 100°K and are
found to obey the power law relationship:

$$\rho(T) - \rho_0 = AT^n$$

where the values A and n are dependent on the composition of the alloy.

The residual resistivity of the dilute NiRu alloy is linear to 5 a/o Ru concentration
(see Figure 1) and the associate coefficient is $K_{NiRu} = 2$ μΩcm/(a/o)Ru. The DMR for
dilute NiRu alloys is large at room temperature.

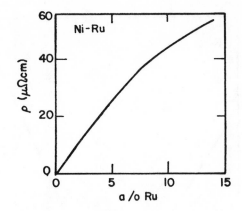

FIGURE 1.   Residual resistivity of N̲iRu alloys
vs. Ru concentration.

REFERENCES

1. Amamou, A., Gautier, F., and Loegel, B., *J. Phys. F*, 5, 1342, 1975.
2. Dorleijn, J. W. F. and Miedema, A. R., *J. Phys. F*, 5, 487, 1975.
3. Dorleijn, J. W. F., *Philips Res. Rep.*, 31, 287, 1976.
4. Durand, J. and Gautier, F., *J. Phys. Chem. Solids*, 31, 2773, 1970.
5. Cadeville, M. C. and Durand, J., *Solid State Commun.* 6, 399, 1968.

Ni-Sb (Nickel-Antimony)

A primary solid solution phase field exists in the Ni end, but not in the Sb end of the Ni-Sb phase diagram. Three narrow intermediate phases and one phase with a large homogeneity range of compositions centered around Ni_1Sb_1 are apparently stable at 25°C.

The DMR has been evaluated[1] for the dilute N̲iSb system and the deviation becomes consistently larger as the temperature increases to 300°K. The resistivity at room temperature has been determined over the homogeneity range of compositions associated with the Ni_1Sb_1 intermediate phase (see Figure 1).

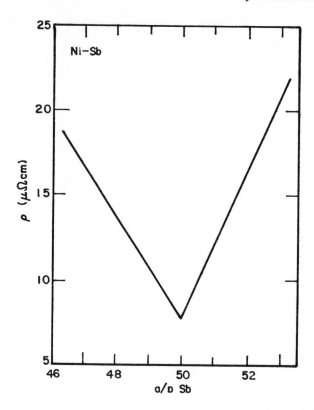

FIGURE 1.   Reistivity vs. Sb content of Ni-Sb alloys.[2]

REFERENCES

1. Ross, R. N., Price, D. C., and Williams, G., *J. Phys. F*, 8, 2367, 1978.
2. Penn, J. and Miller, E., *J. Appl. Phys.*, 44, 177, 1973.
3. Makarov, E. S., *Dokl. Akad. Nauk SSSR*, 40, 191, 1943.

Ni-Si (Nickel-Silicon)

The Ni-Si phase diagram shows that six intermediate phases, Ni_3Si_1, Ni_5Si_2, Ni_2Si_1, Ni_3Si_2, Ni_1Si_1, and Ni_1Si_2, are stable in very restricted composition ranges at 25°C. At 25°C, a primary solid solution region with a 10a/oSi solubility limit exists at the Ni end of the diagram, while no primary solid solution phase appears at the Si end.

The resistivity of the N̲iSi primary phase solid solution alloys has been investigated.[1-4] The concentration coefficient of residual resistivity is reported to be 2.8 $\mu\Omega$cm/(a/o) Si. The temperature dependence of the resistivity for various alloy compositions in the Ni primary solid solution phase field (see Figure 1) shows the expected "knee" in the resistivity at the Curie temperature of each alloy.

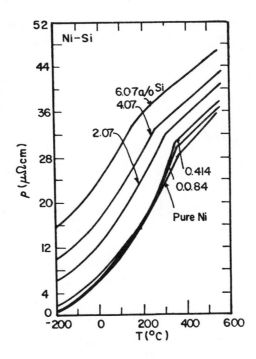

FIGURE 1. Resistivity of Ni-Si alloys. Numbers give a/o Si.[3]

REFERENCES

1. Dorleijn, J. W. F. and Miedema, A. R., *J. Phys. F*, 5, 487, 1975.
2. Dorleijn, J. W. F., *Philips Res. Rep.*, 31, 287, 1976.
3. Dominicali, C. A. and Otter, F. A., *J. Appl. Phys.*, 26, 377, 1955.
4. Arajs, S., *Z. Metallkd.*, 58(6), 263, 1967.

Ni-Sn (Nickel-Tin)

At 25°C, Ni and Sn will form three intermediates phases, Ni_3Sn_1, Ni_3Sn_2, and Ni_3Sn_4. No apparent primary solid solution phases exist at either the Ni or Sn ends of the diagram. The Ni_3Sn_1 and Ni_3Sn_2 phases both order as temperature decreases. The bulk of the Ni-Sn compositions are two-phase alloys at 25°C.

Resistivity measurements have been predominately confined[1-6] to the very dilute NiSn alloy, although some recent work has been reported on Ni_3Sn_2. The concentration coefficient of the residual resistivity is $d\rho/dc = 3$ $\mu\Omega$cm/(a/o)Sn. The DMR for NiSn was found to increase with temperature to 300°K.

The temperature dependence of the resistivity of the intermediate Ni_3Sn_2 phase has been measured after numerous zone melting passes. The resistivities are large and strongly dependent on the number of zone passes (see Figure 1).

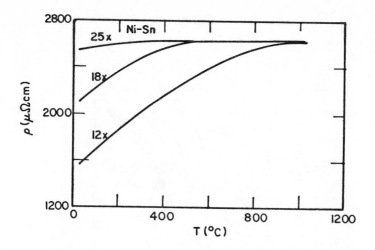

FIGURE 1.   Temperature dependence of the electrical resistivity of zone refined Ni_3Sn_2 as a function of the number of zone passes.[1]

REFERENCES

1. Ross, R. W., Price, D. C., and Williams, G., *J. Phys. F*, 8, 2367, 1978.
2. Dorleijn, J. W. F. and Miedema, A. R., *J. Phys. F*, 5, 487, 1975.
3. Dorleijn, J. W. F., *Philips Res. Rep.*, 31, 287, 1976.
4. van Elst, H. C., *Physica*, 25, 708, 1959.
5. Smit, J., *Physica*, 25, 708, 1959.
5. Smit, J. *Physica*, 21, 877, 1955.
6. Price, D. C. and Williams, G., *J. Phys. F*, 3, 810, 1973.
7. Wever, H. and Wintermann, N., *Z. Metallkd.*, 70, 93, 1979.

Ni-Ta (Nickel-Tantallum)

Ta is slightly soluble (~4% in solid Ni at 25°C and quite soluble (~12%) at 1200°C. An intermediate phase exists at Ta_1Ni_3. The Ta-rich end of the phase diagram is relatively obscure.

Resistivity measurements[1] of the NiTa primary solid solution alloys at 4.2, 77, and 297°K (see Figure 1) yield a concentration coefficient of residual resistivity $d\varrho/dc$ of $5.2 \ \mu\Omega cm/(a/o)Ta$. The DMR for NiTa alloys is relatively small at 297°K.[3]

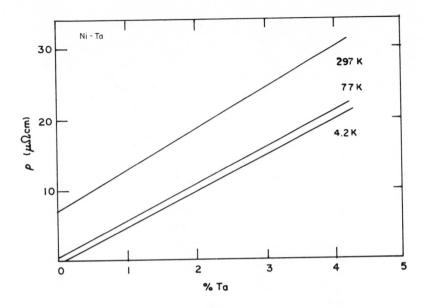

FIGURE 1.   Resistivity of N̲iTa alloys at selected temperatures.[1]

REFERENCES

1. Chessin, H., Arajs, S., and Colvin, R. V., *J. Appl. Phys.*, 35, 2419, 1964.
2. Durand, J., Thesis, Strasbourg, 1973.
3. Durand, J. and Gautier, F., *J. Phys. Chem. Solids*, 31, 2773, 1970.
4. Kornilov, I. I. and Pylaeva, E. N., *Iquest. Sektora, Fiz- Khim. Anal.*, 23, 110, 1953.

Ni-Ti (Nickel-Titanium)

At 500°C, Ni and Ti form two intermediate compounds, Ti_2Ni_1 and Ti_1Ni_3, and one intermediate phase with a moderately wide homogeneity range centered about the equiatomic composition. There is no apparent solid solubility of Ni in Ti, but modest solid solubility (∼8%) of Ti in Ni at 500°C. The resistivity of dilute NiTi alloys between 300 and 700°K (see Figure 1) show the characteristic rapid decline in the resistivity just below the Curie temperature.

The concentration dependence of the residual resistivity[1-5] (see Figure 2), where $d\varrho(FeTi)/dc = 3.0$ $\mu\Omega cm/(a/o)Ti$, and the temperature dependence of the DMR between 300 and 700°K (see Figure 3) have been evaluated for these primary phase NiTi alloys. The peak in the DMR occurs at the Curie temperature and can be employed as a technique for determining the Curie temperature.

The electrical resistivity of the Ti_1Ni_1 intermediate phase has been investigated as a function of temperature. The results (see Figure 4) frequently exhibit a thermal hysteresis that has been attributed to a martensitic transformation in this material.

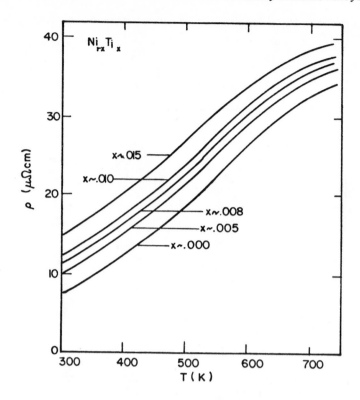

FIGURE 1.   Electrical resistivity of Ni-Ti alloys as a function of temperature between 300 and 750°K.[9]

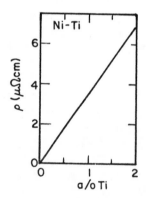

FIGURE 2.   Residual electrical resistivity of Ni-Ti alloys as a function of Ti concentration.[5]

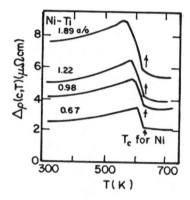

FIGURE 3. $\Delta\varrho(c,T)$ as a function of T for Ni-Ti alloys between 300 and 750°K.[5]

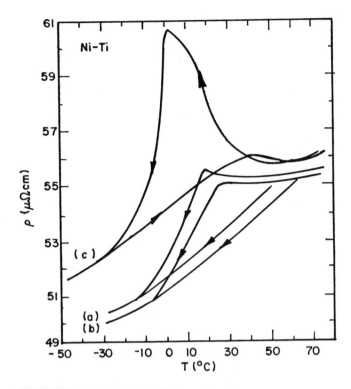

FIGURE 4. Effect of "incomplete" cycles on the electrical resistivity of TiNi (51 a/o Ni) at and around the "martensitic" transition temperature; cycling direction is indicated by arrows. (a) Without "incomplete" cycling; (b) after six "incomplete" cycles; (c) after several hundred "incomplete" cycles.[7]

REFERENCES

1. Chen, C. W., *Phys. Lett.*, 7, 16, 1963.
2. Dorleijn, J. W. F., *Philips Res. Rep.*, 31, 287, 1976.
3. Farrell, T. and Grieg, D., *J. Phys. C*, 1, 1359, 1968.
4. Fert, A. and Campbell, I. A., *Phys. Rev. Lett.*, 21, 1190, 1968; *J. Phys. F*, 6, 849, 1976.
5. Yao, Y. D. and Arajs, S., *Phys. Status Solidi*, 89, K201, 1978.
6. Dorleijn, J. W. F. and Miedema, A. R., *J. Phys. F*, 5, 487, 1975.
7. Wang, F. E., DeSavage, B. F., Beuhler, W. J., and Hosler, W. R., *J. Appl. Phys.*, 39, 2166, 1968.
8. Wayman, C. M., Cornelis, I., and Shimizu, K., *Scr. Metall.*, 6, 115, 1972.
9. Yao, Y. D., Arajs, S., and Rao, K. V., *Chin. J. Phys.*, 15, 1, 1977.

Ni-V (Nickel-Vanadium)

The solid Ni lattice can accomodate an appreciable amount of V ($\sim$14%) at 25°C and a large amount ($\sim$40 a/o) at 1200°C before a second phase precipitates. Several intermediate phases appear to exist in a phase diagram that is not completely established in the V-rich end.

The resistivity of the primary solid solution $\underline{Ni}$V alloys have been determined at low temperatures (<100°K) and were found to be consistent with a power law relationship $\varrho_T = \varrho_o + AT^n$, where A and n are constants that depend on the composition of the alloy.

The resistivity of the dilute $\underline{Ni}$V alloy has been actively investigated,[1-8] and the residual resistivity per unit V concentration is $d\varrho/dc = 4.4$ $\mu\Omega$cm/(a/o) V while the DMR at 25°C is relatively small.

REFERENCES

1. Ammamou, A., Gautier, F., and Loegel, B., *J. Phys. F,* 5, 1342, 1975.
2. Arajs, S., Chessin, H., and Colvin, R. V., *Phys. Status Solidi,* 7, 1009, 1964.
3. Farrel, T. and Grieg, D., *J. Phys. C,* 1, 1359, 1968.
4. Durand, J. and Gautier, F., *J. Phys. Chem. Solids,* 31, 2773, 1970.
5. Chen, C. W., *Phys. Lett.,* 7, 16, 1963.
6. Fert, A. and Campbell, I. A., *J. Phys.,* 32, C1-46-50, 1971; *J. Phys. F,* 6, 849, 1976.
7. Dorleijn, J. W. F., *Philips Res. Rep.,* 31, 287, 1976.
8. Dorleijn, J. W. F. and Miedema, A. R., *J. Phys. F,* 5, 487, 1975.

OSMIUM (Os)

Os-W (Osmium-Tungsten)

Raub and Walter[1] report that the solid solubility of Os in W is of the order of 5 a/o. About 48.5 a/o W is soluble in Os.[2] There exists also an intermediate phase of the composition W_3Os which may be the α-phase.

Alekseyeva et al.[3] studied ϱ of W-Os with up to 5 a/o Os at 77°K. Samples were prepared by arc-melting or plasma-melting. Figure 1 gives the composition dependence of ϱ. It is not quite clear which samples are single-phase systems.

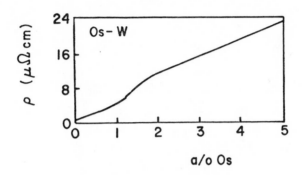

FIGURE 1.   Resistivity of Os-W alloys at 77°K.

REFERENCES

1. Raub, E. and Walter, P., *Festschrift aus Anlass des 100-jahrig Jubilaums der Firma,* W. C. Heraeus, Hanau, 1951.
2. Hansen, M. and Anderko, K., *Constitution of Binary Alloys,* 2nd ed., McGraw-Hill, New York, 1958.
3. Alekseyeva, L. I., Budagovskiy, S. S., Bykov, V. N., Kondakhchan, I. G., Povorova, K. P., Podolyan, N. I., and Savitskiy, Ye. M., *Metall. Metalloved.,* 40(5) 87, 1975.

LEAD (Pb)

Pb-Sb (Lead-Antimony)

The phase diagram[1] of Pb-Sb shows one eutectic reaction at $T = 252°C$. The solubility of Sb in solid Pb at this temperature is 5.8 a/o Pb; a few percent of Pb will dissolve in Sb. Resistivity measurements to study the phase diagram have been given by Hansen.

Aballe et al.[2] measured ϱ of supersaturated PbSb alloys with up to 3 a/o Sb. Samples were prepared from 99.97% pure Pb and 99.99% pure Sb. After casting, samples were soaked for 3 days at 240°C and slowly cooled. Samples were finally extruded between 120 to 180°C. A final anneal followed at 240°C for 1 hr.

Figure 1a shows ϱ(c) of the alloys, as measured at the indicated temperature from 20 to 110°C. The lowest curve gives ϱ(c) at 20°C after samples have been aged for 150°C. Figure 1b gives ϱ(T) for the alloys.

LeBlanc and Schopel[3] measured the conductivity of Pb-Sb alloys over the complete composition range between 20 and 240°C. Dukin and Aleksandrov[4] gave a critical account of pseudopotential calculations and the resistivity increase due to 1 a/o impurity. They gave $[\Delta\varrho/\Delta c]_{exp} = 1.18\ \mu\Omega\text{cm}/(a/o)Sb$ in PbSb.

Pb – Sb

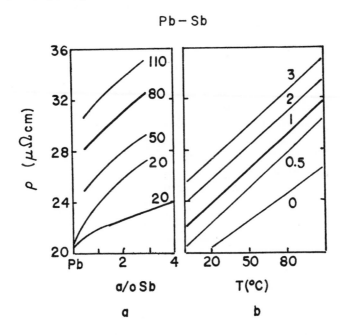

FIGURE 1. Resistivity of Pb-Sb alloys. (a) Number gives temperature in °C; bottom line, sample aged for 150 hr. All other samples in the "as quenched state". (b) Number gives (a/o)Sb.

REFERENCES

1. Raynor, G. V., *Annotated Equilibrium Phase Diagrams*, No. 9, The Institute of Metals, London, 1951.
2. Aballe, M., Regidore, J. J., Sistiaga, J. M.,and Toralba, M., *Z. Metallkd.*, 63(9), 565, 1972.
3. LeBlanc, M. and Schopel, H., *Z. Elektrochem.*, 39(8), 695, 1953.
4. Dukin, V. V. and Aleksandrov, B. N., *Pseudopotential Calculations of the Residual Resistivities of Dilute Solid Alloys Based on Normal Metals, Physics of Low Temperatures*, Akademia Nauk UK SSR Academy of Science, USSR, Kharkov, 1978, 1.

Pb-Sn (Lead-Tin)

The phase diagram[1] of the Sn-Pb system shows one eutectic reaction at 183°C, where Sn with 1.45 a/o Pb and Pb with 29 a/o Sn are in equilibrium with the liquid phase with 26.1 a/o Pb.

Resistivity measurements helped to investigate the phase diagram.[2-8] Cohen et. al[4] determined ϱ of P̲b̲Sn between −196 to +240°C (Pb, 99.9999% pure). $\Delta\varrho/c = 0.26$ $\mu\Omega$cm/(a/o)Sn was independent of temperature.

Pan et al.[10] prepared Sn alloys with 0.36, 0,72, and 1.1a/o Pb. Samples were smelted under Ar and homogenized close to the melting temperature. A quench in water followed. Table 1 gives $R(300°K)/R(4.2°K)$ of the samples. No explanation is given for the result that $R(300°K)/R(4.2°K)$ of the alloy with 1.1 a/o Pb is larger than that for the alloy with 0.72 a/o Pb.

Chuak et al.[11] measured ϱ in the Sn-Pb alloy system from 7 to 300°K. Samples were prepared by vacuum casting, swaging and annealing (1 to 8 hr at 150°), and slow cooling from 99.98% Sn and 99.97% Pb. The residual resistivity ration (RRR) and the resistivity of these samples were

	RRR	$\varrho(295°K)$
Sn pure	187.0 or 392	10.1 or 9.8
$Sn_{85}Pb_{15}$	49.3	11.2
$Sn_{70}Pb_{30}$	59.9	14.5
$Sn_{50}Pb_{50}$	61.5	16.0
$Sn_{30}Pb_{70}$	64.7	16.5
Pb	326.9	17.0

All alloys should be in the two-phase state. $\varrho(T)$ for these alloys is given in Figure 1; the residual resistivity is given in Figure 2. The temperature dependence of electrical resistivity could be given by $\varrho\alpha\ T^5$ for temperatures to 60°K.

Dukin and Aleksandrov[12] gave a critical account of pseudopotential calculations and the resistivity increase due to 1% impurity. They gave $[\Delta\varrho/\Delta c]_{exp} = 0.53$ $\mu\Omega$cm/(a/o)Pb in S̲n̲Pb and $[\Delta\varrho/\Delta c]_{exp} = 0.25$ $\mu\Omega$/(a/o)Sn in P̲b̲Sn.

Table 1

	$R(300°K)/R(4.2°K)$
Sn	1762
Sn + 0.36 a/o Pb	99
Sn + 0.72 a/o Pb	62
Sn + 1.1 a/o Pb	98

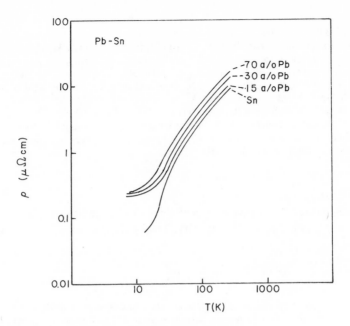

FIGURE 1.   Resistivity of Pb-Sn.

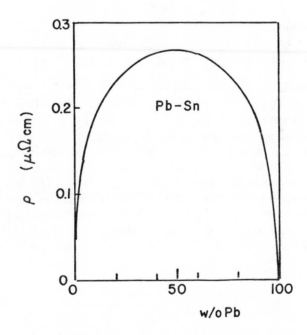

FIGURE 2.   Resistivity of Sn-Pb alloys.

REFERENCES

1. Hansen, M. and Anderko, K., *Constitution of Binary Alloys,* 2nd ed., McGraw-Hill, New York, 1958.
2. Jeffery, F. H., *Trans. Faraday Soc.,* 24, 209, 1928; 26, 588, 1930.
3. Stockburn, A., *J. Inst. Met.,* 66, 33, 1940.
4. Cohen, B. M., Turnbull, D., and Warburton, W. K., *Phys. Rev. B,* 16(6), 2491, 1977.
5. Matuyama, Y., *Sci. Rep. Tohoku Imp. Univ.,* 20, 661, 1931.

6. Parravano, N. and Scortecci, A., *Gazz. Chim. Ital.*, 50(2), 83, 1920.

7. Stockdale, D., *J. Inst. Met.*, 49, 267 (see also discussion), 1932.

8. Borelius, G., Larris, F., and Ohlsson, E., *Ark. Mat. Astron. Fys.*, 31A(10), 1944; Borelius, G., *Trans. AIME*, 191, 477, 1951.

9. Kurzyniec, E. and Wojtaszek, Z., *Bull. Int. Acad. Pol. Sci. Cl. Sci. Math. Nat. Ser. A*, 131, 1951.

10. Pan, V. M., Nikitin, B. G., Korostil, A. M., Nemoshkalenko, V. V., Dogopol, V. P., Dekhtyar, I. Ya., Nishchenko, M. M., and Takzei, G. A., *Sov. Phys. JETP*, 48, (2), 301, 1978.

11. Chuak, D. G. S., Ratnalingam, R., and Seward, R. J., *J. Low Temp. Phys.*, 31(1/2), 153, 1978.

12. Dukin, Y. A. and Aleksandrov, B. N., *Pseudopotential Calculations of the Residual Resistivities of Dilute Solid Alloys Based on Normal Metals, Physics of Low Temperatures,* Akademia Nauk Uk SSR Academy of Science, USSR, Kharkov, 1978, 1.

Pb-Te (Lead-Tellurium)

Dukin and Aleksandrov[1] gave a critical account of pseudopotential calculations and the resistivity increase due to 1% impurity. They gave $[\Delta\varrho/\Delta c]_{exp} = 3.36 \ \mu\Omega cm/(a/o)Te$ in PbTe.

REFERENCES

1. Dukin, V. V. and Aleksandrov, B. N., *Pseudopotential Calculations of the Residual Resistivities of Dilute Solid Alloys Based on Normal Metals, Physics of Low Temperatures,* Akademia Nauk UK SSR Academy of Science, USSR, Kharkov, 1978, 1.

Pb-Tl (Lead-Thallium)

The phase diagram[1] of Pb-Tl shows an extensive range of solid solution, extending from about 13 a/o Pb to pure Pb. Several atomic percent Pb will dissolve in α- and β-Tl.

Predel and Sandig[2] studied the resistivity of solid and liquid Pb-Tl alloys. Pb was of 99.999% purity; Tl was of 99.999% purity. Figure 1 gives $\varrho(T)$ of the alloys. Figure 2 gives $\varrho(c)$ for different temperatures. Included are results by Kurnakow and Zemczuzny[3] and Bridgman.[4] $\varrho(c)$ does not follow Nordheim's rule. $\varrho(c)$ shows a maxima at 20 a/o Tl, between 66 and 70 a/o Tl, and another maxima at 96.5 a/o Tl. $\varrho(c)_{min}$ are found near 25 and 60 a/o Tl for solid solutions and at 30 to 40 and 80 a/o Tl for liquid solutions. Cohen et al.[5] determined ϱ of PbTl between -196 and $+240°C$ (Pb is 99.9999% pure) and found $\Delta\varrho/c = 0.7 \ \mu\Omega cm/(a/o)Tl$, independent of temperature.

Dukin and Aleksandrov[6] gave a critical account of pseudopotential calculations and the resistivity increase due to 1% impurity. They gave $[\Delta\varrho/\Delta c]_{exp} = 1.34 \ \mu\Omega cm/(a/o)Pb$ in TlPb. Room temperature data of ϱ and the thermal coefficient α is given in Table 1.[7]

Pb-Tl

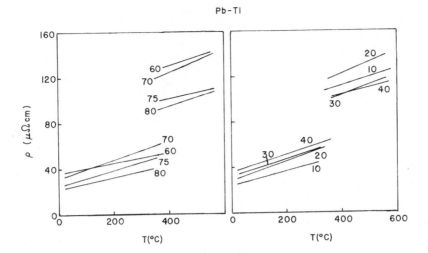

FIGURE 1. Resistivity of Pb-Tl alloys; numbers, (a/o)Tl.

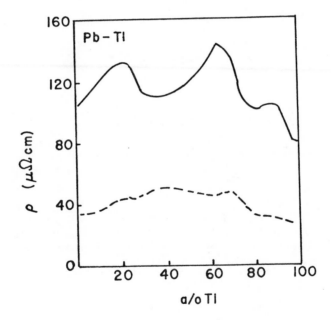

FIGURE 2. Resistivity of Pb-Tl alloys. Dashed line, 200°C; full lines, 500°C.

Table 1

Tl (w/o)	ϱ ($\mu\Omega$cm)	α $10^{-3}/°C$
0	16.6	4.27
3.45	19.15	3.67
4.5	19.26	3.62
7	20.8	3.32
14	26.8	2.51
22	33.35	1.942
32	37.8	1.655
42	40.1	1.523
54	39.2	1.499
63	36.6	1.526
67	34.2	1.629
70	31.9	1.707
75	28.3	1.884
77.5	25.7	2.14
80[a]	23.8	2.30
85[a]	20.8	2.68
89.9[a]	19.2	2.96
90.0[a]	18.9	3.03
93.5[a]	19.04	3.00
95.0[a]	18.13	3.19
96.5[a]	16.81	3.41
98.0[a]	15.50	3.68
99.5[a]	14.23	4.02
100.0[a]	13.80	4.16

[a]   β-form.

REFERENCES

1. Hansen, M. and Anderko, K., *Constitution of Binary Alloys,* 2nd ed., McGraw-Hill, New York, 1958.
2. Predel, B. and Sandig, H., *Z. Metallkd.,* 61, 267, 1970.
3. Kurnakow, N. and Zemczuzny, S., *Z. Anorg. Chem.,* 64, 149, 1909.
4. Bridgman, P. W., *Proc Am. Akad. Arts Sci.,* 84, 1, 1956.
5. Cohen, B. M., Turnbull, D., and Warburton, W. K., *Phys. Rev. B,* 16, 249, 1977.
6. Dukin, V. V. and Aleksandrov, B. N., *Pseudopotential Calculations of the Residual Resistivities of Dilute Solid Alloys Based on Normal Metals, Physics of Low Temperatures,* Akademia Nauk Uk SSR Academy of Science, USSR, Kharkov, 1978, 1.
7. *International Critical Tables,* Vol. 6, McGraw-Hill, New York, 1929, 156.

PALLADIUM (Pd)

Pd-Pr (Palladium-Praseodymium)

Lethuillier[1] melted P̲d̲Pr (Pd, 99.99% pure; Pr, 99.9% pure) in an induction furnace, followed by a rapid quench. Figures 1 and 2 show $\varrho(T)$ of these samples. The different $\varrho(T)$ curves in Figure 1 were obtained on different sample sections, indicating that the sample was quite inhomogeneous. All $\varrho(T)$ curves show a minimum.

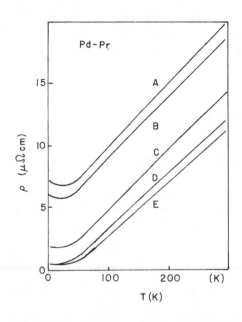

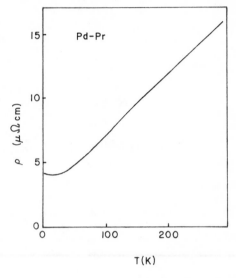

FIGURE 2.   Resistivity of $Pd_{98}Pr_2$.

FIGURE 1.   Resistivity of Pd and $Pd_{97}Pr_3$. (A to D) Pd + 3 a/o Pr; (E) Pd.

REFERENCES

1. Lethuillier, P., *J. Phys.,* 39, 1349, 1978.

Pd-Pt (Palladium-Platinum)

Pd and Pt form a continuous series of solid solutions as, e.g., electrical conductivity and the temperature dependence of the resistivity indicated in early measurements.[1,2]

March[3] prepared Pd-Pt alloys by induction or arc-melting 99.999% pure elements. After rolling and drawing, wires were homogenized at 1000°C and quenched. Figure 1 shows $\varrho(c)$ at 4.2°K and room temperature. The peak in $\varrho(c)$ is found at 50% Pt at 4.2°K, between 45 and 50% Pt at room temperature. The curves follow only approximately the rule that the impurity resistivity is proportional to $c[1 - c]$. The $\varrho(4.2°K)$ data agree well with results obtained by Blood and Greig,[4] who prepared Pd-Pt alloys by arc-melting. After machining, samples were annealed at about 1100°C for a minimum of 24 hr. The temperature dependence of $\varrho(T)$ is given in Table 1. The maximum $\varrho(c)_{4.2}$ value is found at 40 a/o Pt. The resistivity at temperatures below 8°K follows a T^2 dependence; $\varrho(T)$ between 20 and 80°K shows smaller values in the alloys than expected from Matthiessen's rule. Greig and Rowlands[5] reported a resistance minimum in a $Pd_{57} Pt_{43}$ alloy near 3°K, but no evidence for a resistance minimum in a $Pd_{99.5}Pt_{0.5}$ alloy.

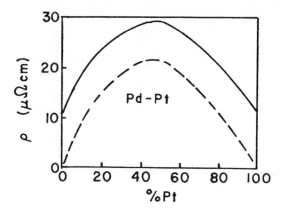

FIGURE 1.   Resistivity of Pd-Pt alloys at room temperature (full line) and 4.2°K (dashed line).

Table 1
THE TEMPERATURE-DEPENDENT RESISTIVITY, $\varrho(T)$

Nominal composition of alloy (a/o)	$\varrho(T)(\mu\Omega cm)$								
	10 K	20 K	30 K	40 K	50 K	60 K	70 K	80 K	90 K
pure Pd	0.004	0.036	0.13	0.32	0.58	0.92	1.30	1.72	2.17
Pd + 0.5 Pt	0.005	0.042	0.14	0.336	0.61	0.99	1.34	1.76	2.21
Pd + 1.0 Pt	0.005	0.042	0.147	0.36	0.64	1.06	1.46	1.85	2.26
Pd + 1.5Pt	0.005	0.046	0.157	0.37	0.65	1.06	1.42	1.82	2.24
Pd + 8.0Pt	0.003	0.034	0.14	0.356	0.65	1.20	1.40	1.79	2.18
Pd + 14Pt	—	0.035	0.13	0.32	0.57	0.90	1.22	1.55	1.89
Pd + 30Pt	—	0.025	0.12	0.29	0.51	0.80	1.08	1.37	1.66
Pd + 43Pt	—	0.020	0.10	0.25	0.48	0.79	1.07	1.36	1.65
Pd + 54Pt	—	0.030	0.10	0.28	0.49	0.80	1.06	1.34	1.62
Pd + 74Pt	—	0.050	0.19	0.43	0.76	1.11	1.51	1.87	2.24
Pd + 87Pt	—	0.059	0.23	0.52	0.88	1.34	1.76	2.16	2.56
Pd + 93Pt	—	0.063	0.22	0.51	0.86	1.30	1.71	2.14	2.56
Pure Pt	0.003	0.036	0.16	0.40	0.72	1.09	1.50	1.91	2.32

REFERENCES

1. Geibel, W., *Z. Anorg. Chem.*, 70, 242, 1911.
2. Schulze, F. A. *Z. Phys.*, 12, 1028, 1911.
3. March, J.-F., *Z. Metallkd.*, 69, 377, 1978.
4. Blood, P. and Greig, D., *J. Phys. F*, 2, 79, 1972.
5. Greig, D. and Rowlands, J. A., *J. Phys. F*, 4, 536, 1974.

Pd-Pu (Palladium-Plutonium)

Nellis and Brodsky[1] measured the resistivity and the magnetic susceptibility of PdPu and suggested that a two-band model cannot explain the resistivity of Kondo-type systems with $d\varrho/dT > 0$ at low temperatures. Their experiments show that $\Delta\varrho/c$ is slightly increasing with increasing temperaturs (see Figure 1).

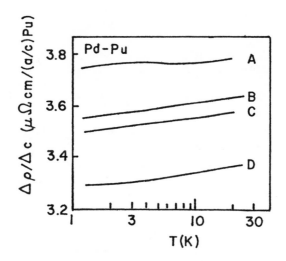

FIGURE 1.   Resistivity of Pd-Pu alloys (a/o Pu). (A)
0.1; (B) 1.1; (C) 0.5; (D) 2.0

REFERENCES

1. Nellis, W. J. and Brodsky, M. B., *Phys. Lett. A*, 32, 267, 1970.

Pd-Ru (Palladium-Ruthenium)

Greig and Rowlands[1] melted appropriate amounts of Pd and Ru in an arc furnace.
After rolling and swaging, samples were annealed for 24 hr at 850°C in vacuum. $\Delta\varrho(T)$
$= \varrho_{meas} - \varrho_o$ of alloys with 0.1 and 0.5 a/o Ru is given in Table 1. ϱ_{meas} of the $Pd_{99.5}Ru_{0.5}$
alloys below 9°K is shown in Figure 1. It shows a resistance minimum. $\varrho(T)$ between
50°mK and 4°K of alloys with 0.1 to 5 a/o Ru has been measured by Schroeder and
Uher.[2]

Table 1

T	19.8	30.1	40.0	66.0	79.7	103.7	199	273	°K
0.1 a/o Ru:$\Delta\varrho(T)$	0.0400	0.157	0.3673	1.204	1.775	2.811	6.562	9.735	($\mu\Omega$cm)
0.5 a/o Ru:$\Delta\varrho(T)$	0.0377	0.1628	0.3867	1.236	1.799	2.807	6.464	9.572	($\mu\Omega$cm)

Note: ϱ_o(0.1 a/o Ru) = 0.773 $\mu\Omega$cm and ϱ_o (0.5 a/o Ru) = 2.440 $\mu\Omega$cm.

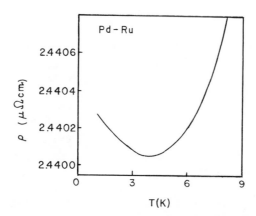

FIGURE 1.   Resistivity of $Pd_{99.5}Ru_{0.5}$.

REFERENCES

1. Greig, P. and Rowlands, J. A., *J. Phys. F*, 4, 536, 1974.
2. Schroeder, P. A. and Uher, C., *Low Temp. Phys.*, 29, 487, 1972.

Pd-Ti (Palladium-Titanium)

Azabar and Williams[1] prepared alloys from 5N purity Pd and 4N purity Ti. After cold rolling, samples were annealed for 24 hr at 650°C. Figure 1 shows the incremental resistivity $\Delta\varrho(T) = \varrho_{alloy}(T) - \varrho_{Pt}$. $\varrho_{Pt} = 0.005 + 15 \times 10^{-6}T^2 + 13 \times 10^{-9}T^5$ ($\mu\Omega$cm). The authors show that the data can be analyzed on the basis of a two-band model and by an emperical scheme by Caplin and Rizzuto.[2]

Pd – Ti

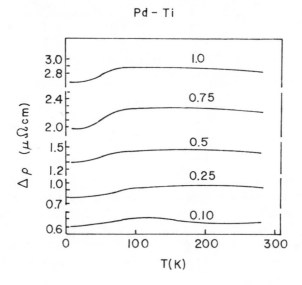

FIGURE 1. $\Delta\varrho = \varrho(T)_{alloy} - \varrho(T)_{Pd}$ of Pd-Ti. Number gives (a/o)Ti.

REFERENCES

1. Azabar, E. K. and Williams, G., *Phys. Rev. B*, 14, 3301, 1976.
2. Caplin, A. D. and Rizzuto, C., *J. Phys. C*, 3, L117, 1970.

Pd-U (Pallidium-Uranium)

Nellis et al.[1] prepared samples from arc-melted Pd-U buttons, rolled into sheet and machined. This was followed by an anneal at 950°C for 18 hr and an oil quench. Figure 1 shows $\varrho(T)$ of alloys with 7 to 11 a/o U. Alloys containing up to 7 a/o U show no anomaly in $\varrho(T)$. The 8 a/o U alloy shows a minimum in $\varrho(T)$ at 25°K. The $\varrho(T)$ minimum for the 9 a/o U alloys is close to room temperature. Further additions of U provide a maxima near 50°K, and $d\varrho/dT < 0$ at 300°K for the 11 and 13 a/o U alloys. $\varrho(4.7°K)$ and $\varrho(300°K)$ are shown in Figure 2. The authors do not associate the minimum in $\varrho(T)$ with the Kondo effect, but with localized spin fluctuations between the spin-up and spin-down U-5f virtual levels.

Andres et al.[2] prepared UPd$_3$ samples by pulling a single crystal from the melt. Its resistivity is anisotropic. It shows a sharp drop near 7 to 8°K (see Figure 3). The authors suggest this may be due to electric quadruple interactions between localized U^{4+}-5f states.

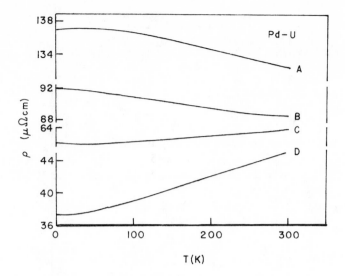

FIGURE 1.   Resistivity of Pd-U alloys (a/o U). (A) 11; (B) 9; (C) 8; (D) 7.

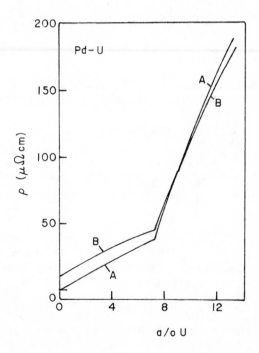

FIGURE 2.   Resistivity of Pd-U alloys (°K). (A) 4.2; (B) 300.

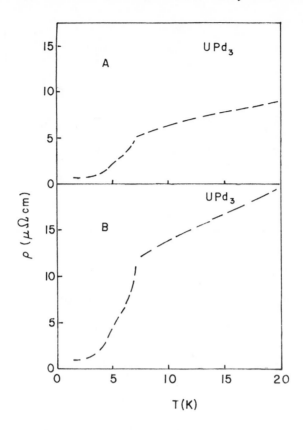

FIGURE 3.   Resistivity of UPd₃. (A) Current parallel to c-axis; (B) ϱ current perpendicular to c-axis.

REFERENCES

1. Nellis, W. J., Brodsky, M. B., Montgomery, H., and Pells, G. P., *Phys. Rev. B*, 2, 4590, 1970.
2. Andres, K., Davidov, D., Dernier, P., Hsu, F., and Reed, W. A., *Solid State Commun.*, 28(5), 405, 1978.

Pd-V (Palladium-Vanadium)

Kao et al.[1] prepared P̲d̲V alloys from 99.999% pure Pd and 99.9% pure V in an arc furnace. Samples were annealed in vacuum at 1000°C for 30 hr. Figure 1 shows $\varrho(T)$. $\Delta\varrho(T = 0)/c = 3.26 \pm 0.13 \ \mu\Omega\text{cm}/(a/o)V$.

The incremental resistivity can be described with an equation of the form $\Delta\varrho(T) = C + D \ \ln[(T^2 + \Theta^2)^{1/2}]$. C and D are approximately proportional to c. The equation is based on a localized spin fluctuation model. Figure 2 shows typical results. Agreement between the equation (line SF) and experiments is good above 40°K. The authors also matched the data to the Appelbaum-Kondo (AK) model, obtaining nearly as good agreement between theory and experiment as with the spin fluctuation (SF) model. The AB theory predicts that $\Delta\varrho(T) = A'[1 + B'[(T/T_K)\ln(T/T_k)]^2]$. Similar results were obtained by Ström-Olsen and Williams.[2]

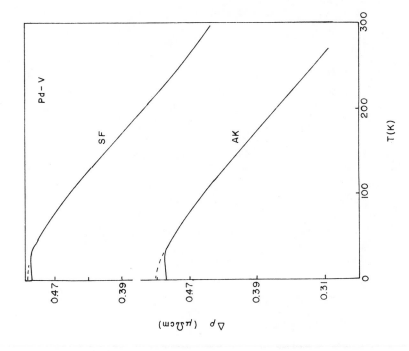

FIGURE 2. Pd = 0.15 a/o V; $\Delta\varrho = \varrho(T)_{alloy} - \varrho(T)_{Pd}$. Full line gives experimental results.

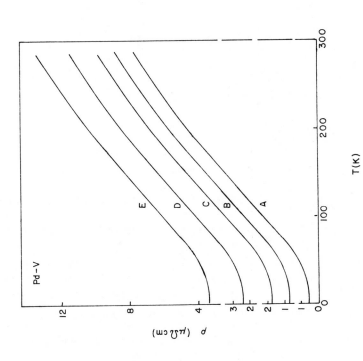

FIGURE 1. Resistivity of Pd-V alloys (a/oV). (A) 0.15; (B) 0.20; (C) 0.50; (D) 0.75; (E) 1.0.

REFERENCES

1. Kao, F. C. C., Kolp, M. E., and Williams, G., *Phys. Rev. B*, 8, 1228, 1973.
2. Ström-Olsen, J. O. and Williams, G., *Phys. Rev. B*, 12, 1986, 1977.

Pd-W (Palladium-Tungsten)

Alloys with up to 14.5 a/o W (equal to 22.6 w/o W) consists of solid solutions of W in Pd.[1] Klyuyeva et al.[2] prepared Pd-W alloys from 99.98% Pd and 99.8% W in a sealed induction furnace. Wires were drawn with intermediate anneals at 900°C in a vacuum. The samples with 2.6 to 12.7 a/o W showed a hysteresis effect as shown in Figure 1.

Lobanov et al.[3] studied the effect of heat treatments on a Pd-20 w/o W alloy. They prepared this alloy from 99.98% Pd and 99.98% W in an induction furnace. Wires were drawn with intermediate vacuum anneals at 900°C. Figure 2 gives $\varrho(T)$ for a variety of heating and cooling procedures. Curve 1 gives $\varrho(T)$ for a healing rate of 3°C/min. An isothermal soaking at 440°C for 15 hr leads to no change in ϱ, however, $\varrho(RT)$ after cooling is 4% higher than the original state. The sample is then heated to 630°C (Curve 3) and soaked for 12 hr, leading to a rise in $\varrho(T)$. The complex temperature dependence is associated with modification in order and K-state formation.[4]

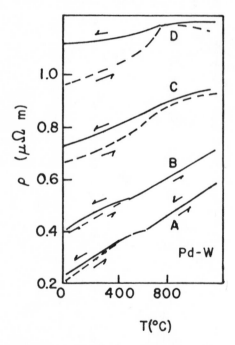

FIGURE 1.   Resistivity of Pd-W alloys (a/o W). (A) 2.6; (B) 6.2; (C) 9.3; (D) 12.7

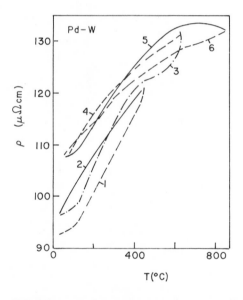

FIGURE 2.   Resistivity of Pd-20W. (1,2) Initial heating and cooling; (3,4) reheating and recooling; (5,6) third heating and cooling.

REFERENCES

1. Hansen, M. and Anderko, K., *Constitution of Binary Alloys*, 2nd ed., McGraw-Hill, New York, 1958.
2. Klyuyeva, I. B., Kuranov, A. A., Chemeriuskaya, L. S., Baranova, Ye. N., Bashkatov, A. N., Syutkin, P. N., Sidorenka, F. A., and Geld, P. V., *Phys. Met. and Metallogr.*, 47(4), 46, 1979.
3. Lobanov, V. V., Klyuyeva, I. B., Kagan, Ye., Ryabov, R. A., Guk, Yu N., and Geld, P. V., *Phys. Met. Metallogr.*, 47(3), 176, 1980.
4. Meskin, V. S., Sergiyenko, R. I., and Popova, L. A., *Fiz. Met. Metalloved*, 3, 127, 1962.

Pd-Y (Palladium-Yttrium)

Fort and Harris,[1] in a study of H diffusion in membrane materials, arc-melted a Pd-Y alloy with 9 a/o Y from 99.99% Pd and 99% Y. The resistivity of the alloys is shown in Figure 1.

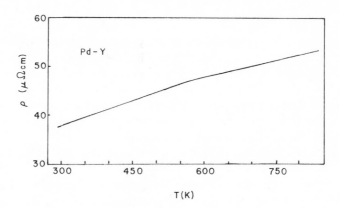

FIGURE 1.   Resistivity of $Pd_{91}Y_9$.

REFERENCES

1. Fort, D. and Harris, I. R., *J. Less Common Met.*, 41, 313, 1975.

PLATINUM (Pt)

Pt-Rh (Platinum-Rhodium)

Pt and Rh form a continuous series of solid solutions. Macfarlane and Collins[1] annealed Pt-Rh wires with 10, 13, and 30 w/o Rh for 2 hr at 900°C followed by a slow cool. Table 1 gives ϱ at 4.2, 77, and 293°K. A value of 6.91 $\mu\Omega$cm for the residual resistivity has been given by Stewart and Huebener[2] for an alloy with 9 a/o Rh (equal to 5 w/o Rh).

Table 1

(w/o)	$\varrho(4.2°K)$ ($\mu\Omega$ cm)	$\varrho(77°K)$ ($\mu\Omega$ cm)	$\varrho(293°K)$ ($\mu\Omega$ cm)
10	10.1	11.9	18.8
13	11.7	12.8	20.3
30	12.3	13.3	18.9

REFERENCES

1. Macfarlane, J. C. and Collins, H. C., *Cryogenics*, 18, 668, 1978.
2. Stewart, R. G. and Huebener, R. P., *Phys. Rev. B*, 1, 3323, 1970.

Pt-Ti (Platinum-Titanium)

Azabar and Williams prepared Pt-Ti alloys from 5N purity Pt and 4N purity Ti. After cold rolling, samples were annealed for 24 hr at 650°C. Figure 1 shows the incremental resistivity $\Delta\varrho(T) = \varrho_{alloy}(T) - \varrho_{Pt}(T)$. ϱ_{Pt} (273°K) = (9.80 ± 0.05) $\mu\Omega$cm. The authors show that the data can be analyzed on the basis of a two-band model and by an emperical scheme by Caplin and Rizzuto.[2]

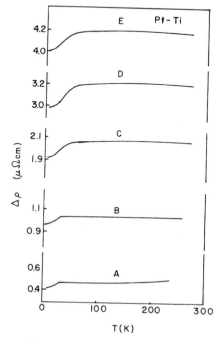

FIGURE 1. $\Delta\varrho = \varrho(T)_{alloy} - \varrho(T)$ Pt of PtTi alloys (a/o Ti). (A) 0.1; (B) 0.25; (C) 0.5; (D) 0.75; (E) 1.

REFERENCES

1. Azabar, E. K. and Williams, G., *Phys. Rev. B,* 14, 3301, 1976.
2. Caplin, A. D. and Rizzuto, C., *J. Phys. C,* 3, L117, 1970.

Pt-V (Platinum-Vanadium)

Azabar and Williams[1] prepared PtV alloys from 5N purity Pt and 3N5 purity V. After cold rolling, samples were annealed for 24 hr at 650°C. Figure 1 shows the incremental resistivity $\Delta\varrho(T) = \varrho_{alloy}(T) - \varrho_{Pt}(T)$. ϱ_{Pt} (273°K) = (9.80 ± 0.05) $\mu\Omega$cm. The authors show that the data can be analyzed on the basis of a two-band model and by an empirical scheme by Caplin and Rizzuto.[2]

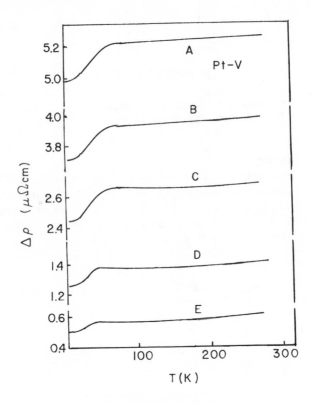

FIGURE 1.   Resistivity of Pt-V alloys (a/o V). (A) 1.00; (B) 0.75; (C) 0.5; (D) 0.25; (E) 0.1

REFERENCES

1. Azabar, E. K. and Williams, G., *Phys. Rev. B.,* 14, 3301, 1976.
2. Caplin, A. D. and Rizzuto, C., *J. Phys. C,* 3, L117, 1970.

PLUTONIUM (Pu)

Pu-Zr (Plutonium-Zirconium)

Elliot and Hill[1] prepared Zr-Pu alloys by arc-casting, then heat treating the samples for 1 week at 600°C. Finally the samples were quenched in the unbroken capsule at room temperature. Alloys with less than 12.5 a/o Pu should be single-phase solid solution.

Figures 1 and 2 give $\varrho(T)$ curves of alloys with up to 16 a/o Pu. The resistivity of the alloys show a minimum similar in appearance to that found in the Kondo effect.

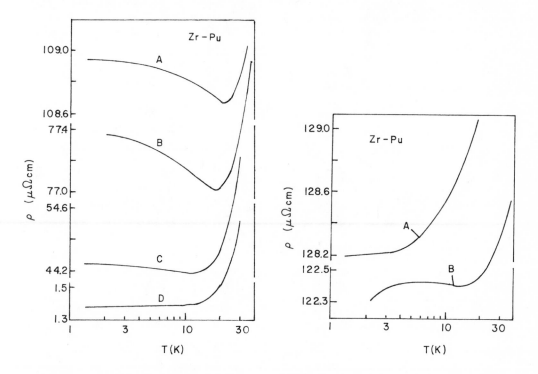

FIGURE 1.   Resistivity of Yu-Zr alloys. (A) 10 a/o Pu; (B) 7.5 a/o Pu; (C) 5.0 a/o Pu; (D) pure Zr.

FIGURE 2.   Resistivity of Zr-Pu alloys (a/o Pu). (A) 16; (B) 12.5

REFERENCES

1. Elliott, R. O. and Hill, H. H., *J. Less Common Met.*, 22, 123, 1970.

RHENIUM (Re)

Re-W (Rhenium-Tungsten)

Zaichenko et al.[1] examined ϱ of W-Re alloys between 300 and 1200°K. These materials should form a solid solution with up to ≈27 a/o Re. However, ϱ(c,T) curves show some anomalies which suggest that some type of ordering may occur in the composition range from 4 to 6 a/o Re and above 12 to 21 a/o Re. Samples were prepared from 99.95% W and electrolytic Re. Impurity levels should be of the order of 10^{-2}%. The room temperature resistivity is shown in Figure 1, ϱ at 77°K is shown in Figure 2, ϱ(c) at 500, 800, and 1200°K is shown in Figure 3, and the temperature dependence in ϱ for several alloys is given in Figure 4. Data follow an equation $\varrho = A + BT$.

Bykov et al.[2] found a similar anomaly in their ϱ(c) curve. They prepared W-Re alloys with up to 25 a/o Re from W and Re powder with a total impurity content of less than 0.02 w/o and 0.05 w/o, respectively. After zone melting the impurity content was less than 0.001 w/o. X-ray studies and metallographic investigations suggested that the samples were single phase.

Vertogradskii and Chekhovskii[3] measured the resistivity of W-Re alloys with 5 and 10 w/o Re at room temperature and between 1200 to 3000°K. The alloys were prepared by powder methods and contained 5 ± 0.25 or 10 ± 0.5 w/o Re. Results are shown in Table 1. The data agree typically within 3% with other measurements.[3-13]

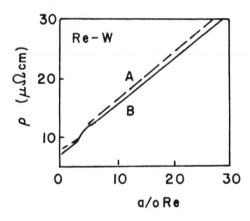

FIGURE 1. Room temperature resistivity of Re-W alloys. (A) Initial state; (B) annealed at 2500°K.

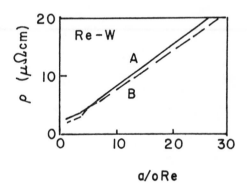

FIGURE 2. Resistivity of Re-W alloys at 77°K. (A) Initial state; (B) annealed at 2500°K.

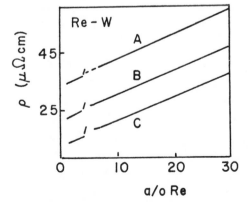

FIGURE 3. Resistivity of Re-W alloys (°K). (A) 1200; (B) 800; (C) 500.

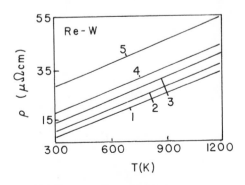

FIGURE 4. Resistivity of Re-W alloys (a/o Re). (1) 1; (2) 5.87; (3) 4.26; (4) 13.03; (5) 27.

Table 1

T	293	1200	140	1600	180	2000	220	2400	260	2800	3000	K
5% Re	0.111	0.370	.429	0.491	.554	.618	.684	0.751	.819	0.889	0.960	μΩm
10% Re	0.172	0.433	.492	0.552	.613	0.674	.736	0.799	.863	0.957	0.992	μΩm

REFERENCES

1. Zaichenko, V. M., Mints, R. G., and Chekhovskoi, V. Ya., *High Temp. USA*, 14(2), 263, 1976.
2. Bykov, V. N., Budagovskiy, S. S., Kondalenko, Yu. V., and Chelnakov, N. P., *Phys. Met. Metallogr.*, 33(2) 30, 1976.
3. Vertogradskii, V. A. and Chekhovskii, V. Ya., *High Temp.*, 11(2), 386, 1973.
4. Nidlender, R. A., Ed., *Metals and Alloys for Electrical Vacuum Instruments*, 1965.
5. Toulovkian, Y. S., Ed., *Thermophysical Properties at High Temp. Solid Materials*, TPRC 2, 1967.
6. Vertogradskii, V. A. and Chekhovskoi, V. Ya., *High Temp. USA*, 9, 399, 1971.
7. Ipatova, S. I., and Pavlova, E. I., in *Materials on Vacuum Technology*, 23, Gosenergoizdat, 1960.
8. Danisheuskii, S. K., Ipatova, S. I., Pavlova, E. I., and Smirnova, N. I., *Zavod. Lab.*, 29, 111, 39, 1963.

RUTHENIUM (Ru)

Ru-V (Ruthenium-Vanadium)

Chu et al.[1] investigated properties of V-Ru alloys near the equiatomic composition. Appropriate amounts of 99.9 w/o V and 99.95 a/o Ru were arc-melted. Samples were annealed at 1000°C for 1 week. This did not affect noticeably the resistivity. $\varrho(T)$ of the alloys is shown in Figure 1. The strong composition dependence of ϱ is associated with a substantial change of the Fermi surface area and the change in the effective number of carriers with the tetragonal distortion. Similar results on alloys with 45, 46, 47, and 48 a/o Ru were obtained by Chu et al.[2]

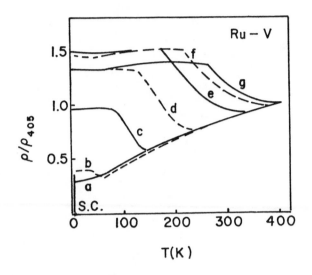

FIGURE 1.   Relative resistivity of Ru-V alloys (a/o Ru). (a) 45; (b) 46; (c) 47; (d) 48; (e) 49; (f) 50; (g) 51.

REFERENCES

1. Chu, C. W., Bucher, E., Cooper, A. S., and Maita, J. P., *Phys. Rev. B*, 4, 320, 1971.
2. Chu, C. W., Huang, S., Smith, T. F., and Corenzwit, E., *Phys. Rev. B*, 11, 1866, 1975.

ANTIMONY (Sb)

Sb-Sn (Antimony-Tin)

The phase diagram[1] of Sb-Sn shows one intermediate phase in two modifications: β and β'. The maximum solubility of Sb in Sn is 10.3 a/o at 246°C. The solubility limit decreases with decreasing temperature. Several percent Sn will dissolve in Sb even at 100°C.

Lane and Dodd[2] prepared Sb-Sn single crystals with up to 3.08 a/o Sn. Figure 1 shows $\varrho(T)$ from room temperature to 4.2°K, and Figure 2 gives the residual resistivity and the mean temperature coefficient between 4.2 and 77.3°K. The residual resistivity is not a linear function of composition, but nearly parabolic: $\Delta\varrho \propto \sqrt{c}$.

Burckbuchler and Reynolds[3] studied deviations from Matthiessen's rule and the orientation dependence of the electrical resistivity of Sn with Sb impurities. Single crystals were grown with the Brideman method from 99.999% pure Sn. The orientation dependence in a tetragonal crystal like Sn can be written as

$$\rho(\Theta) = \rho\perp\left(1 + ([\rho(0°)/\rho(90°)] - 1)\cos^2\Theta\right)$$

with Θ being the angle between current direction and tetragonal axis. $\varrho\perp = \varrho(90°)$ For Sb impurities, one obtains for alloys with up to 1.7 a/o Sb: $\varrho_o\perp/c = 0.63 \pm 0.03$ $\mu\Omega\text{cm}/(\text{a/o})$Sb.

Yurkov et al.[4] prepared SnSb crystals from Sn type OVOOO and Sb type SUOOO. Materials were placed in an evacuated ampoule of fusible glass. Single crystals were prepared by first sealing the alloy in an evacuated refractory glass ampoule, then slowly lowering the ampoule through a furnace. The single crystal had a slight deficit of Sn. The resistivity of the single crystal is given in Figure 3. $\varrho\|$ or $\varrho\perp$ means that the current for the measurement flowed parallel or perpendicular to the cleavage plane of the sample. The resistivity of the polycrystalline sample ϱ_p is $\varrho_p = (2\varrho + \varrho\perp)/3$. The single crystal shows typical metallic properties.

Dukin and Aleksandrov[5] gave a critical account of pseudopotential calculations and the resistivity increase due to 1% impurity. They gave $[\Delta\varrho/\Delta c]_{exp} = 0.70 \ \mu\Omega\text{cm}/(\text{a/o})$Sb in SnSb.

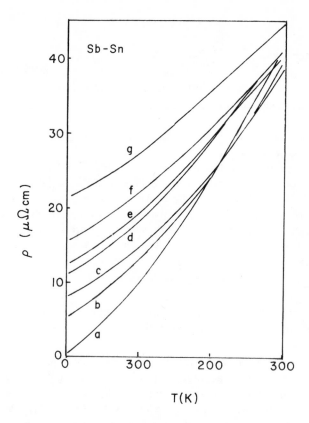

FIGURE 1.   Resistivity of Sb-Sn alloys. (a) 0 Sn; (b) 0.252 a/o Sn; (c) 0.489 a/o Sn; (d) 0.784 a/o Sn; (e) 1.06 a/o Sn; (f) 1.54 a/o Sn; (g) 3.98 a/o Sn.

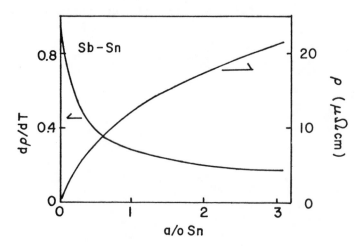

FIGURE 2.   dϱ/dT average in arbitrary units between 4.2 and 77.3°K (left-hand side) and residual resistivity (right-hand side).

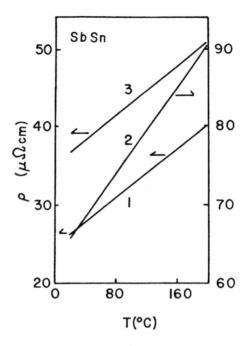

FIGURE 3. SbSn single crystal. (1) $\varrho_{\shortparallel}$ Current parallel to cleavage plan; (2) $\varrho\perp$ current perpendicular cleavage plan; (3) ϱ_p polycrystalline.

REFERENCES

1. Hansen, M. and Anderko, K., *Constitution of Binary Alloys,* 2nd ed., McGraw-Hill, New York, 1958.
2. Lane, C. T. and Dodd, W. A., *Phys. Rev.,* 61, 183, 1942.
3. Burckbuchler, F. V. and Reynolds, C. A., *Phys. Rev.,* 175, 550, 1968.
4. Yurkov, V. A., Yepishchin, I. G., Tugushev, S. Yu., and Soshnikov, V. A., *Phys. Met. Metallogr. (USSR),* 29(1), 111, 1970.
5. Dukin, V. V. and Aleksandrov, B. N., *Pseudopotential Calculations of the Residual Resistivities of Dilute Solid Alloys Based on Normal Metals, Physics of Low Temperatures,* Akademia Nauk Uk SSR Academy of Science, USSR, Kharkov, 1978, 1.

Sb-Tl (Antimony-Thallium)

Dukin and Aleksandrov[1] gave a critical account of pseudopotential calculations and the resistivity increase due to 1% impurity. They gave $[\Delta\varrho/\Delta c]_{exp} = 5.48\ \mu\Omega\text{cm}/(\text{a/o})\text{Sb}$ in TlSb.

REFERENCES

1. Dukin, V. V. and Aleksandrov, B. N., *Pseudopotential Calculations of the Residual Resistivities of Dilute Solid Alloys Based on Normal Metals, Physics of Low Temperatures,* Akademia Nauk Uk SSR Academy of Science, USSR, Kharkov, 1978, 1.

Sb-Zn (Antimony-Zinc)

Burckbuchler and Reynolds[1] studied deviations from Matthiessen's rule in Sn with Zn impurities. Single crystals were grown with the Bridgman method using 99.999% Sn. $\varrho_o\perp/c = 0.82\ \mu\Omega$cm/(a/o)Zn obtained from an alloy with about 0.1 a/o Zn. $\varrho_o\perp$ is the residual resistance measured with current flowing perpendicular to the tetragonal axis.

REFERENCES

1. Burckbuchler, F. V. and Reynolds, C. A., *Phys. Rev.*, 175, 550, 1968.

TIN (Sn)

Sn-Tb (Tin-Terbium)

Singh and Woods[1] measured $\varrho(T)$ of a TbSn crystal of 99.9% purity from 4 to 40°K. $r(T) = R(T)/R(14.5°K)$ is shown in Figure 1; $d\varrho/\varrho(14.5°K)dT$ is shown in Figure 2. The sample is antiferromagnetic below $T_m = 14.5°K$; r-data can be described with a function of the form $[|T-T_m|/T_m]^{-\lambda\pm}$ with $\lambda^+ = -1.0$, and $\lambda^- = -1.3$ for $6 \times 10^{-3} < |(T - T_m)/T_m| < 0.1$. The superscript (+) means $T > T_m$. The superscript (−) means $T < T_m$. This indicates that spin scattering is dominated by short range spin correlations.

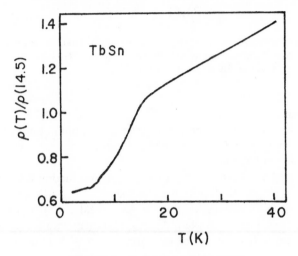

FIGURE 1.   Relative resistivity of TbSn.

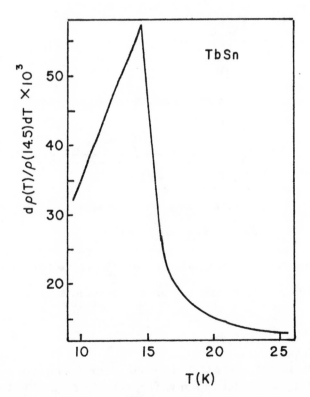

FIGURE 2.   Temperature coefficient of the resistivity.

REFERENCES

1. Singh, R. L. and Woods, S. B., *Phys. Rev. B,* 19, 1555, 1979.

Sn-Te (Tin-Tellurium)

Dukin and Aleksandrov[1] gave a critical account of pseudopotential calculations and the resistivity increase due to 1% impurity. They gave $[\Delta\varrho/\Delta c]_{exp} = 4.05\ \mu\Omega cm/(a/o)Te$ in S̲n̲Te.

REFERENCES

1. Dukin, V. V. and Aleksandrov, B. N., *Pseudopotential Calculations of the Residual Resistivities of Dilute Solid Alloys Based on Normal Metals, Physics of Low Temperatures,* Akademia Nauk Uk SSR Academy of Science, USSR, Kharkov, 1978, 1.

Sn-Ti (Tin-Titanium)

The phase diagram of Sn-Ti has been studied in detail, but not all features are clear.[1] Hansen and Anderko[1] give four intermediate compounds or intermediate phases: Ti_3Sn, Ti_2Sn, Ti_5Sn_3, and Ti_6Sn_5.

Morton et al.[2] studied the resistivity of the superconducting β-W type compound Ti_3Sn between T_c and 300°K. Ingots were prepared in an arc-furnace with starting materials of at least 99.95% metallic purity. An anneal at 950°C for 6 hr followed. The samples were then furnace cooled. They were largely cubic β-W phase. Figure 1 shows $\varrho(T)$. The data follow a curve of the form $\varrho = A + B\ T^r$ between 20 and 50°K, with r = 2.85.

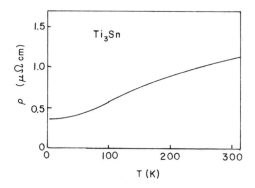

FIGURE 1.   Resistivity of Ti_3Sn alloys.

REFERENCES

1. Hansen, M. and Anderko, K., *Constitution of Binary Alloys,* 2nd ed., McGraw-Hill, New York, 1958.
2. Morton, N., James, B. W., Wostenholm, G. H., and Howard, N. A., *J. Less Common Met.,* 64, 69, 1979.

Sn-Tl (Tin-Thallium)

Dukin and Aleksandrov[1] gave a critical account of pseudopotential calculations and the resistivity increase due to 1% impurity. They gave $[\Delta\varrho/\Delta c]_{exp} = 0.88\ \mu\Omega cm/(a/o)Tl$ in S̲n̲Tl and $[\Delta\varrho/\Delta c]_{exp} = 1.34\ \mu\Omega cm/(a/o)Sn$ in T̲l̲Sn.

REFERENCES

1. Dukin, V. V. and Aleksandrov, B. N., *Pseudopotential Calculations of the Residual Resistivities of Dilute Solid Alloys Based on Normal Metals, Physics of Low Temperatures,* Akademia Nauk Uk SSR Academy of Science, USSR, Kharkov, 1978, 1.

Sn-V (Tin-Vanadium)

Köster and Hauk[1] found that about 10 a/o Sn would dissolve in solid V. The V_3Sn phase was found to be homogeneous around 21.2 a/o Sn, not at the stochiometric composition.

Morton et al.[2] studied the resistivity of the superconducting β-W type compound V_3Sn between T_c and 300°K. Ingots were prepared in an arc furnace with starting materials of at least 99.95% metallic purity. An anneal at 950°C for 6 hr followed. The samples were then furnace cooled. They were largely cubic β-W phase. Figure 1 shows $\varrho(T)$. The data followed a curve of the form $\varrho = A + B\,T^r$ between 20 and 30°K with r = 1.05.

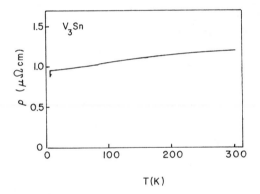

FIGURE 1.   Resistivity of V_3Sn alloys.

REFERENCES

1. Köster, W. and Hauk, K., *Z. Metalldk.,* 48, 327, 1957.
2. Morton, N., James, B. W., Wostenholm, G. H., and Howard, N. A., *J. Less Common Met.,* 64, 69, 1979.

Sn-Zn (Tin-Zinc)

Sn is nearly insoluble in solid Zn. A few percent Zn will dissolve in solid Sn. The phase diagram shows no compounds or intermediate phases, only one eutectic reaction.

Magomedov[2] measured the resistivity of Zn-Sn alloys between 20 and 800°C. Samples were prepared from analytically pure elements. Table 1 gives experimental results.

Dukin and Aleksandrov[3] gave a critical account of pseudopotential calculations and the resistivity increase due to 1% impurity. They gave $[\Delta\varrho/\Delta c]_{exp} = 2.6$ $\mu\Omega$cm/(a/o)Sn in ZnSn and $[\Delta\varrho/\Delta c]_{exp} = 0.78$ $\mu\Omega$cm/(a/o)Zn in SnZn.

Table 1
TEMPERATURE DEPENDENCE OF THE SPECIFIC
RESISTANCE OF Zn-Sn ALLOYS (ϱ, $\mu\Omega$cm)

T(°C)	Zn	w/o Zn (80)	w/o Zn (60)	w/o Zn (40)	w/o Zn (20)	w/o Zn (8)	Sn
20	6.5	7.3	9.5	9.8	10.1	10.6	12.2
40	7.2	7.9	10.3	10.7	11.0	11.5	13.0
60	7.7	8.5	11.2	11.6	12.0	12.4	13.7
80	8.3	9.0	12.0	12.4	12.7	13.2	14.5
100	8.8	9.6	12.7	13.3	13.6	14.1	15.4
120	9.4	10.1	13.6	14.2	14.6	15.0	16.5
140	9.9	10.7	14.4	15.1	15.5	15.9	17.6
160	10.5	11.3	15.3	16.0	16.3	16.8	18.7
180	11.0	11.9	16.0	16.8	17.2	17.7	19.7

REFERENCES

1. Hansen, M. and Anderko, K., *Constitution of Binary Alloys,* 2nd ed., McGraw-Hill, New York, 1958.
2. Magomedov, A. M., *High Temp. USA,* 16(3), 446, 1978.
3. Dukin, V. V. and Aleksandrov, B. N., *Pseudopotential Calculations of the Residual Resistivities of Dilute Solid Alloys Based on Normal Metals, Physics of Low Temperatures,* Akademia Nauk UK SSR Academy of Science, USSR, Kharkov, 1978, 1.

TANTALUM (Ta)

Ta-Ti (Tantalum-Titanium)

Berlincourt and Hake[1] prepared Ta-Ti alloys from elements of usually better than 99.9% purity in an arc furnace. Figure 1 gives the normal state resistivity at 1.2°K, obtained by using very high current densities.

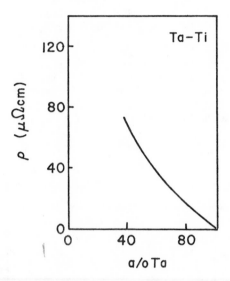

FIGURE 1.   Resistivity of Ta-Ti at 1.2°K.

REFERENCES

1. Berlincourt, T. G. and Hake, K. R., *Phys. Rev.*, 131, 140, 1965.

Ta-W (Tantalum-Tungsten)

Ta and W form a complete series of solid solutions. Thomas[1] prepared Ta-W alloys by powder metallurgical means. Ta powder (metallurgical grade, extra high purity) and W powder (high purity, grade 620 UD, Westinghouse®) as starting materials were mixed, compressed (pressure equal to 30 kg/mm²), and sintered in a vacuum (2400°C, 2 hr, 10^{-5} Torr). The sintered samples were usually once zone melted. The room temperature resistivities are given in Table 1.

The data by Thomas agree within experimental error with data on pure Ta and W.[2] Data on Ta-W alloys by Braun et al.[3] are higher than the data by Thomas. A comparison is only possible if one assumes that the data by Braun et al.[3] were obtained at room temperature.

Cezairliyan[4] developed a high-speed technique to determine $\varrho(T)$. He measured $\varrho(T)$ of a Ta + 9.45 w/o W alloy with less than 0.1% impurities. The experimental results can be fitted to a curve of the form $\varrho = 4.683 + 4.178 \times 10^{-2}T - 2.637 \times 10^{-6}T^2$ $\mu\Omega$m for 1500°K < T < 3200°K (T in °K). Results by Taylor et al.[5] on a sample with 9.45 w/o W can be given by $\varrho = 5.860 \times 10^{-6} + 4.0860 \times 10^{-8} T - 2.549 \times 10^{-12} T^2$ (Ωcm) for sample A (300°K < T < 2400°K) and for sample B $6.362 \times 10^{-6} + 3.9847 \times 10^{-8} T - 2.134 \times 10^{-12}T^2$ (Ωcm) (900°K < T < 2400°K).

Table 1

	W (w/o)	$\varrho(300°K)$ $\mu\Omega cm$
Ta	0	13.0
Ta-10 W	10.5	19.5
20	20.3	20.3
30	29.9	20.3
40	39.9	21.0
50	51.5	19.2
70	70.9	13.6
80	81.4	11.9
90	91.5	9.6
100	100	5.9

Note: $\varrho(T)/\varrho(300°K)$ is given in Figure 1.

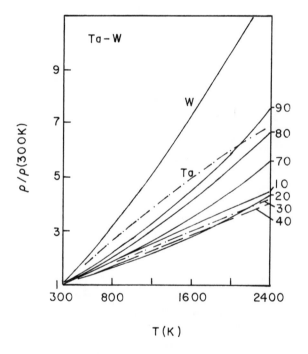

FIGURE 1. ϱ/ϱ (300°K) in Ta-W alloys. Number gives (w/o) W.

REFERENCES

1. Thomas, L., *Z. Metallkd.,* 59, 127, 1968.
2. Malter, C. and Langmuir, D. B., *Phys. Rev.,* 55, 743, 1963.
3. Braun, H., Kieffer, R., and Sedlatschek, K., *Plansee Proc.,* 264, 1958.
4. Cezairliyan, A., *High Temp. High Pressures,* 4, 541, 1972.
5. Taylor, R. E., Kimbrough, W. D., and Powell, R. W., *J. Less Common Metals,* 24, 369, 1971.

TELLURIUM (Te)

Te-Tl (Tellurium-Thallium)

Juodakis and Kannewurf[1] prepared Tl_5Te_3 by direct synthesis of stochiometric proportions of Tl and Te in an evacuated pyrex tube (10^{-3}Torr). It was heated at 500°C for 4 to 12 hr. Single crystals were prepared by vapor transport. The crystal structure is body-centered tetragonal. The resistivity of polycrystalline samples and single crystals is shown in Figure 1. It shows metallic conductivity since $d\varrho/dT > 0$. Samples were superconducting below 2.19°K. Rabenau et al.[2,3] and Akhmedova and Piragas[4] found that the compound Tl_2Te_3 is semiconducting. The same results were found by Flicker and Grass[5] for both TlTe and Tl_2Te_3. They found that alloys with more than 62.5 a/o Tl were metallic. They obtained ϱ(room temperature) = 8.3×10^2 Ωcm of Tl_5Te_3 for an annealed ingot, more than two orders larger than the value obtained in Reference 1. This may be due to the fact that Flicker and Grass prepared their samples in a N_2 atmosphere.

Ikari and Hashimoto[6] melted a stochiometric composition of Tl and Te in an evacuated silica tube (10^{-6} Torr) at 500°C for 24 hr and quenched in water. Then the sample was heated for 500 hr at 260°C. The diffraction pattern was completely that of TlTe. The resistivity of the sample is shown in Figure 2. A sharp kink is found at (172 ± 1)°K. The authors suggest that a phase transition from D_{4h}^{18} (T > 172°K) to D_{2h}^{28} (T < 172°K) takes place. The temperature of ϱ suggests metallic properties, but Flicker and Grass[5] observed semiconducting behavior with an activation energy of 0.15 eV. Cruceanu and Sladara[7] found as Ikari and Hashimoto[6] metallic behavior.

Dukin and Aleksandrov[8] gave a critical account of pseudopotential calculations and the resistivity increase due to 1% impurity. They gave $[\Delta\varrho/\Delta c]_{exp}$ = 14.4 μΩcm/(a/o)Te in TlTe.

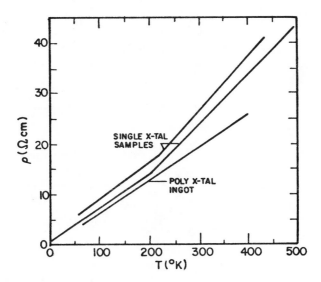

FIGURE 1. Resistivity of Tl_5Te_3 single crystals and a polycrystalline ingot.

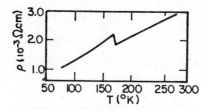

FIGURE 2.   Resistivity of TlTe.

REFERENCES

1. Juodakis, A. and Kannewurf, C. R., *J. Appl. Phys.*, 39, 3003, 1968.
2. Rabenau, A., Stegherr, A., and Eckerlin, P., *Z. Metallkd.*, 51, 295, 1960.
3. Rabenau, A., U.S. Patent, 3096115, 1963.
4. Akhmedova, F. I. and Piragas, L. E., *Izv. Vyssh. Uchebn. Zaved. Fiz.*, 9, 167, 1966.
5. Flicker, P. and Grass, F., *Z. Metallkd.*, 57, 641, 1966.
6. Ikari, T. and Hashimoto, K., *Phys. Status Solidi A*, 31, K115, 1975.
7. Cruceanu, E. and Sladara, St., *J. Mater. Sci.*, 4, 410, 1969.
8. Dukin, V. V. and Aleksandrov, B. N., *Pseudopotential Calculations of the Residual Resistivities of Dilute Solid Alloys Based on Normal Metals, Physics of Low Temperatures,* Akademia Nauk Uk SSR Academy of Science, USSR, Kharkov, 1978, 1.

THALLIUM

Tl-Zn (Thallium-Zinc)

Dukin and Aleksandrov[1] gave a critical account of pseudopotential calculations and the resistivity increase due to 1% impurity. They gave $[\Delta\varrho/\Delta c,]_{exp} = 0.03\ \mu\Omega\text{cm}/$ (a/o)Zn in T̲l̲Zn $[\Delta\varrho/\Delta c]_{exp} = 1.75\ \mu\Omega\text{cm}/$(a/o)Tl in Z̲n̲Tl.

REFERENCES

1. **Dukin, V. V. and Aleksandrov, B. N.,** *Pseudopotential Calculations of the Residual Resistivities of Dilute Solid Alloys Based on Normal Metals, Physics of Low Temperatures,* Akademia Nauk Uk SSR Academy of Science, USSR, Kharkov, 1978, 1.

TITANIUM (Ti)

Ti-V (Titanium-Vanadium)

Collings[1] made precise measurements of the electrical resistivity of single-phase and multiphase Ti-V alloys at 300°K and lower temperatures. Samples were prepared from high-purity ingredients (99.8% pure Ti, 99.95% V) by multiple arc-melting. The ingots were annealed for 1 hr at 1000°C and quenched in iced brine. Coring in samples with more than 50 a/o V was practically eliminated by a second anneal for 8 hr at 1350°C. Samples with less than 10 a/o V were essentially single-phase α'-alloys (hexagonal); alloys with more than 50 a/o V seemed again to be single-phase β-alloys (bcc). In between, two ($\beta + \omega$) or even three phases were found ($\alpha' + \beta + \omega$). The author developed a detailed method to analyze the single-phase and multiphase alloys, as discussed in the introductory chapter.

Berlincourt and Hake[2] prepared alloys from elements of usually better than 99.9% purity in an arc furnace. Figure 1 gives the normal state resistivity at 1.2°K, obtained by using very high current densities. The decrease in $\varrho(1.2°K)$ with increasing impurity concentration is typical for Group IV transition elements with transition elements of Group V as impurities.

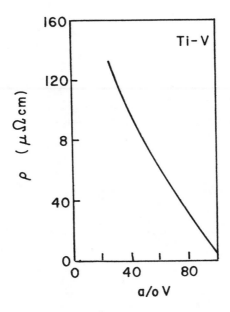

FIGURE 1.   Resistivity of Ti-V at 1.2°K.

REFERENCES

1. Collings, E. W., *Phys. Rev. B,* 9, 3989, 1974.
2. Berlincourt, T. G. and Hake, R. R., *Phys. Rev.,* 131, 140, 1965.

URANIUM (U)

U-Zr (Uranium-Zirconium)

Zr and U form a complete series of solid solutions at high temperatures 1133°C > T > 863°C, but the maximum solubility of Zr and V in each other below 600°C is of the order of a few percent.[1]

Fedorov et al.[2,3] prepared U-Zr samples by arc melting 99.78% pure U and Zr iodide. The samples were annealed at 1000°C in a vacuum of 3×10^{-4} Torr and cooled slowly. The composition dependence of ϱ at 4.2, 77.4, and 295°K is shown in Figure 1. One has to keep in mind that the samples are probably multiphase alloys.

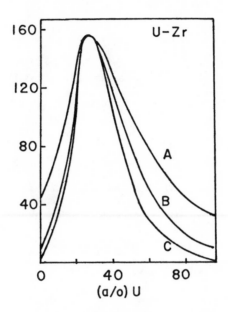

FIGURE 1.   Resistivity isotherms of U-Zr (°K). (A) 295; (B) 74.4; (C) 4.2.

REFERENCES

1. Hansen, M. and Anderko, K., *Constitution of Binary Alloys,* 2nd ed., McGraw-Hill, New York, 1958.
2. Fedorov, G. B., Zuev, M. T., Smirnov, E. A., and Kissil, A. E., *Sov. At. Energy,* 34, 85, 1973.
3. Fedorov, G. B. and Smirnov, E. A., *Sov. At. Energy,* 25, 54, 1968.

VANADIUM (V)

V-Zr (Vanadium-Zirconium)

Hansen and Anderko[1] report that several percent of Zr will dissolve in V. Up to 16.5 a/o V will dissolve in β-Zr, but V is practically insoluble in α-Zr. The Zr_2V compound is stable to 1300°C.

Kimura and Umeda[2] prepared a V_2Zr compound by arc-melting. The purity of the raw material was 99.8%. Figure 1 gives $\varrho(T)/\varrho(300°K)$. V_2Zr undergoes a cubic to rhombohedral transition. The data between 300°K and the phase transition can be given by (T in °K):

$$\rho = 11.4 + 0.044T + 55.1 \exp(-120/T) \quad \mu\Omega cm$$

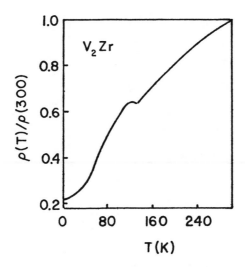

FIGURE 1.   Relative resisitivity of V_2Zr.

REFERENCES

1. Hansen, M. and Anderko, K., *Constitution of Binary Alloys,* 2nd ed., McGraw-Hill, New York, 1958.
2. Kimura, Y. and Umeda, M., *Phys. Status Solidi A,* 47, K111, 1978.

ZINC (Zn)

Zn-Zr (Zinc-Zirconium)

The solubility of Zr in solid Zn is about 0.015 a/o at 400°C. A eutectic and two presumably peritectic reactions were found.[1] The intermediate phase, $ZrZn_2$, with the cubic $MgCu_2$ structure was first reported by Pietrokowsky.[2]

Ogava[3] prepared a $ZrZn_2$ polycrystalline sample by sintering fine powders of high-purity Zr and Zn. A spectrographic analysis shows that the sample contained less than 50 ppm each of Ti, V, Cr, Mn, Co, Ni, and Fe. The sample has a cubic Laves phase structure. $\varrho(T)$ between 1.5 and 300°K is shown in Figure 1. The data are in good agreement with results by Olsen,[4] except for the low-temperature data. Olsen's data are there higher, probably due to grain boundary resistance.

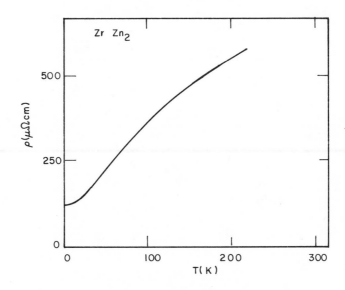

FIGURE 1.   Resistivity of $ZrZn_2$.

REFERENCES

1. Hansen, M. and Anderko, K., *Constitution of Binary Alloys,* 2nd ed., McGraw-Hill, New York, 1958.
2. Pietrokowsky, P., *Trans. AIME,* 200, 219, 1954.
3. Ogava. S., *J. Phys. Soc. Jpn.,* 40, 1007, 1976.
4. Olsen, C. E., *J. Phys. Chem. Solids,* 19, 228, 1961.

Appendix

Appendix

THE ELECTRICAL RESISTIVITY OF RARE EARTH INTERMETALLIC COMPOUNDS

E. Gratz

I. INTRODUCTION

At the beginning of this chapter it is of interest to consider what makes rare earth compounds so interesting that they are described in a separate section. The term rare earth (R) denotes the elements Sc, Y, and La through Lu. The latter are also called lanthanides. Some of them bear a magnetic moment, whereas others do not, as will be discussed in detail. One of the reasons for the growing interest in R-compounds during the last 3 decades is the large number of binary R-compounds which could be prepared and investigated with respect to many different physical properties, such as crystallographic structure, magnetic behavior, transport phenomena, etc. Consulting the review articles by Buschow[1,2] and Kirchmayr and Poldy,[3] more than 800 binary R-systems containing nontransition or transition metals can be identified. Furthermore, a gigantic number of the so-called pseudobinary R-systems in which either the R-element or its partner element is substituted by another element are available. Other reasons for the interest in R-compounds are the large number of exotic magnetic structures which are displayed by numerous systems, depending on the stoichiometry, the temperature, the concentration of the partner element, etc.

Intimately linked with the magnetic structure and the magnetic properties of intermetallic compounds are the transport properties, especially the electrical resistivity. Therefore one can expect that by a careful analysis of the available resistivity data, a more detailed understanding of the question how resistivity is influenced by the different magnetic states can be obtained. Since the transport phenomena, especially the resistivity as well as the magnetic properties of R-intermetallic compounds, are dominated by the electronic structure, the investigation of the mutual influence of both phenomena will undoubtedly help to clarify the basic understanding of the solid state properties, especially the electronic structure.

The subdivision of the material for this chapter is therefore logically based on the presence or absence of a magnetic moment at the respective lattice sites. The simplest case comprises compounds in which neither the R-metal nor the partner element bears a magnetic moment (e.g., YAl_2). In the next group we have compounds in which a magnetic moment is present at the R-site (e.g., $GdAl_2$). Finally compounds are treated where both the rare earth and the partner components carry a magnetic moment (e.g., Gd_4Co_3). The rare earths occurring in the trivalent state present a particularly satisfactory group of elements for investigations of intermetallics since they can be made to replace one another in the pseudobinaries without interfering with the conduction electron concentration. Thus, the pure influence of moment variations can be studied, e.g., in $(La_xGd_{1-x})Ni$. On the other hand, the effet of conduction electron concentration can also be studied by similar substitutions, in this case, however, by replacing one nonrare earth partner element with another of different valency, e.g., $Gd_4(Co,Ni)_3$. It is apparent that the number of possible combinations is unlimited and consequently it is only possible in an article of this length to make a representative selection from the available data.

II. RESISTIVITY OF NONMAGNETIC AND ENHANCED PARAMAGNETIC RARE EARTH COMPOUNDS

A. Theoretical Introduction

The discussion of the temperature dependence of resistivity in nonmagnetic R-compounds (R = Y, Ce, La, Yb, Lu) is based on the assumption of the validity of Matthiessen's rule which states that the various scattering mechanisms contributing to the resistivity may be regarded as independent of one another. They may therefore be summed up at any temperature to give the total resistivity. In the R-compounds in question only defect (or residual) resistivity, ϱ_o, and lattice resistivity, ϱ_{ph}, contributions arise.

The temperature-dependent total resistivity, $\varrho(T)$, is given by

$$\rho(T) = \rho_o + \rho_{ph}(T) \tag{2.1}$$

(The contribution caused by electron-electron collisions to the total resistivity sometimes observed in the lowest temperature range of high-purity metals[4,5] will be neglected.) In the compounds consisting of nonmagnetic R with noble metals, ϱ_o was frequently found to be in the range of a few $\mu\Omega$cm (see Section II.B). It should be pointed out that in most of the cases it is extremely difficult to obtain reliable values for ϱ_o because of the brittleness of most of these R-compounds, especially if the systems crystallize in the MgCu$_2$-structure (e.g., RAl$_2$) or in the CeCu$_2$-structure (e.g., RCu$_2$).[6] The scattering of electrons by vacancies, dislocations, foreign atoms, etc. is assumed to be invariably elastic. As the temperature rises, ϱ_o remains constant, but ϱ_{ph} increases rapidly because of electron scattering by the quantized lattice vibrations (phonons). In nonmagnetic R-compounds with noble metals, one can assume that the influence of d-electrons on physical properties such as susceptibility or transport phenomena is negligible. All the electrons responsible for the charge transport are part of one and the same s-conduction-band and will be scattered by the electron-phonon and electron-impurity interaction into vacant energy states within the same band. The general result for $\varrho_{ph}(T)$ in this case is given by the Bloch-Grüneisen equation:[7]

$$\rho_{ph}(T) = 4C\left(\frac{T}{\Theta_D}\right)^5 \int_0^{\frac{\Theta_D}{T}} \frac{x^5\,dx}{\left(1 - e^{-x}\right)\left(e^x - 1\right)} \tag{2.2}$$

The constant C includes an electron-phonon interaction parameter, the atomic masses of the different sorts of atoms, and a characteristic temperature, Θ_D (Debye temperature). Using the limits of integration as indicated, the standard integral is reduced to an expression proportional to $(\Theta_D/T)^4$ at high temperatures and to a constant equal to 124.4 at low-temperatures. Thus, ϱ_{ph} is proportional to T^5 at low temperature and proportional to T at higher temperatures.

The derivation of Equation 2.2 rests in principle on four assumptions:

1. The electric field E and the current density are described by a semiclassical theory. That means the E-field "accelerates" electrons according to $\dot{k} = -eE/\hbar$ and the current density is caused by the resulting excess of conduction-band electrons with a velocity

$$v = -\frac{1}{\hbar}\, \nabla_k E(k)$$

2. The scattering events are separated spatially by a sufficient number of wave lengths so that an electron "recovers" from a previous collision before experiencing the next one. For free electrons the criterion is $k_F \cdot l \gg 1$ (k_F, Fermi waves vector; l, mean free path).

3. The effective mass approximation is applied.

4. The phonons are in thermal equilibrium and can be described within the Debye model.

A fair number of R-compounds with Al or noble metals was found (see Section II.B) to behave more or less according to Equation 2.2. However, especially in R-compounds with transition metals such as Co, Ru, Rh, etc., serious departures from the Bloch-Grüneisen behavior were frequently found.[8,9] Although these compounds are nonmagnetic, as e.g., YCo_2, the susceptibility is drastically enhanced. Thus, we are speaking of enhanced paramagnetic compounds. For comparison the susceptibility of the isostructural YAl_2 is more than one order of magnitude lower[1,2] than in YCo_2. Summarizing one can say that the ϱ vs. T curves of these enhanced paramagnetic compounds are frequently characterized by two typical types of behavior:

1. A pronounced curvature of the ϱ vs. T curves in the temperature range above 80°K

2. A T^2-dependence instead of the T^5-dependence predicted by the Bloch-Grüneisen law at low temperatures

This behavior is connected with the influence of d-electrons on the temperature dependence of resistivity. Compared with the "nonenhanced" R-compounds, incompletely filled d-bands exist in the enhanced compounds which contribute to resistivity. This contribution can be described in the framework of Mott's s-d scattering mechanism.[10,11] While in the nonenhanced R-compounds, the s-electron scattering by phonons takes place as mentioned above within the same band, in the enhanced R compounds, the s-electrons can additionally be scattered into vacant states of the d-band. Because of the commonly higher density of states in the d-band, this contribution can be of great importance. However, it should be mentioned at this point that this s-d scattering mechanism cannot be the whole story. Recently Allen[12] gave a critical discussion in which he showed that a large number of possible mechanisms can cause a "saturation" in the high-temperature region of resistivity. Not included in those considerations is the influence of spin fluctuations on resistivity. It could be shown[13-15] that spin fluctuations lead to a T^2 increase of the resistivity in the low-temperature range and a curvature towards the temperature axis at high temperatures. Probably these spin fluctuations are of great importance for the discussion of the resistivity behavior in the enhanced R-compounds under consideration.

B. Experimental Results

1. RAl_2 (R = Y, La, Yb, Lu), CeT_2 (T = Co, Rh, Ru, Ir)

All the compounds under consideration crystallize in the cubic $MgCu_2$ structure. The (R-Al) compounds are typical representatives for compounds with negligible d-electron influence on resistivity, thus Equation 2.2 should be applicable. Indeed, in the high-temperature range a nearly linear temperature dependence was found. Figure 2.1 shows the ϱ (T)-curves of RAl_2 compounds.

Figure 2.2 exhibits the temperaure dependence of resistivity for some Ce-transition metal compounds.[8] In all these compounds, Ce is in its nonmagnetic tetravalent state. All the compounds are nonmagnetic, but the susceptibility is about one order of magnitude higher than in the RAl_2 compounds.[1,2] As mentioned in Section II.A, the influ-

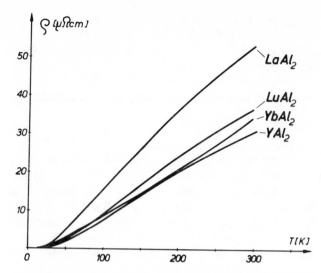

FIGURE 2.1.   The ϱ vs. T curves of some nonmagnetic RAl_2 compounds. (From van Daal, H. J. and Buschow, K. H. J., *Phys. Status Solidi A*, 3, 853, 1970. With permission.)

ence of d-electrons in the enhanced paramagnetic compounds gives rise to a pronounced curvature in the ϱ vs. T curve as can be seen in Figure 2.2.

2. YAl_2, $LuAl_2$, YCu_2, $LuCu_2$, LaNi

In Figure 2.3, the ϱ vs. T curves of the nonmagnetic R-compounds, YAl_2, $LuAl_2$, YCu_2, $LuCu_2$, and LaNi, are presented.[16] A fit procedure of Equation 2.2 to the experimental data was performed and the results are collected in Table 1. Although this fit is possible within about 1%, the obtained Debye temperatures, Θ_R, are considerably smaller than those from specific heat measurements (e.g., Θ_D = 473°K for YAl_2 and 384°K for $LuAl_2$).

3. $La(Ag_{1-x}In_x)$, $Ce(Ag_{1-x}In_x)$

The presence of the 5d-electrons at the Fermi level is of importance, not only with respect to magnetic coupling, but also for crystal structure stability.[17,18] La(AgIn) and Ce(AgIn), both of the CsCl structure, undergo a martensitic type of transition into a tetragonal phase at low temperature. This cubic to tetragonal transition is ascribed to the Jahn-Teller effect. The influence of such a transition on the resistivity in the La(AgIn) pseudobinary system is shown in Figure 2.4.[19] The corresponding measurements in the Ce(AgIn) system are given in Figure 2.5[20] The hysteresis observable in the ϱ vs. T curve is caused by this structural transition. Figure 2.6 represents a survey of the transition temperatures in these systems.[17] (As a transition temperature, the focal point of the loops was taken.) The anomalies in the ϱ behavior can be understood by a dramatic change of the phonon spectrum at this transition.

III. RESISTIVITY OF INTERMETALLIC RARE EARTH COMPOUNDS EXHIBITING LONG-RANGE MAGNETIC ORDER

A. Intermetallic Rare Earth-Nontransition Metal Compounds
1. Theoretical Introduction

The purpose of the following section is to give a detailed description of the resistivity of those rare earth intermetallic compounds whose magnetic properties are due to the localized magnetic moments of the rare earth component only. The collective magnetic

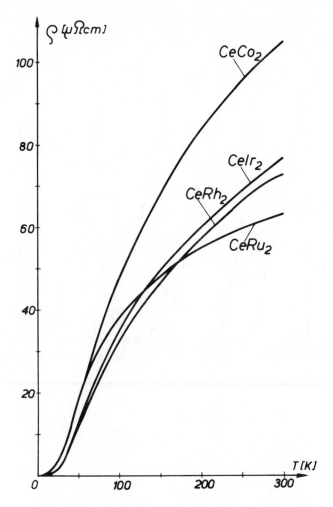

FIGURE 2.2. The ϱ vs. T curves of some CeT_2 compounds (T = Co, Ru, Rh, Ir). (From van Daal, H. J. and Buschow, K. H. J., *Phys. Status Solidi A*, 3, 853, 1970. With permission.)

Table 1
THE DEBYE TEMPERATURES AND MEAN SQUARE DEVIATIONS OBTAINED BY A FIT PROCEDURE OF THE BLOCH-GRÜNEISEN LAW (EQUATION 2.2) TO THE EXPERIMENTAL DATA

Compound	Crystal structure	Θ_R (°K)	Mean square deviation (%)
YAl$_2$	MgCu$_2$ structure	289	0.25
LuAl$_2$	MgCu$_2$ structure	269	0.19
YCu$_2$	CeCu$_2$ structure	169	1.17
LuCu$_2$	CeCu$_2$ structure	177	0.90
LaNi	CrB structure	166	1.28

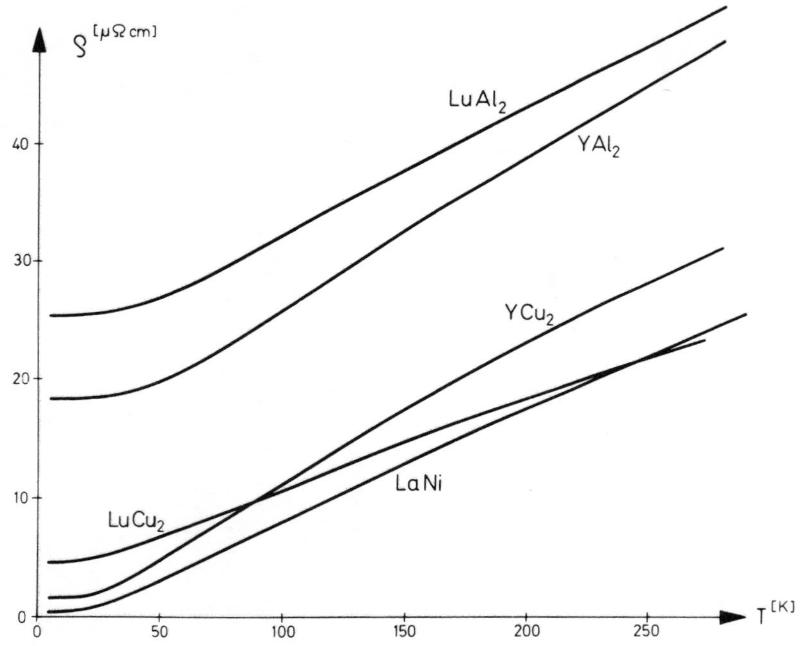

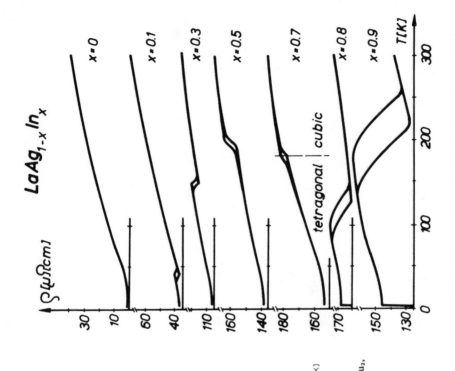

FIGURE 2.4.   The ϱ vs. T curves of the La(Ag₁ ₋ ₓInₓ) pseudobinary system. (From Balster, H., Diplomarbeit, Ruhr Universität Bochum, BRD, Bochum, 1972. With permission.)

FIGURE 2.3.   The experimentally found ϱ vs. T curve of YAl₂, LuAl₂, YCu₂, LuCu₂, and LaNi.

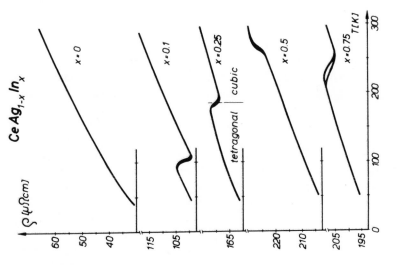

FIGURE 2.6. Transition temperature (T_M) corresponding to the cubic-to-tetragonal transition in La(Ag,In) and Ce(Ag,In). (From Ihrig, H. and Methfessel, S., *Z. Phys. B*, 24, 385, 1976. With permission.)

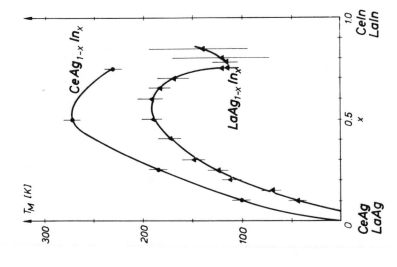

FIGURE 2.5. The ϱ vs. T curves of the Ce(Ag$_{1-x}$In$_x$) pseudobinary system. (From Ihrig, H., Thesis, Ruhr Universität Bochum, BRD, Bochum, 1973. With permission.)

properties of such compounds are based on the polarization of the conduction band by the rare earth moments. The resultant exchange interaction is known as the Ruder-mann-Kittel-Kasuya-Yosida (RKKY) interaction between s- and f-electrons. Since the magnetic properties of these systems have a strong influence on the resistivity, it is appropriate to begin the theoretical discussion with a description of the magnetic moments of free rare earth ions. The subsection continues with a description of the RKKY interaction and with theoretical predictions for the magnetic effect on the electrical resistivity in several temperature regions. The influence of both crystal fields and critical phenomena on the resistivity is also considered.

a. Origin of Magnetic Order

The free ions of the lanthanides usually occur in the trivalent state, and this can also be said of them in the solid state, Exceptions often occur for Ce, Eu, and Yb compounds due to the tendency to adapt a state with either empty, half-full, or full 4f-shells.

In order to select a theoretical model of the R-compounds without d-electron contribution suitable for the description of their magnetic and electrical properties, let us consider the characteristics of the electronic structure of the R-atoms. The external part of an R-atom (apart from the filled shell of the Xe atom) has the following configuration:

$$4f^n 5s^2 5p^6 5d^{0-1} 6s^2 \qquad (n = 1, 2, \ldots, 14)$$

Consequently this configuration consists of

1.   A 4f-shell which is successively filled
2.   A $(5s^2 5p^6)$ shell completely filled and screening the 4f-shell against external fields
3.   A valence shell $5d^{0-1} 6s^2$ which becomes conduction band

It is known from the neutron scattering experiment that the radius of the 4f-electron shell is about 0.35Å[21] which is sufficiently small compared with the inter-rare earth distances to preclude any effects from 4f-4f overlap. Describing the R-compounds with respect to their magnetic behavior, they can be regarded as a collection of trivalent 4f-ions on certain lattice sites immersed in the conduction electron gas. These conduction electrons arise from the 5d and 6s valence electrons of the R-ions and from the valence electrons of the elements Al, Cu, Ag, etc. With respect to magnetism, the components Al, Cu, and Ag (and in some cases Ni) act as a "blank". Each R-ion can be treated — to a first approximation — separately. That means that its magnetic states and energy levels, depending on the number of 4f-electrons, correspond to those of a free trivalent ion having an electron configuration $4f^n$. The conduction electrons are treated within the limits of the band theory as a gas of Fermi particles described by Bloch waves. Then in accordance with the perturbation theory, the exchange interaction between the 4f-shells and the conduction electrons is taken into consideration. The state of the R-ions may be sufficiently well described in the Russell Saunders approximation. The energy levels of the R trivalent ion can be classified according to the quantum numbers L and S of the 4f-shell ion. However, due to the spin orbit interactions, the energy state depends not only on the absolute values of the L and S vectors, but also on their relative orientation. The energy levels for given L and S should be characterized by the value of the quantum number J of the angular momentum operator J = L + S. The energy levels corresponding to different values of J give a so-called multiplet splitting of the level with the given values of L and S. The spin-orbit interaction of electrons in the 4f-shell of the rare earth atoms is so large that the energy difference

(ΔE) between the adjacent levels is of the order of $10^3 cm^{-1}$. Therefore at low temperature, when T$<<$ ΔE/k/$_B$, in general only the ground level of the multiplet is populated. The ground level of the $4f^n$ configuration is determined in accordance with Hund's rules: maximum value of S, then, with this value of S the maximum value of L, and finally

$$J = |L - S| \text{ for the ions with } n < 7$$

$$J = |L + S| \text{ for the ions with } n > 7$$

For a given J, the level of the ion is still $(2J + 1)$-fold degenerated in the direction of the J-vector. This degeneracy can be partly removed by an external magnetic field and by the crystal field. Due to good screening of the 4f-shell by the outer electrons $(5s^25p^6)$, the influence of the crystal field is considerably weaker than in the 3d elements. In most cases the splitting caused by the crystal field is an order of magnitude smaller than the multiplet splitting due to the spin-orbit interaction. Nevertheless, the crystal field essentially influences the magnetic properties and of course the resistivity behavior in R-compounds as will be shown later.

b. The s-f Exchange (RKKY)-Interaction

The interaction of a R-ion having a certain definite value, J, located at the lattice point with position vector, R_j, and an electron belonging to the conduction band with the position vector, r_i, is given by

$$\tilde{H}_{sf} = -2(g - 1) \sum_{ij} G_{sf}(r_i - R_j)s_i \cdot J_j \qquad (3.1)$$

($\tilde{H}_{sf}$ is called the s-f exchange interaction Hamiltonian), where s_i is the spin operator of the conduction electron and $(g - 1)J_j$ describes the projection of the sum spin operator S_j of the 4f-shell. It should be borne in mind that in the case of the rare earths, it is not the ionic spin, S_j, that plays the role of the constant of motion, but the total angular momentum, J_j. The quantity $G_{sf}(r_i - R_j)$ is sometimes called the "exchange integral". Several attempts have been made to obtain this Hamiltonian by means of stricter theoretical computations.[22,23] It has been shown that in the exact s-f exchange Hamiltonian, only the first term of the expansion in powers of a small parameter of the order of $r_{4f}k_F$ (r_{4f} = radius of the 4f-shell, k_F = magnitude of the wave vector of the conduction electron in the vicinity of the Fermi surface) has the form of the spatially isotropic expression in Equation 3.1.

In the following we shall use the simple Hamiltonian (see Equation 3.1), discussing fundamental ideas and results of magnetism and of the influence of localized moments on electrical resistivity. The interaction of the angular momentum, J_j, of the j-th ion with conduction electrons gives rise to a conduction electron spin polarization in the vicinity of this ion. The direction of the polarization depends on the direction of J_j, and the relative magnitude is of the order determined by the ratio G_{sf}/E_F. This is an effect of the first order. One step forward was made by Kasuya[24] and Yosida,[25] who took into account the conduction electron polarization due to their interaction with the angular momentum, J_j, interacting simultaneously with the momentum, $J_{j'}$, of another ion. As a result of this double interaction, the energy of the system pair of angular momentum (J_j and $J_{j'}$) plus conduction electrons becomes a function of the mutual orientation of the moment, J_j and $J_{j'}$. This is a second-order effect. The order of this interaction energy is determined by the ratio G^2_{sf}/E_F. Such an interaction is oscillatory and has long-range character. This theoretical framework is known as the

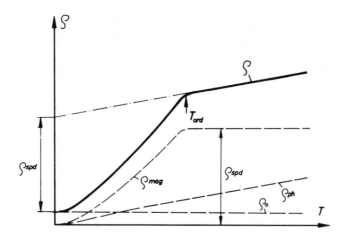

FIGURE 3.1.   Schematic temperature dependence of the electrical resistivity of magnetic R-compounds with negligible d-electron contribution.

RKKY theory.[26-30] In conclusion a great variety of magnetic properties observed in the R-intermetallics under consideration have been explained using the RKKY model.[3]

c. Influence of Magnetism on Resistivity

In Equation 2.1, it was assumed that the temperature dependence of the nonmagnetic R-compounds is determined by the residual resistivity, ϱ_o, and the lattice resistivity, ϱ_{ph}, due to electron-phonon interaction. In R-compounds showing magnetic order, an additional contribution to the resistivity must be taken into consideration. This contribution called ϱ_{mag} describes scattering processes of conduction electrons due to disorder in the arrangement of the magnetic moments.

Assuming the validity of Matthiessen's rule, it follows that the total temperature dependence of the resistivity in these compounds is given by Equation 3.2.

$$\rho(T) = \rho_o + \rho_{ph}(T) + \rho_{mag}(T) \tag{3.2}$$

In Figure 3.1, the temperature dependence of the electrical resistivity of R-compounds without 3d-electron contribution is shown schematically. The contribution, $\varrho_{mag}(T)$, will now be the subject of discussion. $\varrho_{mag}(T)$ is characterized by the following facts:

1.   A temperature independence for $T > T_c$
2.   A pronounced kink at $T = T_c$
3.   A strong decrease for $T < T_c$

In the following a short discussion of the theoretical models describing such behavior will be given.

d. Paramagnetic Regime $(T > T_{ord})$

Using the s-f exchange Hamiltonian given by Equation 3.1, several authors[31-33] solved the conduction electron scattering problem in the paramagnetic regime $(T > T_{ord})$.

The magnitude, $\varrho_{mag} (T > T_c) \equiv \varrho_{spd}$, is given by

$$\rho_{spd} = \frac{3\pi N m^*}{2\hbar e^2 E_F} G^2 (g - 1)^2 J(J + 1) \tag{3.3}$$

Equation 3.3 was derived solving the Boltzmann equation under the relaxation time approximation. The conduction electrons were considered as quasifree particles with an effective mass, m*. The transition probability is obtained in the first Born approximation. Furthermore it is assumed that in the paramagnetic region the local magnetic moments are completely decoupled, possessing $(2J + 1)$ possible states in their orientation to a given direction, z (neglecting the influence of the crystal field at present). For the symbols used in Equation 3.3, see Dekker.[32] Equation 3.3 gives a proportionality of ϱ_{spd} to the so-called de Gennes factor $(g - 1)^2 J(J + 1)$. ϱ_{spd} provides a large constant addition to the total resistivity as was shown in the schematic picture (see Figure 3.1).

Assuming that the linear increase of the total $\varrho(T)$ curve is caused by electron-phonon scattering, in accordance to the Bloch-Gruneisen law (Equation 2.2), an extrapolation procedure of the high-temperature parts to absolute zero temperature indicates the magnitude of ϱ_{spd} given by Equation 3.3. For a binary system consisting of two rare earths $(A_x B_{1-x})$, it could be shown that Equation 3.3 may be written as [32]

$$\rho_{spd}(x) = \frac{3\pi N m^*}{2\hbar e^2 E_F} \left\{ xG_a^2(g_a - 1)^2 J_a(J_a + 1) + (1 - x)G_b^2(g_b - 1)^2 J_b(J_b + 1) - \right.$$
$$\left. x(1 - x) [G_a(g_a - 1)J_a - G_b(g_b - 1)J_b]^2 \right\} \qquad (3.4)$$

where G_a and G_b are the exchange coupling constants for the alloy components and g_a and g_b are the respective Lande g-factors. Equation 3.4 is derived under the assumption that the domains of ferromagnetically aligned spins are of a small size compared with the conduction electron mean free path. When the system contains only one magnetic rare earth compound, e.g., in $(Gd_x Y_{1-x})$, Equation 3.4 is reduced to

$$\rho_{spd}(x) = \frac{3\pi N m^*}{2\hbar e^2 E_F} (g - 1)^2 G^2 Jx(1 + Jx) \qquad (3.5)$$

Assuming that the mean free path of the conduction electrons is much larger than the size of magnetic regions, Dekker[32] proposed the following formula for $\varrho_o(x)$:

$$\rho_0(x) = \frac{3\pi N m^*}{2\hbar e^2 E_F} x (1 - x)[V_{ab}^2 + G^2 (g - 1)^2 J^2] \qquad (3.6)$$

for a system in which only one component has a magnetic moment. $V_{ab} (= V_a - V_b)$ denotes the difference in the Coulomb potential of the two components in a binary system.

e. Influences of Crystal-Fields on Spin Disorder Resistivity

In Section III.A.1.a, it was pointed out that in rare earths and rare earth intermetallic compounds the 4f-electrons are well shielded by the surrounding electrons $(5s^2 5p^6)$ and therefore do not play a significant role in chemical bonding. The metallic rare earth systems, with a few exceptions, may be treated as an assemblage of tripositive ions located in a sea of conduction electrons. The 4f-electrons possess large angular momentum and are therefore responsible for the interesting variety of magnetic properties. As was mentioned above, the exchange interaction given by the Hamiltonian (Equation 3.1) is responsible for the coupling between 4f-moments.

An array of charged R-ions in a crystal produces an electric field at any ion, the so-called crystalline electric field. The presence of this field caused a splitting of the free ion energy level which is in the absence of this field $(2J + 1)$-fold degenerated. This splitting results in a substantial modification of magnetic, electrical, and thermal properties of these materials. The reader may refer to the following citations for information on crystal field theory.[34-36]

Taking the origin of the coordinate system at the nucleus of the rare earth, the following electrostatic potential at a point (r, Θ, Φ) (near the origin of the nucleus) exists due to the surrounding k-ions

$$V(r,\Theta,\Phi) = \sum_k \frac{q_k}{|R_k - r|} \tag{3.7}$$

If the rare earth ion has the charge q_i at (r_i, Θ_i, Φ_i), then the crystalline potential will be

$$H_{CF} = \sum_i q_i V(r_i) = \sum_{i,k} \frac{q_i q_k}{|R_k - r_i|} \tag{3.8}$$

where the summation with respect to "i" is only taken over electrons in unfilled shells as the crystal field affects closed shells only in a high-order perturbation. The perturbation potential of Equation 3.8 lifts the (2J + 1) degeneracy of the ground multiplet of the free rare earth ion. The crystalline potential, Equation 3.8, may be calculated in Cartesian coordinates or directly in terms of spherical harmonics.[35] The exchange scattering of conduction electrons by localized moments neglecting the crystal field influence gives rise to a contribution to the electrical resistivity given by Equation 3.3. This is valid at temperatures sufficiently high in comparison to the crystal field splitting of the rare earth ion. In this "high-temperature" approximation, one obtains temperature independence of ϱ_{spd}. However, at lower temperatures, there can be a very marked influence of the crystal field on ϱ_{spd}, as was first shown for CeAl$_2$.[36]

The method proposed essentially takes into account (1) the distribution of the rare earth ions among the crystal field states using the Boltzmann factor and (2) the contribution to the resistivity from inelastic scattering processes where the rare earth ion undergoes a transition from one crystal field state to another and the conduction electron gains or loses the corresponding amount of energy. The expression for the resistivity at a temperature T becomes

$$\rho_{spd}(T) = \frac{3\pi N m^*}{\hbar e^2 E_F} G^2 (g-1)^2 \sum_{\substack{m_s, m_s' \\ i, i'}} |<m_s' i'|s.J \qquad i' \tag{3.9}$$

where m_s and m_s' represent the spins of the conduction electrons in the initial and final states, and i and i′ are the initial and final rare earth crystal field states. p_i is the Boltzmann probability that the rare earth ion is in the crystal field state of energy, E_i, given by

$$p_i = \frac{e^{(-E_i/k_B T)}}{\sum_j e^{(-E_j/k_B T)}} \tag{3.10}$$

E_i depends on the extent of the crystal field splitting, and $f_{ii'}$ is given by

$$f_{ii'} = \frac{2}{1 + e^{(E_{i'} - E_i)/k_B T}}$$ (3.11)

Rao and Wallace[36] applied Equation 3.9 to the experimental result of $CeAl_2$ especially to explain the unusually broad "knee" in the ϱ vs. T curve at about 70°K. They ascribed this behavior to the influence of the crystal field on ϱ_{spd} given by Equation 3.9. One of the important consequences of the influence of the crystal field on resistivity in the paramagnetic state is that ϱ_{spd} now becomes temperature dependent.

f. Resistance Minimum—Kondo Effect

The occurrence of minima in the temperature dependence of the electrical resistivity of metallic systems has been known for a long time. Traditionally the metallic systems consist of elements such as Cu, Ag, or Au as a solvent with very small amounts ($\leqslant 0.1$ atomic %) of transition metals such as Cr, Mn, and Fe as a solute.[37] A theoretical explanation of this effect was presented by Kondo.[38] Kondo has shown that the spin disorder resistivity is not a constant (as given by Equation 3.3), but increases towards low temperature under certain circumstances. This result follows from second-order perturbation theory using the Hamiltonian given by Equation 3.1. In this model it is assumed that the conduction electrons can also be scattered from an initial state to a final state via intermediate states. The efficiency of this additional "channel" is large in the low-temperature range and gives rise to the observed increase of the spin disorder resistivity. The appearance of a minimum in the ϱ vs. T curve follows from the combined effect of the increasing spin disorder resistivity and the decreasing lattice scattering resistivity on cooling down the sample. Assuming a simple s-S exchange interaction, the expression derived by Kondo reads

$$\rho_{Kondo}(T) = \rho_{spd}\left[1 + 2N(E_F)G \ln(T/T_F)\right]$$ (3.12)

where ϱ_{spd} is given by Equation 3.3 and $(g - 1)^2 J(J + 1)$ is replaced by $S(S + 1)$. T_F is the temperature corresponding to the Fermi energy. G is the exchange coupling constant as in Equation 3.3. $N(E_F)$ denotes the density of states at the Fermi energy.

Summarizing one can say that the Kondo anomaly should be observable if

1. Localized magnetic moments exist in a metallic matrix.
2. These local moments are decoupled and flips of these moments can take place.
3. The sign of G is negative, where G denotes the exchange coupling constant between the local moments and the conduction electrons.

As it was mentioned above, initially the minimum in the ϱ vs. T curve was observed in noble metal systems containing small quantities of 3d-atoms. The magnetic moment of the 3d-atoms is considerably smaller than that of the R-compounds. Therefore one would expect that in dilute R-compounds very pronounced minima should be observable, remembering that ϱ_{spd}, the prefactor in Equation 3.12 (given by Equation 3.3), is large due to the large magnetic moment of the R-ions. However, contrary to this expectation of the 13 R-ions with partially filled 4f-shells, Kondo-like behavior has been observed only in a few Ce, Pr, Sm, Eu, Tm, and Yb alloys.[39] The reason why most of the R-compounds exhibit no Kondo anomalies is explained by the fact that the exchange constant G has a positive sign.[40] A critical discussion is given in a number of investigations.[8,41-43]

g. Critical Phenomena at $T \sim T_c$ or T_N

As was discussed at the beginning of this section, the $\varrho(T)$ curves in the R-compounds in question show pronounced kinks at $T = T_c$ or T_N. These phenomena are obviously caused by the influence of magnetism on electrical resistivity as shown schematically in Figure 3.1. The Curie points (T_c) in ferromagnets and the Neel points (T_N) in antiferromagnets are examples of critical points similar to the well-known critical points in liquid-vapor systems. Order-disorder phase transitions involving some kinds of long-range order parameter take place at these magnetic critical points. For the ferromagnetic case, this parameter is the spontaneous magnetization, M_o. At $T = 0$, the ferromagnetic spin system of a pure crystal is completely ordered; with increasing temperature, M_o decreases continuously. The breakdown of the magnetic long-range order is initially slow because when most of the spins are well aligned, a rather large amount of energy is required to turn a spin in the presence of the exchange field, but as the process continues and the temperature rises, the breakdown of long-range order increases rapidly. Consequently disordering becomes progressively easier, until at the Curie point the last vestiges of long-range order just vanish. Nevertheless, the so-called critical fluctuations still maintain a limited order among the spins above T_c in small regions of the lattice, so small as to contain comparatively few spins above T_c. When the temperature is sufficiently high $(T \gg T_c)$, complete disorder can be assumed, but as the Curie point is approached on lowering the temperature, the exchange interaction is able to initiate some limited amount of order over small volumes of the crystal. In this way small assemblies of parallel spins appear in different parts of the crystal. The total net spontaneous magnetization remains, however, practically zero.

The process is a dynamic one and time dependent so that at a given temperature small ordered spin assemblies, each having a local net spontaneous magnetization, are continuously forming and dissolving in different parts of the crystal. It is because of this situation that the expression "fluctuation" has come into use. As T_c is further approached, the spontaneous magnetization grows both in magnitude and extension, until ordering spreads throughout the whole crystal.

Many physical properties are thereby influenced by these critical phenomena, including equilibrium properties such as specific heat and magnetic susceptibility and also nonequilibrium properties such as the transport properties. Here we distinguish between two classes of ferromagnetic materials. One is the class which can be described in terms of localized spin, represented by the rare earth compounds. The other class comprises the weak ferromagnetic materials whose magnetic properties can satisfactorily be explained using the band model. According to the degree of localization of the magnetic spins, the kink at T_c is more or less pronounced. In rare earth compounds containing Al, Ni, Cu, Ag, Au, etc., the kink in the ϱ vs. T curve in the vicinity of T_c or T_N is found to be extremely sharp, whereas in weak ferromagnetic substances like $ZrZn_2$, practically no anomaly could be found at T_c in the ϱ vs. T curves.[44] De Gennes and Friedel[31] examined the effect of the above-mentioned fluctuations of the short-range order on the resistivity for ferromagnetic metals of the localized spin type with the exchange Hamiltonian given in Equation 3.1. The calculation was carried out in the first Born approximation using the Ornstein-Zernicke method for evaluating the static spin-spin correlation function which is needed to derive $\varrho_{mag}(T)$. In this paper, $\varrho_{mag}(T)$ close to T_c is given by

$$\rho_{mag}(T) = \rho(T_c) - b \left| \frac{T - T_c}{T_c} \right| \left| \ln \left| \frac{T - T_c}{T_c} \right| \right|^{-1} \tag{3.13}$$

with $b > 0$.

Fisher and Langer[45] introduced a modified correlation function which gave a theoretical expression for the divergence in $d\varrho_{mag}/dT$ similar to that of the expression for the specific heat in the temperature range just above T_c. Richard and Geldart[46] extended this model and showed that this is equally true at temperatures below T_c. In antiferromagnetic materials, $d\varrho_{mag}/dT$ at T_c is also influenced by the existence of superzones.[47,48]

h. Intermediate-Temperature Range ($0.2\, T_{ord} \leqslant T \leqslant T_{ord}$)

Over this temperature range, ϱ_{mag} is temperature dependent. Cooling through T_c, the localized spins begin to order so that the conduction electron scattering sharply decreases causing the pronounced kink at the ordering temperature. As it was first demonstrated by Kasuya,[33] one can use the molecular field approximation instead of the s-f exchange interaction given in Equation 3.1. This model neglects the correlations between the individual localized spins. Only the interaction of a localized spin with the field produced by the neighbors is taken into consideration. The scattering of the conduction electrons from a particular local moment at $\mathbf{R}_j$ is due to the difference between S and $<S>_T$ the average ionic spins. The appropriate Hamiltonian for the molecular field model is given by

$$\tilde{H} = -G\mathbf{s} \cdot \left(\mathbf{S} - <\mathbf{S}>_T \right) \qquad (3.14)$$

in the case where the R-ion only carries a spin moment. Finally the result obtained reads

$$\rho_{mag}(T) = \rho_{spd} \left[1 - \frac{<S>_T^2}{S(S+1)} \right] \qquad (3.15)$$

where ϱ_{spd} is given by Equation 3.3 for an S-state ion such as Gd.

The computation of $<S>_T$ as a function of temperature results in a temperature dependence of ϱ_{mag} in this intermediate range. At one extreme, above T_c, $<S>_T$ is zero and spin disorder is complete. That means ϱ_{mag} is at its maximum value and is constant as discussed in Section III.A.1.d. At the other extreme, at absolute zero, all the localized spins are perfectly ordered and ϱ_{mag} becomes zero.

i. Low-Temperature Regime ($T \ll T_{ord}$)

In the lowest-temperature regime, the magnetic properties of the R-compounds discussed in this section can be described satisfactorily using the spin wave-model. At absolute zero, it is assumed that the localized spins are perfectly aligned in a hypothetical magnetic single crystal. With increasing temperature collective vibrations in this coupled system of 4f-moments will be created which are called spin waves. Therefore at low temperature, the conduction electrons become scattered, not only by phonons, but also by spin waves. In most of the magnetic materials at low temperature, electron-spin wave scattering gives a large contribution to the total resistivity. Several authors derived a T^2 dependence for this spin wave contribution to the total resistivity in ferromagnetic materials.[5,49,50] For a simple antiferromagnetic material, a T^4 dependence was predicted.[5]

For the calculation of the temperature dependence of resistivity due to spin wave scattering, the shape of the spin wave spectrum, $\omega(q)$, at low q values is very important. Assuming that $\omega(q) \propto q^2$ for ferromagnetic materials, the above-mentioned T^2 dependence of the spin wave resistivity is obtained. The assumption of $\omega(q) \propto q$ for the spin waves in antiferromagnetic materials gives a T^4 temperature dependence. This predicted $\varrho \propto T^4$ law seems to be questionable because on the one hand more sophisticated calculations for the resistivity of antiferromagnetic materials at low temperatures gives[51]

$$\rho(T) = \begin{cases} \propto T^5 & \text{for } T_N \gg T \gg D \\ \propto Te^{(-\sqrt{T_N D}/T)} & \text{for } T_N D \gg T \end{cases} \qquad (3.16)$$

where T_N is the Neel temperature and D is the temperature corresponding to the anisotropy energy of the spin system. On the other hand, to our knowledge no experimental verification on any of the calculated temperature dependencies could be given in the case of antiferromagnetic order.

As it was shown,[52,53] a gap in the spin wave dispersion relation caused by the magnetic anisotropy of a material leads to a modification of the T^2 law given by

$$\rho(T) = \propto T^2 \, e^{(-\Delta/k_B T)} \qquad (3.17)$$

Δ is the gap energy which represents the minimum energy required to excite a spin wave in the anisotropy field.

2. Experimental Results
a. RAl₂ (R = Y, La. Pr, Nd, Sm, Gd, Tb, Dy, Ho, Er, Tm)

In Figure 3.2, the temperature dependence of the resistivity of the heavy RAl_2 compounds (R = Gd, Tb, Ho, Er, Tm) is shown. Note the similarity of these ϱ vs. T curves to the behavior predicted by the schematic picture (see Figure 3.1). The dependence of ϱ_{spd} on the de Gennes factor $(g - 1)^2 J(J + 1)$ for these heavy RAl_2 compounds is shown in Figure 3.3. In this figure, the ϱ_{spd} value obtained by van Daal and Buschow[54] and that obtained by the author[55] are presented. The agreement with the theoretical model discussed in Section III.A.1.d and given by Equation 3.3 is satisfactory. The ϱ vs. T curves of some of the light RAl_2 compounds (R = La, Pr, Nd, Sm) are shown in Figure 3.4.

b. CeAl₂, CeAl₃, Ce₃Al₁₁, and CeIn₃

Figure 3.5 shows the temperature dependence of the resistivity of $CeAl_2$, $CeAl_3$, Ce_3Al_{11}, and $CeIn_3$.[1] In each case, a minimum in the corresponding ϱ vs. T curve is observable. The shape of the resistivity curve of $CeAl_2$ deviates drastically from that of the other RAl_2 compounds shown in Figures 3.2 and 3.4, although $CeAl_2$ crystallizes in the same structure ($MgCu_2$ structure type). The sharp maximum at about 5°K indicates the onset of antiferromagnetism ($T_N = 4.5°K$).[56] The minimum at about 15°K is interpreted as due to the Kondo effect.[54,57,58] The strong curvature occurring at about 70°K was attributed to the influence of the crystal field effect on the spin disorder resistivity given by Equation 3.9. The resistivity of the nonmagnetic $CeAl_3$ compound (Ni_3Sn structure) has a broad minimum at room temperature and a huge maximum near 35°K. It has proved impossible to explain this resistivity behavior of $CeAl_3$ in terms of a crystal field influence, taking into account thermal repopulation of the three doublet levels of the Ce ions. A model has been put forward in which the resistivity anomalies are explained by also taking into account Kondo exchange scattering of the conduction electrons[42,59] or by considering conduction electron scattering on virtual bound 4f-states situated partly below and partly above E_F, with a width narrower than the crystal field splitting.[60] Magnetic investigations performed on Ce_3Al_{11}[61] show the onset of ferromagntism at about 6°K, followed by a further magnetic transition at

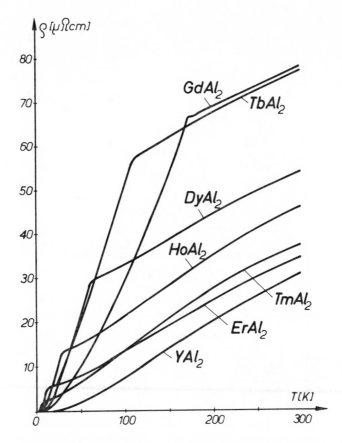

FIGURE 3.2. The ρ vs. T curves of heavy RAl₂ compounds. (From van Daal, H. J. and Buschow, K. H. J., *Solid State Commun.*, 7, 217, 1969. With permission.)

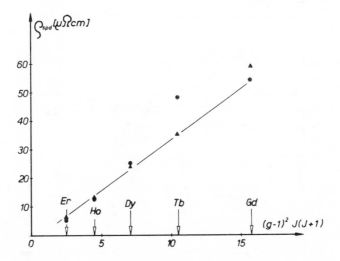

FIGURE 3.3. The dependence of ϱ_{spd} on the de Gennes factor $(g - 1)^2 J(J + 1)$ of some heavy RAl₂ compounds (●, van Daal et al.[54]; ▲, obtained by the author).

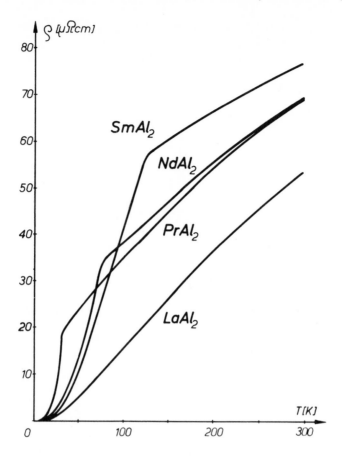

FIGURE 3.4.   The ϱ vs. T curves of some light RAl₂ compounds. (From van Daal, H. J. and Buschow, K. H. J., *Solid State Commun.*, 7, 217, 1969. With permission.)

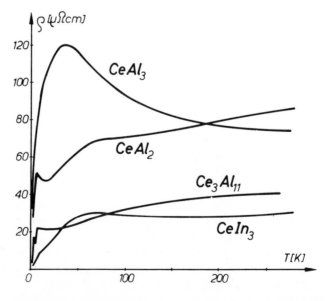

FIGURE 3.5.   The ϱ vs. T curves of CeAl₂, CeAl₃, Ce₃Al₁₁, and CeIn₃. (From Buschow, K. H. J., *Rep. Prog. Phys.*, 42, 1373, 1979. Copyright The Institute of Physics. With permission.)

about 3°K. Both transitions can clearly be seen in the ϱ vs. T curve. Ce$_3$Al$_{11}$ crystallizes in the orthorhombic La$_3$Al$_{11}$ structure.[62] The occurrence of the minimum at about 20°K in the resistivity is attributed to the Kondo effect. Another example of Kondo-like ϱ-behavior in the presence of magnetic ordering effects has been found in the cubic CeIn$_3$ compound. The sharp change in the temperature dependence of ϱ at 9°K is due to magnetic ordering.[63] A comparison of the crystal field parameter A$_4$<r>^{4} and A$_6$<r>^{6} indicates an increase of the energy separation between the 4f-level and the conduction band in going from CeIn$_3$ to NdIn$_3$.[64] A small separation between the 4f-level and the Fermi level can explain the observation of the resistivity minimum in this compound.[1] At this point it should be mentioned that a number of pseudobinary R-compounds are known in which the evidence of a Kondo-like behavior was found.[51,61,63,65-67]

c. PrAl$_3$

PrAl$_3$ can be considered as a classical example to demonstrate the crystal field influence on a local R-moment in the paramagnetic state. In this compound, the results obtained from heat capacity, susceptibility, and spin disorder resistivity provide a unique description of the crystal field influence on the Pr ion.[68] The nonmagnetic PrAl$_3$ crystallizes in the hexagonal Ni$_3$Sn structure. A comparison of the experimental results and the corresponding calculations with respect to the susceptibility (see Figure 3.6a), specific heat (see Figure 3.6b), and resistivity (see Figure 3.6c) of PrAl$_3$ is shown. For all calculations the same crystal field parameters have been used.[68] Figure 3.6a includes also the corresponding energy level scheme. The resistivity is given in a reduced representation; ϱ_{spd}(T) is given by Equation 3.9, and ϱ_{spd}(T = ∞) is given by Equation 3.3.

d. Gd(Al$_{1-x}$Cu$_x$)$_2$

The substitution of Al by Cu in Gd(Al$_{1-x}$Cu$_x$)$_2$ was investigated in order to study the influence of the variation of the conduction electron concentration on the ordering temperature and on resistivity.[69] The ϱ vs. T curves of several samples of this cubic pseudobinary system are given in Figure 3.7. The concentration dependence of ϱ_{spd} together with the Curie temperatures are shown in Figure 3.8. ϱ_{spd}(x) and T$_c$ (x) vary with increasing Cu concentration in a very similar way. The change in these two quantities is explained by the variation of the exchange coupling constant, G, as a function of the conduction electron concentration.[69]

e. R(Ni$_x$Cu$_{1-x}$)$_2$ (R = Tb, Gd)

The concentration dependence of ϱ_{spd} in R(Ni$_x$Cu$_{1-x}$)$_2$ (R = Tb,Gd) (0 ≤ × ≤ 0.4) has been investigated.[6,70] In both of these orthorhombic pseudobinaries (showing CeCu$_2$ structure) depending on the Ni concentration, a transition from an antiferromagnetic to a ferromagnetic ordered state was found. This critical concentration is about 13% Ni in the Gd system and about 8% Ni in the Tb system. The temperature dependence of resistivity in the Tb(Ni,Cu)$_2$ system is given in Figure 3.9. The corresponding concentration dependence of ϱ_{spd} for both pseudobinaries is given in Figure 3.10. In both cases, steps were found at these critical concentrations apparently in connection with the change of the magnetic structure. It follows clearly that these anomalies can be explained by the influence of superzone boundaries on resistivity.[6,70] Superzone boundaries exist if the static arrangement of localized moments is not colinear (e.g., antiferromagnetism or spin spiral structure). This noncolinearity causes an exchange field at the conduction electrons with a lower symmetry than that of the lattice.[71] This superzone effect makes the use of ϱ_{spd} in Equations 3.3 or 3.5 for the calculation of the parameter G and m* meaningless.[3,6,70] This fact is the reason for the step obtained in the (Gd,Y) system as will be seen later, (see Figure 3.23).

In the antiferromagnetically ordered samples of the Tb(Ni,Cu)$_2$ system, no sharp

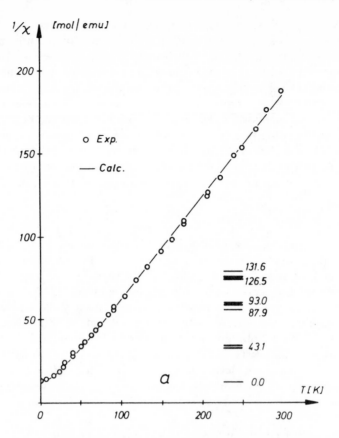

FIGURE 3.6a.   The experimental and calculated inverse susceptibility of PrAl₃ as a function of the temperature, according to the shown energy level scheme. (From Wallace, W. E., Sankar, S. G., and Rao, V. U. S., *Structure and Bonding,* Vol. 33, Springer-Verlag, New York, 1977, 1. With permission.)

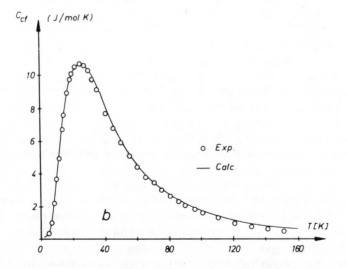

FIGURE 3.6b.   The experimental and calculated crystal field heat capacity of PrAl₃ as a function of temperature. (From Wallace, W. E., Sankar, S. G., and Rao, V. U. S., *Structure and Bonding,* Vol. 33, Springer-Verlag, New York, 1977, 1. With permission.)

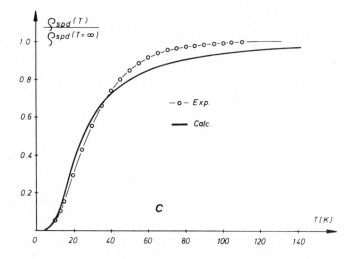

FIGURE 3.6c. The comparison of experimental and calculated ϱ_{spd} data in a reduced representation. $\varrho_{spd}(T)$ is given by Equation 3.9; $\varrho_{spd}(T = \infty)$ is given by Equation 3.3. (From Wallace, W. E., Sankar, S. G., and Rao, V. U. S., *Structure and Bonding*, Vol. 33, Springer-Verlag, New York, 1977, 1. With permission.)

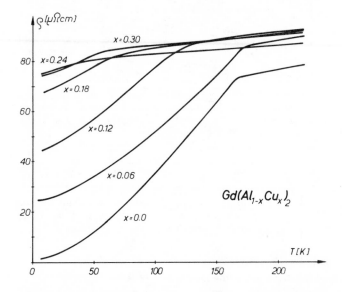

FIGURE 3.7. The ϱ vs. T curves of several concenrations of the Gd(Al$_{1-x}$Cu$_x$)$_2$ pseudobinary system. (From Sakurai, J., Ishimasa, T., and Komura, Y., *J. Phys. Soc. Jpn.*, 43, 5, 1977. With permission.)

kink at T_N is observable, instead a pronounced increase in the ϱ vs. T curves at about 30°K was found, as can be seen in Figure 3.9. Magnetic measurements[6] give evidence that in the samples with x = 0.05, 0.065, and 0.08 in this Tb(Ni$_x$Cu$_{1-x}$)$_2$ system, a change of the magnetic structure appears at these temperatures. This kind of order-to-order transition is typical of most of the heavy RA$_2$ compounds (A = Cu, Ag, Au), as will be shown for the RAg$_2$ and RAu$_2$ systems (see Sections III.A.2.f and III.A.2.g).

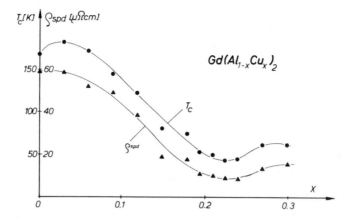

FIGURE 3.8. ϱ_{spd} and T_c as a function of the Cu concentration in Gd(Al$_{1-x}$Cu$_x$)$_2$. (From Sakurai, J., Ishimasa, T., and Komura, Y., *J. Phys. Soc. Jpn.*, 43, 5, 1977. With permission.)

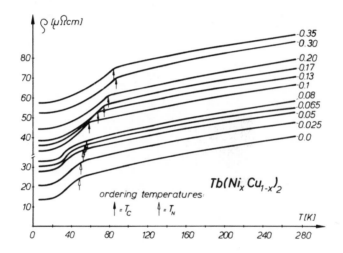

FIGURE 3.9.   The ϱ vs. T curves of several concentrations of the Tb(Ni$_x$Cu$_{1-x}$)$_2$ pseudobinary system. (From Poldy, C. A. and Gratz, E., *J. Mag. Mag. Mat.*, 8, 223, 1978. With permission.)

f. RAg$_2$ (R = Gd, Tb, Dy, Ho, Er)

Magnetic investigations on some of the heavy RAg$_2$ compounds (crystallizing in the MoSi$_2$-structure) yield Neel temperatures in the range of 5°K for ErAg$_2$ and 25°K for GdAg$_2$.[72] The temperature dependence of resistivity[73] is shown in Figure 3.11. In some of the compounds (R = Dy, Ho, Er) below the ordering temperature, T_N, an abrupt increase in resistivity with increasing temperature was observed (as it is indicated by arrows in Figure 3.11). However, no such anomalies in resistivity were observed for GdAg$_2$ and TbAg$_2$; this also applies to the magnetic properties.[3] These sharp increases indicate an order-to-order transition[74] similar to those found in the RAu$_2$ system (see Section III.A.2.g).

Using the experimental data for ϱ_{spd} in Equation 3.3, |G| was calculated to be about 0.08 eV. These results are considerably smaller than those values for |G| obtained from magnetic investigations.[72,73] This circumstance can be understood by the fact that the substitution of ϱ_{spd} in Equation 3.3 by the experimentally obtained value is not applicable for samples which show antiferromagnetic order.[70]

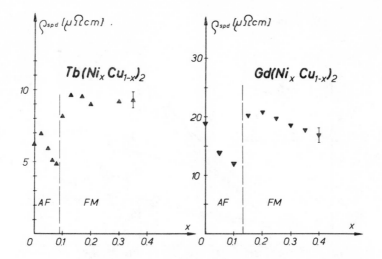

FIGURE 3.10.   Concentration dependence of ϱ_{spd} in $R(Ni_xCu_{1-x})_2$ (R = Gd, Tb) pseudobinaries. (AF = antiferromagnetism, FM = ferromagnetism.) (From Poldy, C. A. and Gratz, E., *J. Mag. Mag. Mat.*, 8, 223, 1978. With permission.)

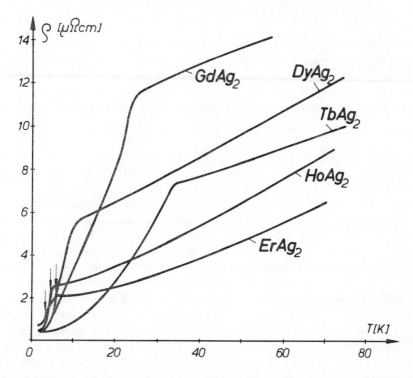

FIGURE 3.11.   The ϱ vs. T curves of some RAg_2 compounds (R = Gd, Tb, Dy, Ho, and Er). (The arrows indicate the transitions in the ordered state in $ErAg_2$, $HoAg_2$, and $DyAg_2$.) (From Ohashi, M., Kaneko, T., and Miura, S., *J. Phys. Soc. Jpn.*, 38, 588, 1975. With permission.)

g. RAu_2 (R = Gd, Tb, Dy, Ho, Er)

The RAu_2 compounds crystallize in the tetragonal $MoSi_2$-structure.[75] The neutron diffraction investigations[74] have shown that most of these compounds exhibit a layered antiferromagnetic spin structure below T_N (β) and a linear-transverse wave spin-structure between $T_N(\beta)$ and $T_N(\alpha)$. The temperature dependence of resistivity in these com-

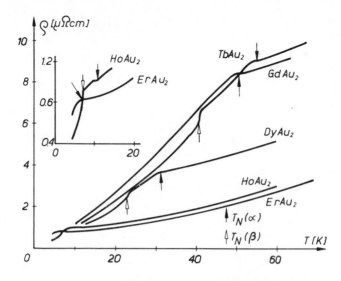

FIGURE 3.12.   The ϱ vs. T curves of some RAu_2 compounds (R = Gd, Tb, Dy, Ho, and Er). (From Kaneko, T., Miura, S., Ohashi, M., and Yamauchi, H., *Proc. ICM-73 TOM-V*, Nauka, Moscow, 1974, 370. With permission.)

pounds is shown in Figure 3.12.[76] The step in the ϱ vs. T curve of $TbAu_2$, $DyAu_2$, and $HoAu_2$ is caused by the magnetic order-to-order transition discussed above. In contrast to the RAl_2 system, no proportionality between ϱ_{spd} and the de Gennes factor in this RAu_2 system is observable. The experimentally defined ϱ_{spd} values were used for the calculations of |G| (exchange coupling constant),[76] as in the RAg_2 system (see Section III.A.2.f).

h. RA (R = Gd, Tb, Dy, Ho, Er, Tm; A = Cu, Ag, Au)

The electrical resistivity of a number of CsCl structure-type binary systems are shown in Figures 3.13 (RCu), 3.14 (RAg), and 3.15 (RAu) (R = Gd, Tb, Dy, Ho, Er, Tm).[77] In all the cases the residual resistivity is subtracted. The shape of these ϱ vs. T curves is in agreement with those which one expects for materials possessing local moments only, represented by the schematic diagram in Figure 3.1. However, it should be pointed out that in the case of very low magnetic ordering temperatures, the ϱ vs. T curves just above the transition temperature are not linear (e.g., ErCu in the inset of Figure 3.13). This is caused by the fact that the phonon contribution ϱ_{ph} is not a linear function of temperature in this temperature range. The experimental data were, however, analyzed to obtain ϱ_{spd} using the usual extrapolation method[77] (see Figure 3.1) which for low-ordering temperatures is not applicable and resulted in a negative sign of the ϱ_{spd}-values. A more applicable method is either to subtract the ϱ-value from the ϱ-value just above T_N or as Pierre[78] did it, to subtract the ϱ-value of the isostructural nonmagnetic Y compound. This latter technique avoids the unwanted influence of superzone boundaries upon the evaluation of ϱ_{spd}. In any case, the proportionality between ϱ_{spd} and the de Gennes factor as in RAl_2 (see Figure 3.3) is not obtained for these CsCl-type compounds.

i. RCd (R = Sm, La)

These compounds crystallize in the CsCl structure. Figure 3.16 shows the temperature dependence of resistance (R) of the magnetic SmCd (T_c = 194°K) and the nonmagnetic LaCd which shows superconductivity below 3°K.[79] The shape of the R vs. T curve of SmCd is typical of a material possessing well-localized magnetic moments.

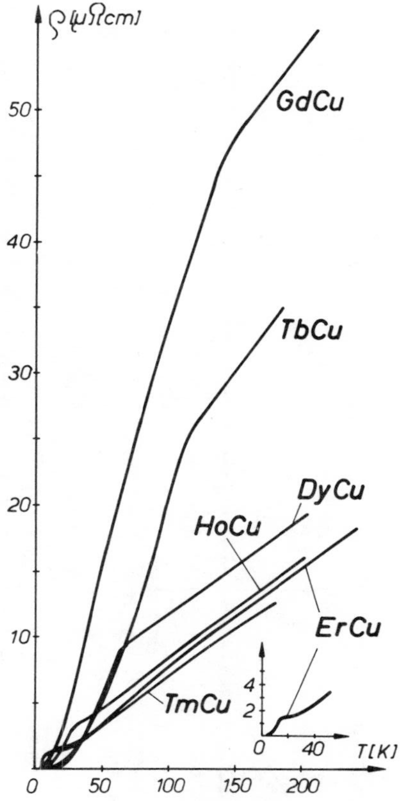

411

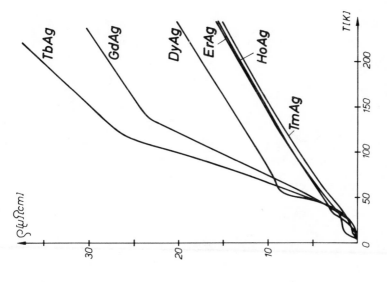

FIGURE 3.13. The ρ vs. T curves of some magnetic RCu compounds. (From Chao Chang-Chih, *J. Appl. Phys.*, 37, 2081, 1966. With permission.)

FIGURE 3.14. The ρ vs. T curves of some magnetic RAg compounds. (From Chao Chang-Chih, *J. Appl. Phys.*, 37, 2081, 1966. With permission.)

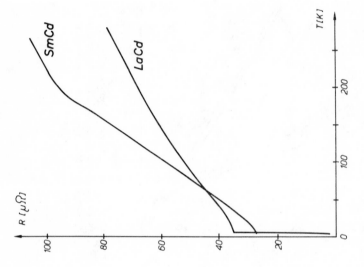

FIGURE 3.16.   The ϱ vs. T curves of the magnetic SmCd (T_c = 194°K) and the nonmagnetic LaCd. (The step in the LaCd curve indicates the superconducting transition). (From Stewart, A. M., Costa, G., and Olcese, G., *Aust. J. Phys.*, 27, 383, 1974. With permission.)

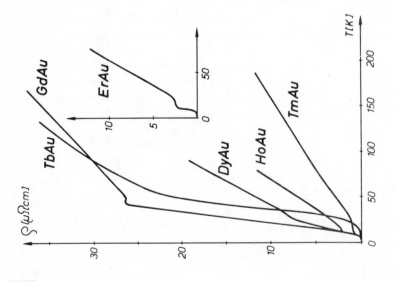

FIGURE 3.15.   The ϱ vs. T curves of some magnetic RAu compounds. (From Chao Chang-Chih, *J. Appl. Phys.*, 37, 2081, 1966. With permission.)

Magnetization measurements performed on SmCd at 77°K exhibit extremely small magnetic moments for the Sm ions (0.05 μ_B per Sm).[79] One possible explanation is that the small bulk magnetization is due to the magnetic order being of noncolinear type, e.g., a conical spiral, that does not allow the Sm moments to become parallel until a very large external magnetic field is applied.[79] Alternatively a crystal field influence reducing the Sm magnetic moment in the ordered state was discussed.[79] From the shape of the resistance curve (pronounced kink at T_c and large ϱ_{spd}), one would expect a much bigger magnetic moment. Apart from the step in the resistance curve of LaCd at about 3°K indicating the onset of superconductivity, no anomaly was reported.[79] In another investigation,[20] a small step in the resistivity curve at 70°K was measured; it is referred to a structural transition appearing at this temperature in LaCd.[80] The influence of such structural transitions on resistivity is described in the La(Ag,In) and Ce(Ag,In) compounds in Section II.B.3.

j. RA_2 (R = Gd, Dy; A = Al, Ni, Co, Rh, Pt)

The critical phenomena at T_c for six ferromagnetic R-cubic Laves phase compounds (GdAl$_2$, DyAl$_2$, GdNi$_2$, GdCo$_2$, GdRh$_2$, and GdPt$_2$) were investigated.[81] All samples are characterized by pronounced anomalies in the $d\varrho/dT$ vs. T dependence in the vicinity of the magnetic ordering temperature, T_c. In Figure 3.17, the results of three of them are presented. Both the electrical resistivity and the computed $d\varrho/dT$ for GdNi$_2$, GdRh$_2$, and GdPt$_2$ are shown from 4.2°K to room temperature. All three $d\varrho/dT$-curves show a broad maximum at high temperatures well above T_c and a minimum as T is decreased which is further followed by a sharp rise towards a well-defined maximum. The temperature corresponding to the maximum for each system is associated with the magnetic ordering temperature. The interpretation of this behavior in these systems is based on the assumption of s-f exchange interaction. Close to T_c the long-range part of the critical fluctuations in the short-range order contributes significantly to the temperature dependence of the scattering. According to the interpretation given in this paper,[81] the shape of the $d\varrho/dT$ curve can be explained using the spin-spin correlation function.[31,82] The application of this correlation function leads to a cusp in $\varrho(T)$ and an infinite discontinuity in $d\varrho/dT$ with positive values of $d\varrho/dT$ for $T < T_c$ and negative ones for $T > T_c$.

Figure 3.18 shows the $\varrho(T)$ and the computed $d\varrho/dT$ behavior for GdCo$_2$, GdAl$_2$, and DyAl$_2$ around the transition temperatures. As one moves from DyAl$_2$ to the GdAl$_2$ on to the GdCo$_2$, one can see a more pronounced peak in $d\varrho/dT$. This behavior was referred in part to the more linear temperature dependence of the phonon contribution which can be expected at higher temperatures.[81] In DyAl$_2$ the weakness of the maximum shows the limit on how well one can determine T_c from the resistivity data. Conversely GdAl$_2$ shows a sharp and well-defined maximum in $d\varrho/dT$ at about 160°K which is associated with the Curie temperature. The behavior of GdCo$_2$ is to a certain extent comparable to the behavior of pure Ni.[83] As will be discussed in Section III.B.2.c, the R-Co coupling plays a significant role in the magnetic properties in these compounds. The moments on the Co atom sites aligned antiparallel to the R-moments are not so well localized as they are on the Gd atom sites. Probably the existing 3d-electron influence in this compound gives the similarity of $d\varrho/dT$ to that of Ni.

k. GdSb

The temperature dependence of resistance (R) and its temperature coefficient dR/dT in the vicinity of T_N (22.8°K) of the antiferromagnetic GdSb compound is shown by Figure 3.19.[84] GdSb crystallizes in the rocksalt structure. In this compound it could be demonstrated that the steep change in dR/dT above T_n is quantitatively comparable to that observed in the ferromagnetic Ni.[83] This is shown by using the second derivation d^2R/dT^2.[84]

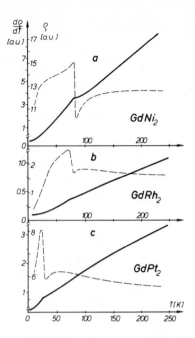

FIGURE 3.17.   The ϱ vs. T curves and the temperature dependence of $d\varrho/dT$. (a) GdNi$_2$; (b) GdRh$_2$; and (c) GdPt$_2$. (-----, $d\varrho/dT$;———, $\varrho(T)$). (From Kawatra, M. P., Mydosh, J. A., and Budnick, J. I., *Phys. Rev. B,* 2, 665, 1970. With permission.)

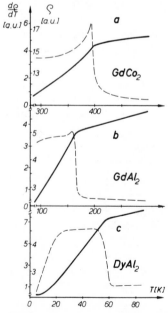

FIGURE 3.18.   The ϱ vs. T curves and the temperature dependence of $d\varrho/dT$. (a) GdCo$_2$; (b) GdAl$_2$; and (c) DyAl$_2$ (-----, $d\varrho/dT$;———, $\varrho(T)$). (From Kawatra, M. P., Mydosh, J. A., and Budnick, J. I., *Phys. Rev. B,* 2, 665, 1970. With permission.)

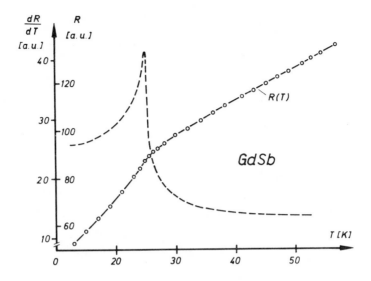

FIGURE 3.19. The resistance vs. temperature (R vs. T) and the temperature dependence of dR/dT of the antiferromagnetic GdSb compound. (From Taub, H. and Williamson, S. J., *Solid State Commun.*, 13, 1021, 1973. With permission.)

l. (La$_x$Gd$_{1-x}$)Ni

The influence of dilution of the magnetic R-component was studied in the orthorhombic (CrB structure) (La$_x$Gd$_{1-x}$)Ni pseudobinary.[85] Figure 3.20 shows a survey of the ϱ vs. T curves in this system. The arrows indicate the Curie temperatures T_c obtained by initial susceptibility measurements. With increasing La concentration, both ϱ_{spd} and T_c decrease together. $\varrho_{spd}(x)$ is given in Figure 3.21. A determination of ϱ_{spd} from experimental data in the high La concentration region was not possible using the previously mentioned extrapolation procedure (see Figure 3.1). Figure 3.21 also shows the concentration dependence of the residual resistivity ϱ_o. $\varrho_o(x)$ deviates significantly from the parabolic shape according to the Nordheim's rule. The asymmetric shape of the $\varrho_o(x)$-curve skewed to the Gd-rich side indicates a growing influence of spin dependent scattering processes on the residual resistivity for increasing Gd content.

In Figure 3.22, the temperature dependence of resistivity in the range from 4.2 to about 20°K for the two boundary compounds GdNi and LaNi is presented in a double logarithmic diagram. LaNi represents a nonmagnetic R-compound. The experimentally observed T^5 dependence (given by the slope of the straight line in Figure 3.22) is in agreement with the temperature dependence predicted by the Bloch-Grüneisen law (see Equation 2.2). The ferromagnetic GdNi compound shows a T^2 dependence as expected from the spin wave contribution to the resistivity. For comparison the resistivity results obtained for the antiferromagnetic TbCu compound are also presented in this diagram. This TbCu measurement shows a rather large deviation from the theoretical behavior predicted by Equation 3.16.

m. Gd$_x$Y$_{1-x}$

Popplewell et al.[86] used Equation 3.5 in Section III.A.1.d to calculate the exchange coupling constant |G| for the Gd$_x$Y$_{1-x}$ binary system based on experimental data for ϱ_{spd}. It is assumed that E_F and m*, the Fermi energy and the effective mass of the conduction electrons, respectively, are unchanged in the whole binary system. An amended estimation of |G| is given by Arnold and Popplewell[87] taking into account the variation of m* using the formula for the paramagnetic Curie temperature in the

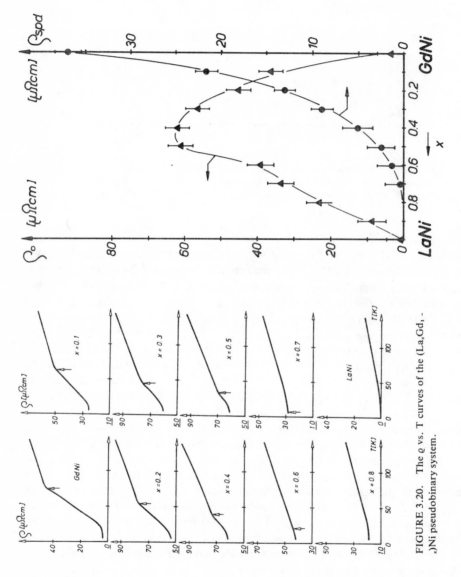

FIGURE 3.21. ϱ_{spd} and ϱ_o vs. concentration in the (La$_x$Gd$_{1-x}$)Ni pseudobinary system.

FIGURE 3.20.   The ϱ vs. T curves of the (La$_x$Gd$_{1-x}$)Ni pseudobinary system.

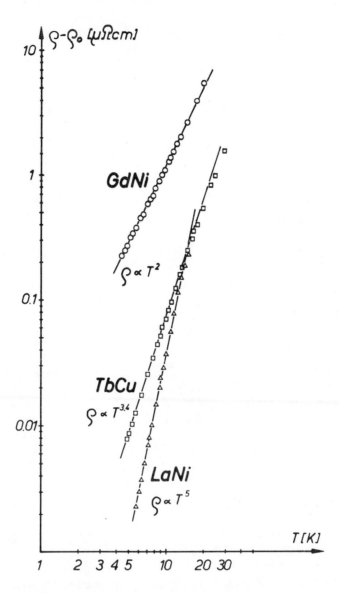

FIGURE 3.22. The electrical resistivity behavior at low temperatures of the ferromagnetic GdNi, the antiferromagnetic TbCu, and the nonmagnetic LaNi.

molecular field model. Figure 3.23 shows the concentration dependence of |G| in the Gd-Y system given by Arnold and Popplewell.[87] The pronounced step at x = 0.6 is caused by the transition from ferromagnetism (x > 0.6) to antiferromagnetism (x < 0.6). This concentration-dependent transition is clearly shown by Ito.[88] In this careful investigation performed on single crystals, it is shown that ϱ_{spd} shows anomalous behavior only for the c-direction. This behavior can be explained by the influence of superzone boundaries on the magnetic contribution to resistivity[6,71] below the ordering temperature.

n. $R_x R'_{1-x}$ (R, R' are Different Rare Earths)

The temperature dependence of electrical resistivity in the range of about 4 to 300°K was investigated for the following light-heavy R-binaries, $Tb_x Pr_{1-x}$, $Dy_x Nd_{1-x}$, $Nd_x Y_{1-x}$, and $Dy_x La_{1-x}$.[89] The discussion of the resistivity is based on the room

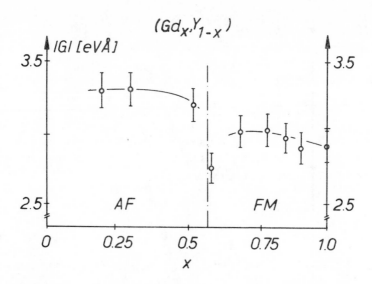

FIGURE 3.23.   The variation of /G/ (exchange coupling constant) with composition for the Gd_xY_{1-x} binary system (AF = antiferromagnetism; FM = ferromagnetism). (From Arnold, P. G. and Popplewell, J., *J. Phys. F,* 3, 1985, 1973. Copyright The Institute of Physics. With permission.)

temperature crystallographic structure of the corresponding system, i.e., the d-hex., the Sm-like, and the hcp structure. The Tb_xPr_{1-x} binary system, for example, exhibits the following structure depending on the concentration $1.0 \geq x \geq 0.7$ — hcp structure, $0.67 \geq x \geq 0.45$ — Sm structure, and $0.3 \geq x \geq 0$ — d-hex. structure. In some compounds a temperature dependent change of the structure was observed similar to that found in pure Dy.

In the discussion of the ϱ vs. T curves the following equation is used.[89]

$$\rho(T) = \rho_0 + \rho_{ph}(T) + \rho_{sf}(T) + \rho_{sr}(T) \tag{3.18}$$

ϱ_o and ϱ_{ph} represent the residual resistivity and the phonon contribution, respectively. ϱ_{sf} has the same meaning as ϱ_{spd} in Equation 3.3. $\varrho_{sr}(T)$ is a contribution to the resistivity due to the presence of short-range order between free magnetic moments above the ordering temperature. This contribution is suggested to be responsible for the curvature of the ϱ vs. T curve in the paramagnetic range observable in some of these compounds.[89] Examples of the behavior expressed by Equation 3.18 are the Tb-Pr and the Dy-Nd systems. For describing the Nd-Y system, $\varrho_{sr}(T)$ was replaced by $\varrho_{cf}(T)$ describing the crystal field influence on the local moments of Nd.

B. Intermetallic Rare Earth-Transition Metal Compounds
1. Theoretical Introduction

In Section III.A intermetallic R-compounds have been discussed consisting of R-elements with local moments and nontransition metals. As could be shown in Section III.A.1, a comparatively broad understanding of the physics of resistivity based on an unsophisticated indirect exchange mechanism between the R-ions has been developed.

Once R-compounds are formed with moment carrying partner elements, such as the transition metals Mn, Fe, Co, etc., we must expect the appearance of much more complex magnetic and transport behavior. The interaction between the different ions in these compounds are in principle of three different kinds, namely the R-R, R-d-element, and the d-element-d-element interaction. It is assumed that the last two interactions are mainly responsible for the physical properties found in compounds of this type.[2,3] In the R-3d intermetallics there exist a large number of isotypic compounds in which the replacement of one d-component by another can lead to mixed crystal series without change of the crystal structure type.[3]

Because of the complex nature of the magnetic interactions, one can only expect a phenomenological description of the resistivity behavior in these compounds under consideration. The contribution, due to magnetism, to the temperature-dependent resistivity ϱ_{mag} is now given by the local R-moment part and in addition by the itinerant magnetism of the d-electrons. In contrast to the well-localized R-moments, the magnetism of the transition metal partner is caused by the split d-bands. The influence of band magnetism or itinerant magnetism on resistivity and in general on all of the transport properties in these compounds is not fully understood yet. However, one can say that the ϱ vs. T curves are characterized by the following attributes:

1. The resistivity is in general much higher in comparison to R-nontransition metal compounds (see Section III.A.2). In some of the R-transition metal pseudobinary compounds in the middle concentration range, even a negative slope of the ϱ vs. T curve was found (e.g., in the $R_6(Mn,Fe)_{23}$ system).

2. Pronounced curvatures towards the temperature axis below and above T_c were found (e.g., $(Gd,Y)_4Co_3$).

3. The kink at T_c is frequently smeared out.

2. Experimental Results
a. $(Gd_xY_{1-x})_4Co_3$

In Figure 3.24 the ϱ vs. T curves for various concentrations of the pseudobinary $(Gd,Y)_4Co_3$ system are shown.[90] This system exhibits a solid solution in the hexagonal Ho_4Co_3 structure. This series was selected to investigate the effect of dilution of the magnetic Gd by nonmagnetic Y. The Y ions can be considered as a "blank" with respect to magnetism. Gd was chosen because of its S-state to avoid the influence of the crystal field on the results. The electrical resistivity has two essential contributions, arising from the localized 4f-moments and the itinerant 3d-electrons.[91] The former invariably leads to kinks in the ϱ vs. T curves at T_c. The latter gives rise to a more or less pronounced strong negative curvature in the ϱ vs. T curve and is given to a first approximation by the ϱ vs. T curve of Y_4Co_3, where the magnetic properties are defined only by the Co ions.[92] Referring to the discussion in Section II.A, the negative curvature caused by the influence of the Co 3d-electrons can be explained as arising from a density of state feature[93] (s-d scattering mechanism) or from spin fluctuations.[15]

b. $Gd_4(Co_{1-x}Ni_x)_3$

The influence of the substitution of Co by Ni in Gd_4Co_3 on resistivity is given in Figure 3.25.[94] The solid solubility limit is reached at 20% Ni concentration. In contrast to the previously described $(Gd,Y)_4Co_3$ system where the ordering temperatures are practically proportional to Gd content, the $Gd_4(Co,Ni)_3$ system shows hardly any change of T_c with Ni content. Although the replacement of Co by Ni is expected to raise the Fermi level, the 3d-band appears not to become totally filled, as indicated by the remaining convexity of the ϱ vs. T curves and as supported by magnetic investigations.[94]

c. RCo_2 (R = Nd, Tb, Dy, Ho, Er)

In a large number of investigations concerning the RCo_2 compounds, it was shown that in $DyCo_2$, $HoCo_2$, and $ErCo_2$ the magnetic transition is of the first-order type, in contrast to the other RCo_2 compounds which are of second-order type. In the case of

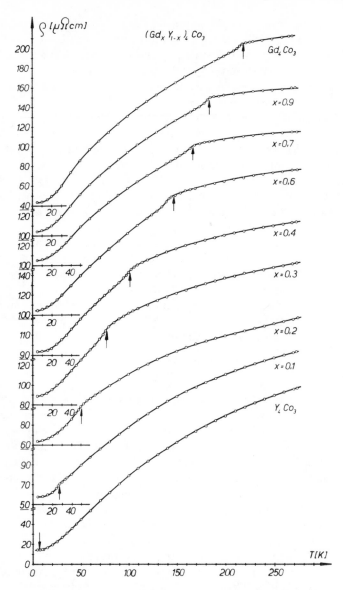

FIGURE 3.24.   The ϱ vs. T curves for various concentrations of the pseudobinary $(Gd_xY_{1-x})_4Co_3$. The ordering temperatures are indicated by the arrows. (From Gratz, E. et al., *J. Phys. F,* 10, 2819, 1980. Copyright The Institute of Physics. With permission.)

a first-order transition, most of the physical parameters such as lattice constants,[95] magnetization,[96] specific heat,[97] and electrical resistivity[98] show an extremely strong change at this transition temperature. Figure 3.26 shows the temperature dependence of the resistivity of the RCo_2 compounds in question. The discontinuities observable in the ϱ vs. T curve of the Er, Ho, Dy compounds represent the influence of this first-order transition on resistivity. Note that, e.g., in $ErCo_2$ at 32°K, the magnitude of the electrical resistivity increases by about 600% on heating up the sample by a few tenths of a degree. Precise investigations of the resistivity in the vicinity of T_c for $ErCo_2$ show a hysteresis in temperature of about 0.8°K.[99] The magnitude of this jump in the ϱ vs. T curve decreases with increasing ordering temperature ($ErCo_2$ T_c = 32°K, $HoCo_2$ T_c = 78°K, and $DyCo_2$ T_c = 135°K). This is brought in connection with the increasing

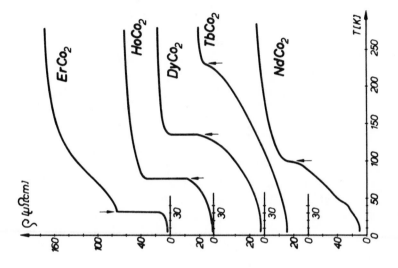

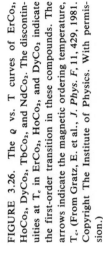

FIGURE 3.26. The ϱ vs. T curves of ErCo$_2$, HoCo$_2$, DyCo$_2$, TbCo$_2$, and NdCo$_2$. The discontinuities at T_c in ErCo$_2$, HoCo$_2$, and DyCo$_2$ indicate the first-order transition in these compounds. The arrows indicate the magnetic ordering temperature, T_c. (From Gratz, E. et al., *J. Phys. F*, 11, 429, 1981. Copyright The Institute of Physics. With permission.)

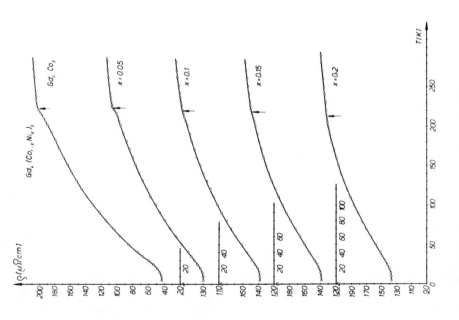

FIGURE 3.25. The ϱ vs. T curves for various concentrations of the pseudobinary Gd$_4$(Co$_{1-x}$Ni$_x$)$_3$ system. The ordering temperatures are indicated by the arrows.

influence of temperature.[98] The discontinuities at T_c are referred to the sudden appearance of a high degree of order in rare earth moments immediately below T_c (in contrast to a second-order transition where the alignment in the moments increases continuously below T_c). Coupled with the ordering of the R-moments is a splitting of the Co-3d-band and the consequent appearing of an induced 3d-moment of about 1 μ_B which escalates the ordering process. The extremely flat increase of the ϱ vs. T curve in the high-temperature region experimentally found in all the RCo_2 compounds (YCo_2 included) is explained as caused by the s-d scattering processes of conduction electrons into available empty d-states (see Section II.A).

d. $(Ho_xY_{1-x})Co_2$

In Figure 3.27, the ϱ vs. T curves of some concentrations of the $(Ho,Y)Co_2$ pseudo-binary system are presented.[100] It appears that the discontinuity found in $HoCo_2$ at T_c exists for up to 30% Y substitution. The cause of the minimum in the resistivity curves in the intermediate concentration range is still an open question. From magnetic measurements in $HoCo_2$ a reorientation of the easy axis of magnetization at 14°K was found.[101] The small discontinuities observable for $0.7 \leqslant x \leqslant 1.0$ around 14°K (shown in the inset in Figure 3.27) are connected with this reorientation.[100]

e. $Y(A_{1-x}Fe_x)_2 (A = Co, Ir)$

The magnetic and the resistivity behavior in three pseudobinary cubic Laves phases were investigated.[102] The measurements were performed in the magnetically dilute region, i.e., in the low Fe concentration range. The ϱ vs. T curves measured for several Fe concentrations of $Y(Co_{1-x}Fe_x)_2$ are collected together in Figure 3.28. Note that the substitution of only 10% Fe causes an extremely high increase of the residual resistivity connected with a nearly temperature independent behavior shown by the ϱ vs. T curves. For Fe concentrations $x > 0.12$, "kink"-like anomalies were observed, indicating the ordering temperatures which are given by the arrows in this figure.

In the upper part of Figure 3.28, the ϱ vs. T curves of two $Y(Ir_{1-x}Fe_x)_2$ compounds are shown. In these samples $d\varrho/dT < 0$ was detected. It was reported that for the $Hf(Co,Fe)_2$ system no such behavior was found.[102]

f. $R(Al_{0.9}T_{0.1})_2 (R = Y, Gd, Dy, Er; T = Fe, Co)$

Figure 3.29 shows the influence of a small substitution of Al by transition metals in RAl_2.[55] In some selected RAl_2 compounds, 10% of Al was substituted by Fe or Co. Two facts are apparent: (1) a dramatic increase of the residual resistivity ϱ_o and (2) a pronounced decrease of ϱ_{spd} in the substituted samples.

These effects could not be observed when the R-ions are substituted by Y (as in $(Dy_{0.9}Y_{0.1})Al_2$, see Figure 3.29) or Al is substituted by Cu.[69] In Table 2 these facts are summarized. The ratio r in Table 2 is given by the following formula:

$$r = \frac{\rho_{spd}[R(Al_{0.9}T_{0.1})_2]}{\rho_{spd}(RAl_2)} \qquad (3.19)$$

The extremely strong influence of a transition metal substitution is referred to structural and magnetic local distortions.[55] X-ray investigations of the $R(Al_{0.9}T_{0.1})_2$ (T = Co, Fe) shows an extremely strong broadening of the X-ray reflexions. Furthermore, it has been found that the pseudobinary compounds with Fe or Co are extremely hard to saturate in magnetic fields.

g. $R_6(Fe_xMn_{1-x})_{23} (R = Y, Er)$

Unusually interesting magnetic behavior was found in the R-transition metal compounds exhibiting the Th_6Mn_{23} structure.[103,104] The temperature dependence of the re-

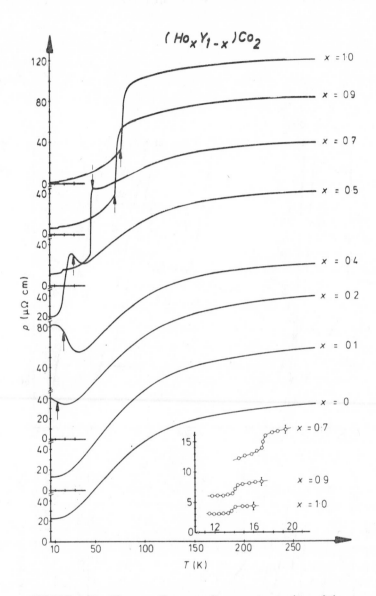

FIGURE 3.27. The ϱ vs. T curves of some concentrations of the $(Ho_xY_{1-x})Co_2$ pseudobinary system. The magnetic ordering temperatures, T_c, are given by the arrows. The inset shows the effect of the reorientation of the easy axis of magnetization on an expanded scale. (From Steiner, W., Gratz, E., Ortbauer, H., and Camen, H. W., *J. Phys. F*, 8, 1525, 1978. Copyright The Institute of Physics. With permission.)

sistivity of two $R_6(Fe,Mn)_{23}$[105] pseudobinary series is shown in Figure 3.30 $(Y_6(Fe,Mn)_{23})$ and Figure 3.31 $(Er_6(Fe,Mn)_{23})$. The substitution of only a small fraction of Fe or Mn in the boundary compounds R_6Fe_{23} or R_6Mn_{23} causes an extremely strong increase of the residual resistivity ϱ_0. In Figure 3.32 the concentration dependence of ϱ_0 for both systems is given. In the middle concentration range, the ϱ vs. T curves are marked by $d\varrho/dT < 0$ over the whole temperature range, irrespective of whether the R-ion carries a local moment (as in the Er compounds) or not (Y compounds). This behavior referred to the very complex crystallographic structure of these compounds[105] which suggests the application of Mooij's rule,[106] stating that if the mean free path of

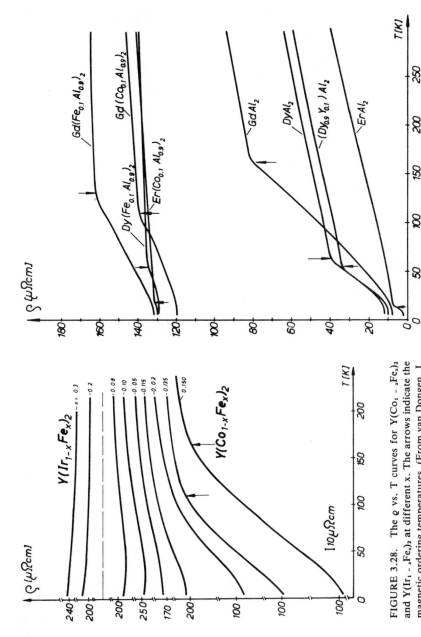

FIGURE 3.28. The ϱ vs. T curves for Y(Co₁₋ₓFeₓ)₂ and Y(Ir₁₋ₓFeₓ)₂ at different x. The arrows indicate the magnetic ordering temperatures. (From van Dongen, J. C. M., Nieuwenhuys, G. J., Mydosh, J. A., van der Kraan, A. H., and Buschow, K. H. J., *Proc. Phys. of Transition Metals*, Leeds, 1980, Inst. Phys. Conf. Ser. No. 55, Chap. 4. Copyright The Institute of Physics. With permission.)

FIGURE 3.29. The effect of a 10% substitution of Al by Fe or Co (and Dy by Y) on the ϱ vs. T curves in RAl₂ compounds (R = Gd, Dy, Er).

Table 2
ELECTRICAL RESISTIVITY FEATURES
OF MATERIALS BASED ON GdAl₂,
DyAl₂, AND ErAl₂

Compound	ϱ_o ($\mu\Omega$cm)	ϱ_{spd} ($\mu\Omega$cm)	r
GdAl$_2$	8	59	(1)
Gd(Al$_{0.9}$Fe$_{0.1}$)$_2$	132	31	0.52
Gd(Al$_{0.9}$Co$_{0.1}$)$_2$	120	20	0.33
Gd(Al$_{0.9}$Cu$_{0.1}$)$_2$	42	46	0.78
DyAl$_2$	10	24	(1)
Dy(Al$_{0.9}$Fe$_{0.1}$)$_2$	129	5.3	0.22
(Dy$_{0.9}$Y$_{0.1}$)Al$_2$	12	17	0.77
ErAl$_2$	1.5	5	(1)
Er(Al$_{0.9}$Co$_{0.1}$)$_2$	130	1	0.2

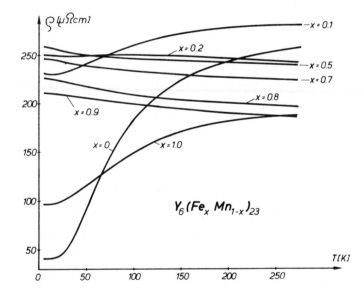

FIGURE 3.30. The ϱ vs. T curves of the Y$_6$(Fe$_x$Mn$_{1-x}$)$_{23}$ pseudobinary system. (From Gratz, E. and Kirchmayr, H. R., *J. Mag. Mag. Mat.*, 2, 187, 1976. With permission.)

the conduction electrons is reduced to the size of the atomic distances, the ϱ vs. T curves become temperature independent. Furthermore, the s-d scattering mechanism (see Section II.A) was also taken into consideration. Practically, no anomalies could be found at T_c in these highly resistive crystalline systems.

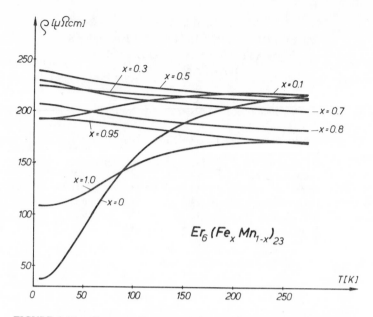

FIGURE 3.31.   The ϱ vs. T curves of the $Er_6(Fe_xMn_{1-x})_{23}$ pseudobinary system. (From Gratz, E. and Kirchmayr, H. R., *J. Mag. Mag. Mat.*, 2, 187, 1976. With permission.)

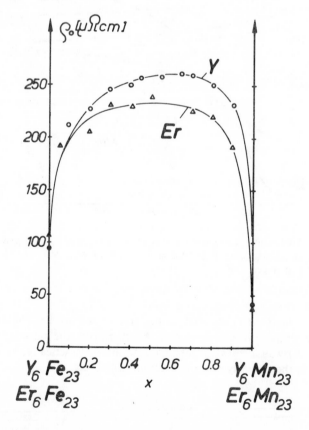

FIGURE 3.32.   The concentration dependence of the residual resistivity of $Y_6(Fe_xMn_{1-x})_{23}$ and $Er_6(Fe_xMn_{1-x})_{23}$. (From Gratz, E. and Kirchmayr, H. R., *J. Mag. Mag. Mat.*, 2, 187, 1976. With permission.)

IV. SUMMARY

Summarizing we have found a large variety of different forms of temperature dependences of resistivity in R-compounds. In principle four main types can be discussed.

1. Starting with the simplest case as exemplified by YAl_2, we have a pure phonon influence which can be approximated by the Bloch-Grüneisen law, characterized by a T^5 increase in the lowest temperature range and a linear increase at high temperatures.

2. The next case is distinguished by a curvature towards the temperature axis for higher temperatures; in addition we also have deviations from the T^5 dependence at low temperatures. Examples of this group are $CeRu_2$, $CeCo_2$, YCo_2, etc. This behavior arises from the d-character of the partner elements. Such compounds are commonly called enhanced paramagnetic materials.

3. Replacing now the nonmagnetic R-component with a magnetic rare earth, we observe the behavior shown in Figure 3.1. The contribution caused by magnetic ordering leads to a more or less sharp kink at the ordering temperature. In the paramagnetic region, the magnetic contribution is a constant superposed on the linear increase due to the phonon contribution. This behavior is found in local moment compounds, such as RAl_2, RAg_2, etc.

4. When both the rare earth and the partner d-elements are magnetic (e.g., Gd_4Co_3), we have a combination of the convexity of (2) and the kink shown by (3). In this complex case no satisfactory theory has been put forward in contrast to the relatively "simple" models for case (3). In general the resistivity even at low temperature is relatively high which can lead in some cases, e.g., in $R_6(Fe,Mn)_{23}$, to a negative slope. Apart from these main categories we can now note a number of exotic types of behavior.

Anomalies in the ϱ vs. T curves with hysteresis caused by crystal structure change are found in R(Ag,In). Similar to this behavior is the step-like anomaly found in some RAg_2 and RAu_2-compounds as a result of an order-order transition in magnetic structure. Here we can also note the anomalies caused by superzone boundaries, e.g., $Tb(Ni,Cu)_2$.

Another kind of anomaly found in compounds belonging to the above-mentioned group (3) is caused by the influence of the crystal field on the local moments which gives rise to a pronounced temperature-dependent ϱ_{spd} in the paramagnetic state. This is demonstrated by the case of $PrAl_3$ which shows no ordering temperatures. If the magnetic transition is of a first-order type, extremely pronounced steps are observed at the ordering temperatures, e.g., $ErCo_2$.

Finally one can hope that as a result of the investigations of the electrical resistivity in the R-compounds a more detailed knowledge with regard to the question of how magnetism influences the electrical resistivity can be expected.

ACKNOWLEDGMENTS

I am indebted to Professor H. Kirchmayr for giving me the possibility to perform this work and for fruitful discussions. I am especially grateful to Dr. C. Poldy for many helpful discussions and for the reading of the manuscript. My sincere thanks also go to Professor M. Zuckermann for many comments on this work. I would like to thank Dipl. Ing. K. Haidinger and A. Zuser for their help during the preparation of the manuscript.

REFERENCES

1. Buschow, K. H. J., *Rep. Prog. Phys.*, 42, 1373, 1979.
2. Buschow, K. H. J., *Rep. Prog. Phys.*, 40, 1179, 1977.
3. Kirchmayr, H. R. and Poldy, C. A., *Handbook on the Physics and Chemistry of Rare Earths*, Vol. 2, Gschneider, K. A., Jr. and Eyring, L., Eds., North-Holland, Amsterdam, 1979, 57.
4. Ziman, J. M., *Electrons and Phonons*, Clarendon Press, Oxford 1960, 170.
5. Volkenshtein, N. V., Dyakina, V. P., and Startsew, V. E., *Phys. Status Solidi B*, 57, 9, 1973.
6. Poldy, C. A. and Gratz, E., *J. Mag. Mag. Mat.*, 8, 223, 1978.
7. Blatt, F. J., *Physics of Electronic Conduction in Solids*, McGraw-Hill, New York, 1968, 189.
8. van Daal, H. J. and Buschow, K. H. J., *Phys. Status Solidi A*, 3, 853, 1970.
9. Meaden, G. T., *Contemp. Phys.*, 12, 313, 1971.
10. Mott, N. F., *Proc. R. Soc. London Ser. A*, 156, 368, 1936.
11. Jones, H., *Handbuch der Physik*, Vol 14, Springer-Verlag, Berlin, 1956, 268.
12. Allen, P. B., *Superconductivity in d- and f-Band Metals*, Suhl, H. and Maple, M. B., Eds., Academic Press, New York, 1980, 291.
13. Mathon, J., *Proc. R. Soc. London Ser. A*, 306, 355, 1968.
14. Kaiser, A. B. and Doniach, S., *Int. J. Magn.*, 1, 11, 1970.
15. Ueda, K. and Moriya, T., *J. Phys. Soc. Jpn.*, 39, 605, 1975.
16. Nowotny, H. and Gratz, E., to be published.
17. Ihrig, H. and Methfessel, S., *Z. Phys. B*, 24, 381, 1976.
18. Ihrig, H. and Methfessel, S., *Z. Phys. B*, 24, 285, 1976.
19. Balster, H., Diplomarbeit, Ruhr Universität Bochum, BRD, Bochum, 1972.
20. Ihrig, H., Thesis, Ruhr Universität Bochum, BRD, Bochum, 1973.
21. Koehler, W. C. and Wollan, E. O., *Phys. Rev.*, 92, 1380, 1953.
22. Liu, S. H., *Phys. Rev.*, 121, 451, 1961.
23. Kondo, J., *Prog. Theor. Phys.*, 27, 772, 1962.
24. Kasuya, T., *Prog. Theor. Phys. (Kyoto)*, 16, 45, 1956.
25. Yosida, K., *Phys. Rev.*, 106, 893, 1957.
26. Rudermann, M. A. and Kittel, C., *Phys. Rev.*, 96, 99, 1954.
27. de Gennes, P. G., *Radium*, 29, 510, 1962.
28. Kasuya, T., *Magnetism*, Vol. 2B, Rado, G. T. and Suhl, H., Eds., Academic Press, New York, 1966, 215.
29. Rocher, Y. A., *Philos. Mag. Suppl.*, 11, 233, 1962.
30. Kittel, C., *Solid State Physics*, Vol. 22, Seitz, F. and Turnball, H., Eds., Academic Press, New York, 1968, 1.
31. de Gennes, P. G. and Friedel, J., *J. Phys. Chem. Solids*, 4, 71, 1958.
32. Dekker, A. J., *J. Appl. Phys.*, 36, 906, 1965.
33. Kasuya, T., *Prog. Theor. Phys.*, 16, 58, 1956.
34. Stevens, K. W. H., *Proc. Phys. Soc. London Sect. A*, 218, 553, 1953.
35. Hutchings, M. T., *Solid State Physics*, Vol. 16, Seitz, F. and Turnball, H., Eds., Academic Press, New York, 1966, 227.
36. Rao, V. U. S. and Wallace, W. E., *Phys. Rev. B*, 2, 4613, 1970.
37. van den Berg, G. J., *Prog. Low Temp. Phys.*, 4, 194, 1964.
38. Kondo, J., *Prog. Theor. Phys. (Kyoto)*, 32, 37, 1964.
39. Maple, M. B., De Long, L. E., and Sales, B. C., *Handbook on the Physics and Chemistry of Rare Earths*, Vol. 1, Gschneider, K. A., Jr. and Eyring, L., Eds., North-Holland, Amsterdam, 1978, 797.
40. Coqblin, B. and Blandin, A., *Adv. Phys.*, 17, 281, 1968.
41. Coqblin, B. and Schriefer, J. R., *Phys. Rev.*, 185, 847, 1969.
42. Maranzana, F. E., *Phys. Rev. Lett.*, 25, 239, 1970.
43. Buschow, K. H. J., van Daal, H. J., Maranzana, F. E., and van Aken, P. B., *Phys. Rev. B*, 3, 1662, 1971.
44. Ogawa, S., *J. Phys. Soc. Jpn.*, 40, 1007, 1976.
45. Fisher, M. E. and Langer, J. S., *Phys. Rev. Lett.*, 20, 665, 1968.
46. Richard, T. G. and Geldart, D. J. W., *Phys. Rev. Lett.*, 30, 290, 1973.
47. Kasuya, T. and Kondo, A., *Solid State Commun.*, 14, 249, 1974.
48. Ausloos, M., *Physica*, 86, 338, 1977.
49. Kasuya, T., *Prog. Theor. Phys. (Kyoto)*, 22, 227, 1959.
50. Mannari, J., *Prog. Theor. Phys. (Kyoto)*, 22, 335, 1959.
51. Yamada, H. and Takada, S., *Prog. Theor. Phys.*, 52, 1077, 1974.
52. Miwa, H., *Prog. Theor. Phys.*, 28, 208, 1962.
53. Mackintosh, A. R., *Phys. Lett.*, 4, 140, 1963.
54. van Daal, H. J. and Buschow, K. H. J., *Solid State Commun.*, 7, 217, 1969.

55. Gratz, E., Grössinger, R., Österreicher, H., and Parker, F. T., *Phys. Rev.,* 23, 2542, 1981.
56. Walker, E., Purwins, H. G., Landolt, M., and Hullinger, F., *J. Less Common. Metals,* 33, 203, 1973.
57. van Daal, H. J. and Buschow, K. H. J., *Phys. Rev. Lett.,* 23, 408, 1969.
58. Maple, M. B., Thesis, University of California, San Diego, 1969.
59. Coqblin, B., Bhattacharjee, A. K., Cornut, B., Gonzales-Jimenez, F., Iglesias-Sicardi, J. R., and Jullien, R., *J. Mag. Mag. Mat.,* 3, 67, 1976.
60. Andres, K., Graebner, J. E., and Ott, H. R., *Phys. Rev. Lett.,* 35, 1779, 1975.
61. van Daal, H. J. and Buschow, K. H. J., *Phys. Lett. A,* 31, 103, 1969.
62. Gomes de Mesquita, A. H. and Buschow, K. H. J., *Acta Crystallogr.,* 22, 497, 1967.
63. Elenbaas, R. A., Schinkel, C. J., and van Deudekom, C. J. M., *J. Mag. Mag. Mat.,* 15, 979, 1980.
64. Lethuillier, P. and Chaussy, J., *J. Phys.,* 37, 123, 1976.
65. Buschow, K. H. J. and van Daal, H. J., *Solid State Commun.,* 8, 363, 1970.
66. Bakanowski, S., Crow, J. E., and Mihalisin, T., *Solid State Commun.,* 22, 241, 1977.
67. Rao, V. U. S., Hutchens, R. D., and Greedan, J. E., *J. Phys. Chem. Solids,* 32, 2755, 1971.
68. Wallace, W. E., Sankar, S. G., and Rao, V. U. S., *Structure and Bonding,* Vol. 33, Springer-Verlag, New York, 1977, 1.
69. Sakurai, J., Ishimasa, T., and Komura, Y., *J. Phys. Soc. Jpn.,* 43, 5, 1977.
70. Gratz, E. and Poldy, C. A., *Phys. Status Solidi B,* 82, 159, 1977.
71. Elliot, R. J. and Wedgwood, F. A., *Proc. Phys. Soc.,* 81, 846, 1963.
72. Miura, S., Kaneko, T., Ohashi, M., and Yamauchi, H., *J. Phys. Soc. Jpn.,* 37, 1464, 1974.
73. Ohashi, M., Kaneko, T. and Miura, S., *J. Phys. Soc. Jpn.,* 38, 588, 1975.
74. Atoji, M., *J. Chem. Phys.,* 51, 3877, 1969.
75. Conner, R. A., *Acta Crystallogr.,* 22, 745, 1967.
76. Kaneko, T., Miura, S., Ohashi, M., and Yamauchi, H., *Proc. ICM-73 TOM V,* Nauka, Moscow, 1974, 370.
77. Chao Chang-Chih, *J. Appl. Phys.,* 37, 2081, 1966.
78. Pierre, J., *Solid State Commun.,* 7, 165, 1969.
79. Stewart, A. M., Costa, G., and Olcese, G., *Aust. J. Phys.,* 27, 383, 1974.
80. Ihrig, H., Vigren, D. T., Kubler, J., and Methfessel, S., *Phys. Rev. B,* 8, 4525, 1973.
81. Kawatra, M. P., Mydosh, J. A., and Budnick, J. I., *Phys. Rev. B,* 2, 665, 1970.
82. Kim, D. J., *Prog. Theor. Phys.,* 31, 921, 1964.
83. Craig, P. P., Goldberg W. I., Kitchens, T. A., and Budnick, J. I., *Phys. Rev. Lett.,* 19, 1334, 1967.
84. Taub, H. and Williamson, S. J., *Solid State Commun.,* 13, 1021, 1973.
85. Gratz, E., Hilscher, G., Sechovsky, V., and Sassik, H., to be published.
86. Popplewell, J., Arnold, P. G., and Davies, P. M., *Proc. Phys. Soc.,* 92, 177, 1967.
87. Arnold, P. G. and Popplewell, J., *J. Phys. F,* 3, 1985, 1973.
88. Ito, T., *J. Sci. Hiroshima Univ. Ser. A,* 37, 107, 1973.
89. Krizek, H., Taylor, K. N. R., and Corner, W. D., *Phys. Status Solidi A,* 41, 251, 1977.
90. Gratz, E., Sechovsky, V., Wohlfarth, E. P., and Kirchmayr, H. R., *J. Phys. F,* 10, 2819, 1980.
91. Gratz, E. and Wohlfarth, E. P., *J. Mag. Mag. Mat.,* 15, 903, 1980.
92. Gratz, E., Kirchmayr, H. R., Sechovsky, V., and Wohlfarth, E. P., *J. Mag. Mag. Mat.,* 21, 191, 1980.
93. Hilscher, G. and Gratz, E., *Phys. Status Solidi A,* 48, 473, 1978.
94. Gratz, E., Hilscher, G., Kirchmayr, H. R., and Sassik, H., *The Rare Earth in Modern Science and Technology,* Vol. 2, Rhyne, J., McCarthy, G., and Silver, H., Eds., Plenum Press, New York, 1980, 327.
95. Lee, E. W. and Pourarian, F., *Phys. Status Solidi A,* 33, 483, 1976.
96. Lemaire, R., *Cobalt,* 33, 201, 1966.
97. Voiron, J., Berton, A., and Chaussy, J., *Phys. Lett. A,* 50, 17, 1974.
98. Gratz, E., Sassik, H., and Nowotny, H., *J. Phys. F,* 11, 429, 1981.
99. Gratz, E., Sassik, H., Nowotny, H., Steiner, W. and Mair, G., *J. Phys. Collaq.,* 40(Suppl. C5), 186, 1979.
100. Steiner, W., Gratz, E., Ortbauer, H., and Camen, H. W., *J. Phys. F,* 8, 1525, 1978.
101. Lee, E. W. and Pourarian, F., *Phys. Status Solidi A,* 34, 383, 1976.
102. van Dongen, J. C. M., Nieuwenhuys, G. J., Mydosh, J. A., van der Kraan, A. H., and Buschow, K. H. J., *Proc. Phys. of Transition Metals,* Leeds, 1980, in press.
103. Hilscher, G. and Rais, H., *J. Phys. F,* 8, 511, 1978.
104. Hardmann, K., James, W. J., Long, G. L., Yelon, W. B., and Kebe, B., *The Rare Earth in Modern Science and Technology,* Rhyne, J., McCarthy, G., and Silver, H., Eds., Plenum Press, New York, 1980, 315.
105. Gratz, E. and Kirchmayr, H. R., *J. Mag. Mag. Mat.,* 2, 187, 1976.
106. Mooij, J. H., *Phys. Status Solidi A,* 17, 521, 1973.

Index

INDEX

D